PROBLEMS BOOK AND STUDY GUIDE

Nancy L. Pruitt
Colgate University

Cell and Molecular Biology

Concepts and Experiments

Gerald Karp
Formerly of the University of Florida, Gainesville

JOHN WILEY & SONS, INC.
New York • Chichester • Brisbane • Toronto • Singapore

ISBN 0-471-14287-5

Printed in the United States of America

10 9 8 7 6 5 4 3 2 1

TO THE STUDENT

This companion to *Cell Biology: Concepts and Experiments* by Gerald Karp, like the textbook, is about learning by doing and learning by seeing what others have done. We want to challenge you to apply the concepts you read about in the text, and to stretch your imagination in ways that memorization cannot do. To do that, I have written a workbook that asks you to apply the principles presented in your text to novel situations, and even derive new concepts from those you've learned from your reading. You will find that this type of conceptual learning is more challenging than memorization– and more rewarding.

How to Use This Study Guide

After you've read each chapter of the text, remind yourself of the content using the outline called **Reviewing the Chapter**. You can glance only at the major headings in bold type, or allow your eye to take in more detail by reading subheadings. One of my favorite ways of testing my understanding of a concept is to try to explain it to a friend. Practice explaining each of the **Learning Objectives** to someone, anticipating their questions. Can you answer them? If not, go back to the book or to the outline in **Reviewing the Chapter**.

Next, glance down the list of **Key Terms and Phrases**. They are arranged roughly in the order in which they are presented in the chapters so related terms will be close together. Do the terms conjure up mental images or ideas? If not, better check the text or **Reviewing the Chapter**.

The **Key Figure** was chosen from the text to represent an important concept from each chapter. Use the **Questions for Thought** as a catalyst for discussion with a classmate, your professor, or teaching assistant, or just to ponder on your own.

Now get out your calculator and sharpen your pencils, you are ready to tackle the **Review Problems**. If you consistently find that your answers to the Short Answer section exceed the space that is allotted, perhaps you need to rethink your approach. Remember that a few carefully constructed, concise phrases are better than long-winded explanations in most cases. Some of the Multiple Choice questions are simple recall questions. Others will make you think. Lastly, the Problems and Essays will ask you to articulate clear, scientific explanations, interpret data, design experiments, or solve problems relating to each chapter. Do your calculations right in the book and don't cheat! Give yourself ample opportunity to find the answers on your own before looking in the **Appendix**.

Remember that your most important learning resource is not a study guide, a textbook, or even your professor. Your most important resource is YOU! Commit a certain number of hours every day to the study of cell biology; *don't* let yourself get behind. This study guide, the *Cell Biology* textbook, your professor, and all the resources at your university are there to help *you* to take charge of your education.

Acknowledgments

Many thanks to Olga Naidenko and Mark Baustian for their help in outlining the chapters. My friend Marilyn Graber did a superb job finding my errors. Any that remain are entirely my responsibility. Gerry Karp, whose earlier *Cell Biology* textbook sparked my interest as a student, provided a wonderful book from which to work and much encouragement while I prepared this study guide. Finally, Jonathan, Nathan, and Danny fended for themselves many evenings and weekends, allowing me the time to write this book. To them, I am most grateful of all.

TABLE OF CONTENTS

CHAPTER ONE

Introduction to the Study of Cell Biology

Learning Objectives

When you have finished this chapter, you should be able to:

1. Give a brief history of ideas about the cellular nature of life.
2. List the basic properties of cells.
3. Distinguish between prokaryotic and eukaryotic cells, and describe the characteristics of each, including:
   a. the major subkingdoms of prokaryotic cells.
   b. the process of differentiation and specialization in eukaryotes, including examples.
4. Appreciate the relative sizes of cells and their components, and the factors that limit the size of cells.
5. Understand the nature of viruses, their origins, and their mechanisms of infection.
6. Become familiar with other noncellular pathogens, including virions and prions.

Key Terms and Phrases

Robert Hooke	Anton van Leeuwenhoek	cell theory
George Gey	Schleiden, Schwann and Virchow	HeLa cells
ATP	metabolism	Hans Driesch
organelles	prokaryote	eukaryote
cell wall	cell surface receptors	nucleoid
nucleus	nuclear envelope	mesosomes
mitochondria	endoplasmic reticulum	Golgi complex
chloroplast	cytoskeleton	ribosome
bacterial conjugation	flagella	Archaebacteria
Eubacteria	methanogens	halophile
thermoacidophile	mycoplasma	cyanobacteria
nitrogen fixation	Dictyostelium	differentiation
micrometer	nanometer	virus
virion	Wendell Stanley	capsid
provirus	viroid	prion

Reviewing the Chapter

I. Introduction
   A. Cells are the topic of intense study.
   B. The study of cells requires creative instruments and techniques.
   C. The study of cells lies on a spectrum of human endeavor, from subatomic physics to astronomy.

II. The Discovery of Cells

A. The discovery of cells followed from the invention of the microscope by Robert Hooke, and its refinement by Anton van Leeuwenhoek.

B. Cell theory was articulated in the mid-1800s by Schleiden, Schwann and Virchow.
   1. All organisms are composed of one or more cells.
   2. The cell is the structural unit of life.
   3. Cells arise from preexisting cells by division.

III. Basic Properties of Cells.

A. Life is the most basic property of cells.

B. Cells can grow and reproduce in culture for extended periods.
   1. HeLa cells are cultiured tumor cells isolated from a cancer patient named Henrietta Lacks by George Gey in 1951.
   2. Cultured cells, including HeLa cells, are an essential tool of cell biologists.

C. Cells are highly complex and organized.
   1. Cellular processes are highly regulated.
   2. Cells from different sources share similar structure, composition and metabolic features.

D. Cells have a genetic program.
   1. Genes encode information to build each cell, and the organism.
   2. Genes encode information for cellular reproduction.

E. Cells reproduce, and each daughter cell receives a complete set of genetic instructions.

F. Cells acquire and utilize energy.
   1. Photosynthesis provides fuel for all living organisms.
   2. Animal cells derive energy from the products of photosynthesis, often in the form of glucose.
   3. Cells can convert glucose into ATP— a substance with readily available energy.

G. Cells engage in numerous mechanical activities, including locomotion.

H. Cells are able to respond to stimuli via surface receptors that sense changes in the chemical environment.

I. Cells are capable of self-regulation.

IV. Two Fundamentally Different Classes of Cells

A. Prokaryotes are all bacteria. They arose 3.5 billion years ago.

B. Eukaryotes are protists, fungi, plants and animals. The first eukaryotes arose 1.5 billion years ago.

C. Characteristics that distinguish prokaryotic and eukaryotic cells:
   1. Complexity: Prokaryotes are relatively simple; eukaryotes are more complex.
   2. Genetic material:
      a. Packaging: Prokaryotes have a nucleoid region whereas eukaryotes have a true, membrane-bound nucleus.
      b. Amount: Eukaryotes have several orders of magnitude more genetic material than prokaryotes.
      c. Form: Eukaryotes have many chromosomes that are made of both DNA and protein whereas prokaryotes have a single DNA chromosome.
   3. Cytoplasm: Eukaryotes have membrane-bound organelles and cytoskeletal proteins; prokaryotes have neither. Both have ribosomes, although they differ in size.
   4. Cellular reproduction: Eukaryotes divide by mitosis; prokaryotes divide by simple fission.
   5. Locomotion: Eukaryotes use both cytoplasmic movement, and cilia and flagella; prokaryotes have flagella, but they differ in both form and mechanism from eukaryotic flagella.

V. Types of Prokaryotic Cells: Two Subkingdoms

A. Subkingdom Archaebacteria:
1. Methanogens
2. Halophiles
3. Thermoacidophiles

B. Subkingdom Eubacteria:
1. Includes the smallest known cells— the mycoplasma.
2. Includes cyanobacteria— the photosynthetic bacteria that gave rise to green plants and an oxygen-rich atmosphere. Some cyanobacteria are nitrogen fixers, making nitrogen available for use by other organisms.

VI. Types of Eukaryotic Cells: Cell Specialization

A. Unicellular eukaryotes are complex single-celled organisms.

B. Multicellular eukaryotes have different cell types for different functions.
1. The slime mold Dictyostelium is an example of unspecialized cells becoming specialized via differentiation during the life cycle.
2. Differentiation occurs during embryonic development in other multicellular organisms.
3. Numbers and arrangements of organelles relate to the function of the cell.
4. Despite cellular differentiation, eukaryotic cells have many features in common.

VII. The Size of Cells and Their Components

A. Cells are commonly measured in units of micrometers (1 $\mu m = 10^{-6}$ meter) and nanometers (1 nm = 10^{-9} meter).

B. Cells are typically small; cell size is limited:
1. by the volume of cytoplasm that can be supported by the genes in the nucleus.
2. by the volume of cytoplasm that can be supported by exchange of nutrients across a limited surface area, i.e., the surface area/volume ratio.
3. by the distance over which substances can efficiently travel through the cytoplasm via diffusion.

VIII. Viruses

A. Viruses are pathogens first described in the late 1800s.

B. Viruses are obligatory intracellular parasites.

C. A virion is a virus particle outside the host cell.

D. Viral structure:
1. The genetic material can be single- or double-stranded DNA or RNA.
2. The protein capsid surrounds the genetic material in the virion.
3. The capsid may be polyhedral in some viruses.
4. A lipid envelope may surround the capsid in some viruses; the lipids may contain viral proteins.

E. Virus and host:
1. Viruses have surface proteins that bind to the surface of the host cell.
2. Viral specificity for a certain host is determined by the virus's surface proteins.

F. Viruses are not considered living.

G. Viral infection types:
1. Lytic infection— the virus redirects the host into making more virus particles, the host cell ruptures and releases the viruses.
2. Integrated infection— the virus integrates its DNA (called the provirus) into the host cell's chromosomes.
   a. The infected host may behave normally until an external stimulus activates the provirus, leading to lysis and release of viral progeny.
   b. The host may give rise to viral progeny by bidding, as in HIV.
   c. The host may become malignant.

H. Viral origins:
    1. Viruses had to arise after their hosts evolved— they cannot survive without their hosts.
    2. Viruses probably arose as fragments of host chromosomes that became somewhat autonomous.

I. Viruses are good models for studying mechanisms of genetic expression, and good vectors for introducing foreign genes into cells.

IX. Viroids

A. Viroids are pathogens, each consisting of a small, naked RNA molecule.

B. Viroids cause disease by interfering with gene expression in host cells.

X. Prions (Experimental Pathways)

A. New evidence links two degenerative brain diseases to single proteins called prions (proteinaceous infectious particles).

B. The mechanism of infection by a protein is not well understood, and somewhat controversial.

Key Figure

Figure 1.9. The structure of cells.

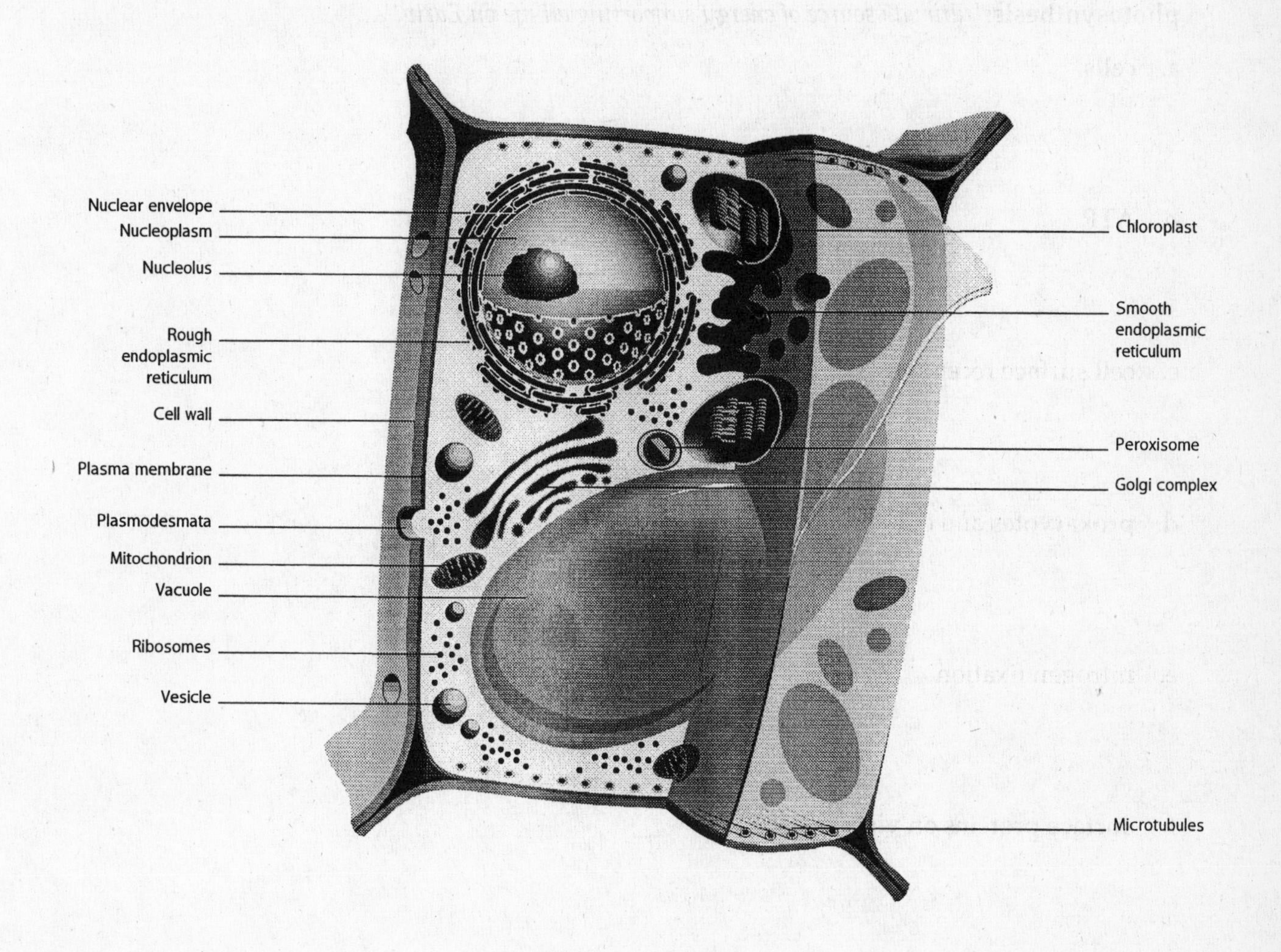

Questions for Thought:

1. What structures immediately identify this as a plant cell?

2. What structures shown here do plant cells, animal cells and prokaryotic cells have in common?

3. What is the largest intracellular structure in the plant cell? The animal cell? The bacterial cell?

4. This is a "generalized" cell, shown to give you an overview of what kinds of structures occur in this cell type. How might the numbers of the different organelles differ in cells with specific functions?

Review Problems

Short Answer

1. Give the significance (not the definition) of the following terms or phrases. Say what they do or why they are important. For example:
   photosynthesis: *ultimate source of energy supporting all life on Earth.*

   a. cells

   b. ATP

   c. cell surface receptors

   d. prokaryotes and eukaryotes share the same genetic "language"

   e. nitrogen fixation

   f. surface proteins on viruses

   g. genetic similarity between genes of viruses and their hosts

   h. life cycle of Dictyostelium

2. Compare and contrast the following:
   a. prokaryote vs. eukaryote

b. endoplasmic reticulum vs. mesosomes

c. virus vs. virion

d. lytic infection vs. integrative infection

Multiple Choice

1. All of the following individuals contributed to cell theory except:
   - a. Robert Hooke.
   - b. Matthias Schleiden.
   - c. Theodor Schwann.
   - d. Rudolf Virchow.

2. Of the following structures, which is the smallest?
   - a. viroid
   - b. hydrogen atom
   - c. bacterium
   - d. mitochondrion

3. Of the following, which is the most primitive?
   - a. virus
   - b. eukaryote
   - c. Archaebacteria
   - d. mitochondria

4. Cell theory includes all of the following except:
   - a. All organisms are composed of one or more cells.
   - b. The cell is the most primitive form of life.
   - c. The cell is the structural unit of life.
   - d. Cells arise by division of preexisting cells.

5. All of the following are basic properties of cells except:
   - a. Cells have nuclei and mitochondria.
   - b. Cells have a genetic program and the means to use it.
   - c. Cells are capable of producing more of themselves.
   - d. Cells are able to respond to stimuli.

6. The Archaebacteria include all of the following except:
   - a. methanogens.
   - b. halophiles.
   - c. thermoacidophiles.
   - d. cyanobacteria.

7. Evolutionary relationships between groups of organisms are determined using which of the following types of information?
   a. comparisons of nucleotide sequences
   b. comparisons of biochemical pathways
   c. comparisons of structural features
   d. all of the above

8. All of the following are features of prokaryotes except:
   a. nitrogen fixation.
   b. photosynthesis.
   c. sexual reproduction.
   d. locomotion.

9. Which of the following may account for the small size of cells?
   a. the rate of diffusion
   b. the surface area/volume ratio
   c. the number of mRNAs that can be produced by the nucleus
   d. all of the above

10. Which of the following statements is not true of viruses:
   a. Viruses have been successfully grown in pure cultures in test tubes.
   b. All viruses are obligatory intracellular parasites.
   c. All viruses have either DNA or RNA as their genetic material.
   d. Viruses probably arose from small fragments of cellular chromosomes.

11. If you were to study the sequences of nucleic acids in a variety of viruses and viral hosts, you would probably find more similarities:
   a. among different viruses than between viruses and their hosts.
   b. among different viral hosts than among different viruses.
   c. among different viral hosts than between viruses and their hosts.
   d. between viruses and their hosts than among different viruses.

Problems and Essays

1. Mitochondria and chloroplasts can be isolated from broken cells and, given the proper nutrients, can undergo respiration and photosynthesis in a test tube for several hours. Why, then, are these organelles not considered the basic unit of life?

2. Given that all the cells of the slime mold Dictyostelium have identical genetic information, why do some develop into stalk cells and others into spore cells?

3. Given that cell structure reflects cell function, what structural features would you predict in the following?
   a. cells that line the digestive tract and take up nutrients across the wall of the intestine into the blood

   b. cells that synthesize and secrete protein, such as an insulin-producing cell from the pancreas or an antibody-producing lymphocyte

   c. a leaf cell

   d. a contractile cell, such as a muscle cell

4. Of viruses that infect bacteria or viruses that infect humans, which type is older and why?

5. Consider a large, roughly cuboidal cell that measures 100 µm on a side.
   a. What is its surface area/volume ratio?

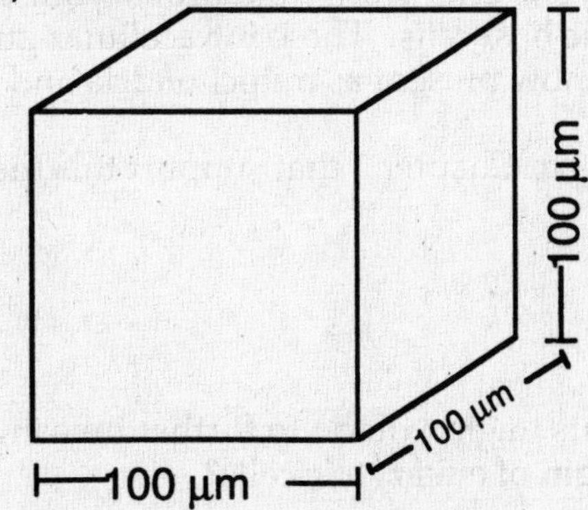

b. Assume that a cell requires a surface area/volume ratio of at least 3 to survive. Would dividing this cell into 125 cells which together had a volume of 1,000,000 μm3 ensure survival?

6. Given that some antibiotic drugs kill pathogens by interfering with replication of bacterial chromosomes, why can't we use these drugs to cure the common cold?

7. Why does the HIV virus lack genes for synthesizing its own lipid envelope?

8. What biochemical evidence could be used to show that modern green plants evolved from cyanobacterial ancestors and not from other photosynthetic prokaryotes?

9. Biologists believe that eukaryotic cells arose from prokaryotic ancestors that took up residence in other prokaryotic host cells. These intracellular guests eventually lost the ability to live on their own, and are now present as mitochondria and chloroplasts in modern eukaryotes.
   a. Find three facts presented in Chapter 1 that support this theory of eukaryotic origins.

   b. What kinds of experiments might you do to further investigate whether this explanation could account for the origin of eukaryotic cells?

CHAPTER TWO

The Chemical Basis of Life

Learning Objectives

When you have finished this chapter, you should be able to:

1. Explain why an understanding of chemistry is essential to the study of cell biology.
2. Describe the nature of covalent bonds, and why some bonds have unequal distributions of electrical charge.
3. Identify polar and nonpolar molecules, and understand the consequences of polarity in cellular molecules.
4. List the types of weak bonds and explain why they are important in cells.
5. Explain why water is ideally suited to life.
6. Define and understand the importance of pH in cells.
7. List the four cellular macromolecules and their building blocks. Describe how the building blocks are assembled into macromolecules.
8. Describe the levels of protein structure, and the factors that determine the structure of proteins in cells.
9. Give examples of some important cellular proteins.
10. Recognize some of the potential of biotechnology– and some of the dangers.

Key Terms and Phrases

molecule
electronegativity
ionization
free radical
hydrogen bond (H-bond)
van der Waals forces
carbonyl group
stereoisomer (enantiomer)
glycosidic bond
glycogen
amylopectin
amphoteric
K_W
inorganic molecule
saturated hydrocarbon
functional group
macromolecule
polysaccharide
monomer
cellulose
cellulase
compound
polar
anion
weak bond
hydrophilic
metabolic intermediate
ketose
asymmetric carbon
disaccharide
starch
dissociation
acid
buffer
biochemical
unsaturated hydrocarbon
ester bond
protein
sugar
nucleic acid
chitin
lipid
covalent bond
nonpolar
cation
ionic bond
hydrophobic
carbohydrate
aldose
pyranose
oligosaccharide
amylose
pH
base
organic molecule
hydrocarbon
structural isomer
amide bond
amino acid
polymer
nucleotide
glycosaminoglycan
fatty acid

fat	oil	triacylglycerol
amphipathic	adipocyte	steroid
atherosclerosis	phospholipid	diacylglycerol
polypeptide chain	peptide bond	C-terminus
N-terminus	R group (side chain)	disulfide bridge
conjugated protein	nucleoprotein	lipoprotein
glycoprotein	primary structure	secondary structure
tertiary structure	quarternary structure	sickle cell anemia
conformation	alpha helix (α helix)	beta sheet (ß sheet)
fibrous protein	globular protein	keratin
collagen	x-ray crystallography	myoglobin
prosthetic group	protein domain	protein motif
homodimer	heterodimer	subunit
antibody (immunoglobulin)	heavy chain	light chain
site-directed mutagenesis	deoxyribonucleic acid (DNA)	ribonucleic acid (RNA)
purine	pyrimidine	adenine
guanine	cytosine	uracil
thymine	adenosine triphosphate (ATP)	self-assembly
heat shock response	molecular chaperone	

Reviewing the Chapter

I. Introduction

A. Understanding cellular function requires knowledge of structure.
B. Structure and function in cells is closely related to the structure of molecules and atoms.
C. The study of chemistry is essential for understanding cell biology.

II. Covalent Bonds

A. Bonds between atoms with shared pairs of electrons are called covalent bonds.
   1. Molecules are stable combinations of atoms held together by covalent bonds.
   2. Compounds are molecules with more than one type of atom.
B. Atoms tend to fill their outer electron shell by sharing electrons with other atoms.
   1. Hydrogen forms a single covalent bond by sharing its one unpaired electron.
   2. Oxygen forms two covalent bonds by sharing two unpaired electrons.
   3. Water results when oxygen bonds with two hydrogens.
C. It requires 80 to 100 kilocalories to break a mole of covalent bonds.
D. Multiple bonds are formed when more than one pair of electrons are shared by two atoms.
E. Shared electrons stay closest to the nucleus with the highest electronegativity.
   1. The electronegativity depends upon the number of protons in the nucleus.
   2. The electronegativity also depends upon the distance of the shared electrons from the nucleus.

III. Polar and Nonpolar Molecules

A. Polar molecules have asymmetric distributions of electrical charge.
B. Nonpolar molecules lack polarized bonds.
C. Some biological molecules have both polar and nonpolar regions, e.g., some proteins and phospholipids, and are called "amphipathic" molecules.

IV. Ionization

A. Ions result when strongly electronegative nuclei capture electrons.
B. Anions have extra electrons; cations have lost electrons.

C. Free radicals are unstable atoms or molecules with unpaired electrons.
   1. Free radicals, and the inability to neutralize them, have been implicated in disease.
   2. Aging has been tentatively attributed to cumulative cellular damage by free radicals.

V. Noncovalent Bonds

A. Noncovalent bonds are weaker than covalent bonds (about 1 to 5 kcal/mole).
B. Weak bonds are readily broken and re-formed.
C. Weak bonds play an important role in dynamic cellular processes.
D. Types of noncovalent bonds
   1. Ionic bonds: Attraction between charged atoms:
      a. Ionic bonds are weakened in the presence of water.
      b. They may be quite important within large biological molecules.
   2. Hydrogen bonds:
      a. Hydrogen bonds occur when covalently bound hydrogen has a partial positive charge and attracts electrons of a second atom.
      b. H-bonds determine the structure and properties of water.
      c. H-bonds occur in biological molecules, notably between the strands of DNA.
   3. Hydrophobic interactions and van der Waals forces:
      a. These occur when nonpolar molecules associate and minimize their exposure to polar molecules.
      b. Van der Waals forces, or attractions between nonpolar molecules, are due to transient dipole formation.

VI. The Life Supporting Properties of Water

A. The structure of water is suitable for sustaining life.
   1. The molecule is asymmetric– both H atoms are on one side.
   2. Both covalent O–H bonds are highly polarized.
   3. All three atoms readily form hydrogen bonds.
B. The properties of water result from its structure.
   1. Water requires a lot of heat energy to raise its temperature.
   2. Water requires a lot of heat to evaporate it.
   3. Water is an excellent solvent for many substances.
   4. Water determines the interactions between many biological solutes.

VII. Acids, Bases and Buffers

A. Acids release protons; bases accept protons.
B. Amphoteric molecules can act as either acids or bases; water is an example of an amphoteric molecule.
C. Acidity is measured using the pH scale.
   1. $pH = -\log[H^+]$
   2. The ion product constant of water, $K_W = [H^+][OH^-] = 10^{-14}$ at 25 °C.
   3. As $[H^+]$ increases, then $[OH^-]$ decreases such that the product of the two always equals 10^{-14}.
D. Biological processes are sensitive to pH.
   1. Changes in pH affect the ion state and function of proteins.
   2. Buffers in living systems, such as bicarbonate, resist changes in pH.

VIII. The Nature of Biological Molecules

A. Carbon is central to the chemistry of life.
   1. Carbon forms four covalent bonds, with itself or other atoms.
   2. Several hundred thousand carbon-containing (organic) molecules have been identified.

B. Hydrocarbons:
  1. Hydrocarbons contain only carbon and hydrogen.
  2. Hydrocarbons vary in the number of carbons, and the number of double and triple bonds between carbons.
  3. Structural isomers have the same chemical formulae, but the atoms are arranged differently.

C. Functional groups are groups of atoms that behave as units.

D. Functional classification of biological molecules:
  1. Macromolecules:
     a. Macromolecules are large structural and functional molecules in cells.
     b. Macromolecules include proteins, nucleic acids, polysaccharides and lipids.
  2. Building blocks of macromolecules include amino acids for proteins, nucleotides for nucleic acids, sugars for polysaccharides, and fatty acids for lipids.
  3. Metabolic intermediates are compounds formed in metabolic pathways.
  4. Miscellaneous others include vitamins, hormones, ATP, and metabolic waste products.

IX. Four Families of Biological Molecules

A. Carbohydrates include simple sugars and sugar polymers.
  1. Carbohydrates serve as energy storage molecules.
  2. Structure:
     a. Chemical formula is $(CH_2O)_n$.
     b. Ketose sugars have a carbonyl on an internal carbon.
     c. Aldose sugars have a carbonyl on a terminal carbon.
     d. Sugars can be linear, but usually form ring structures called pyranoses.
  3. Stereoisomerism:
     a. Asymmetric carbons bond to four different groups.
     b. Molecules with asymmetric carbons can exist in two mirror-image configurations called enantiomers or stereoisomers.
     c. Enantiomers are designated either D- or L-isomers, depending upon whether the more electronegative group extends to the right or left of the asymmetric carbon.
     d. Sugars can have many asymmetric carbons, but are designated D- or L- according to the arrangement around the carbon farthest from the carbonyl group.
  4. Linking sugars together:
     a. Glycosidic bonds are C–O–C links between sugars.
     b. Disaccharides are used as a source of readily available energy.
     c. Oligosaccharides are found bound to cell surface proteins and lipids, and may be used for cell recognition.
  5. Polysaccharide are polymers of sugars joined by glycosidic bonds.
     a. Glycogen and starch: nutritional polysaccharides.
        (1) Glycogen is an animal product made of branched glucose polymers.
        (2) Starch is a plant product made of both branched and unbranched glucose polymers.
     b. Cellulose, chitin, and glycosaminoglycans: structural polysaccharides.
        (1) Cellulose is a plant product made of unbranched glucose polymers that most animals cannot digest.
        (2) Chitin is a component of invertebrate exoskeleton made of unbranched N-acetylglucosamine polymers.

B. Lipids are a diverse group of nonpolar molecules.
  1. Fats are made of glycerol molecules esterified to fatty acids.
     a. Fatty acids are unbranched hydrocarbons with one terminal carboxyl group; they are amphipathic.
     b. Saturated fatty acids lack C=C double bonds and are solids at room temperature.
     c. Unsaturated fatty acids have one or more C=C double bonds and are liquid at room temperature.
     d. In animals, fats are stored for energy in adipocytes.
  2. Steroids are four-ringed animal lipids that have been implicated in atherosclerosis.
  3. Phospholipids are amphipathic lipids that are a major component of cell membranes.

C. Proteins are polymers of amino acids and form a diverse group of macromolecules with a variety of cellular functions.
  1. Amino acids all have an alpha carbon, an amine group, a carboxyl group, and a variable R group.
  2. Amino acids in nature occur as the L stereoisomer.
  3. Amino acids are linked together by peptide bonds into polypeptide chains to make proteins.
  4. Peptide bonds form between the α-carboxyl and the α-amino of participating amino acids.
  5. Amino acids differ in the R group attached to one of the bonds of the α-carbon.
  6. Properties of R Groups:
     a. R groups may be polar and charged.
     b. R groups may be polar and uncharged.
     c. R groups may be nonpolar.
     d. R groups may have other properties.
        (1) Glycine has only -H as its R group and is small.
        (2) Proline's α-carbon is part of a ring, creating kinks in the protein.
        (3) Cysteine forms disulfide bonds with other cysteines.
     e. The nature of the R groups determines the function of the protein.
  7. Conjugated proteins have nonprotein constituents.
  8. Protein structure:
     a. Primary structure, the sequence of amino acids in the polymer, is critical to protein function.
     b. Secondary structure refers to the conformation of adjacent amino acids into α-helix, ß-sheet, kinks. loops or turns.
     c. Tertiary structure is the conformation of the entire polymer, and describes the manner in which the helical and nonhelical portions fold back on themselves.
     d. Quarternary structure applies to proteins with subunits, and refers to the manner in which subunits interact.
  9. Examples of protein structure
     a. Collagen: An Abundant Fibrous Protein:
        (1) Collagen is abundant in connective tissue.
        (2) Collagen strands are three helices intertwined.
        (3) Collagen has glycine as every third amino acid, and hydroxylated prolines and lysines.
     b. Myoglobin: The first globular protein whose tertiary structure was determined:
        (1) Myoglobin stores oxygen in muscle cells.
        (2) Myoglobin has a heme prosthetic groups that binds oxygen.

(3) Myoglobin structure was derived using x-ray crystallography, and found to be predominantly α-helix.

10. Protein domains and motifs:
    a. Domains occur when large proteins are composed of more than one distinct functional region.
    b. Motifs are recurring substructures in different proteins that may be related to similar evolutionary histories, or similar functions.
11. Dynamic changes within a protein may occur with protein activity.
12. Multiprotein complexes occur when groups of proteins associate to perform a specific function.
13. Antibody structure and function:
    a. Antibodies, or immunoglobulins, are proteins that recognize and bind to foreign materials called antigens.
    b. Antibodies can be highly specific for the antigens to which they bind.
    c. Each antibody is a tetramer with four protein chains: two light chains and two heavy chains.
    d. There are five classes of antibodies that differ in the amino acid sequence of their heavy chains.
    e. There are two classes of light chains, kappa and lambda.
    f. The model system for studying antibodies is malignant lymphocytes that produce copious amounts of one antibody.
    g. Both heavy chains and light chains have a constant region and a variable region.
    h. Hypervariable regions of both heavy and light chains play a role in antigen binding.
    i. Different combinations of light chains and heavy chains are partially responsible for the variety of antibodies an individual can produce.
    j. Constant regions have effector functions, such as activation of the complement system, cell anchoring, and stimulation of lymphocyte proliferation.

D. Protein engineering:
1. Current technology enables us to make artificial genes that code for proteins of specified amino acid sequences.
    a. Peptides with simple secondary and tertiary structures have been produced.
    b. More complex peptides have proven more difficult to produce.
2. Knowledge of a protein's amino acid sequence rarely permits us to predict a protein's higher-order structure.
3. Site-directed mutagenesis enables researchers to make alterations in single amino acids of a protein by altering the DNA that encodes it.
    a. This technique is a powerful tool in research.
    b. The role of single amino acids in protein structure and function can be determined.
4. Tissue plasminogen activator (tPA) is a natural protein that is used therapeutically to dissolve blood clots in coronary patients.
    a. Synthetic tPAs, better "clot-busters," have been produced using biotechnology.
    b. Dangers may be associated with the use of synthetic tPAs.

E. Nucleic acids are polymers of nucleotides that store and transmit genetic information.
1. DNA holds the genetic information in all cellular organisms and some viruses.
2. RNA holds the genetic information in other viruses.
3. Nucleotides consist of a five-carbon sugar (ribose in RNA), a phosphate group and a nitrogenous base.

4. Nucleotides are connected via 3′-5′ phosphodiester bonds between the phosphate of one nucleotide and the 3′ carbon of the next.
5. The nitrogenous bases are either purines or pyrimidines.
    a. The purines are adenine and guanine in both DNA and RNA.
    b. The pyrimidines are cytosine and uracil in RNA and cytosine and thymine in DNA.
6. RNA is single stranded and DNA is double stranded.
    a. RNA may fold back on itself to form complex three dimensional structures, as in ribosomes.
    b. RNA may have catalytic properties; such RNAs are called ribozymes.
7. ATP is a ribonucleotide with three phosphates, and plays an important role in cellular metabolism.

F. An RNA world:
1. RNA was formerly thought to be only a messenger molecule.
2. Recent discoveries of ribozymes have lead researchers to believe that RNA may have been the precursor to both DNA and enzymes in early evolution.

X. The Formation of Complex Molecular Structures

A. Complex proteins and other molecular aggregates may self-assemble spontaneously.

B. The assembly of tobacco mosaic virus particles and ribosomal subunits:
1. Experiments in the 1950s showed that TMV particles are capable of self-assembly.
2. Prokaryotic ribosomal subunits were shown in the 1960s and 1970s to assemble spontaneously in a step-by-step manner in vitro.
3. The components of the mature eukaryotic ribosomal subunits require accessory proteins to assemble.

C. Cells may use accessory proteins, called chaperones, to affect assembly of molecular structures.

XI. Building Protein Structure (Experimental Pathways)

A. Some proteins, for example ribonuclease, reassemble into the active conformation after denaturation.

B. Other proteins require molecular chaperones for proper folding.
1. Molecular chaperones may also act to protect protein structure during increases in temperature.
2. The heat shock response involves synthesis of new proteins called heat shock proteins, that prevent denaturation of existing proteins.
3. Heat shock proteins and molecular chaperones function by preventing aggregation of denatured or newly synthesized proteins.
4. The protein GroEL is synthesized in E. coli and is essential for the proper folding of other cellular proteins.

Key Figure

Figure 2.49. Nucleotide Strands.

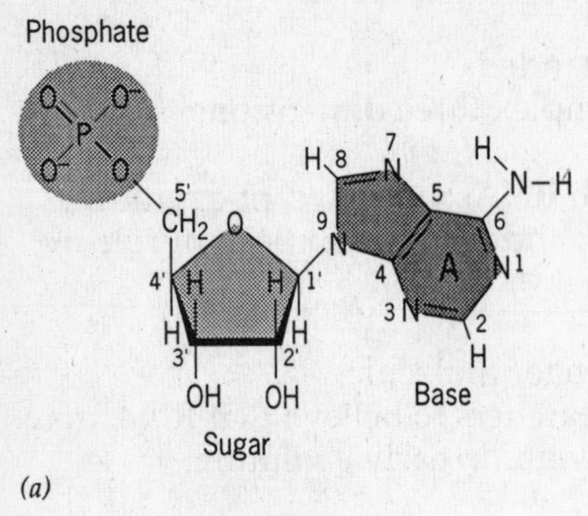

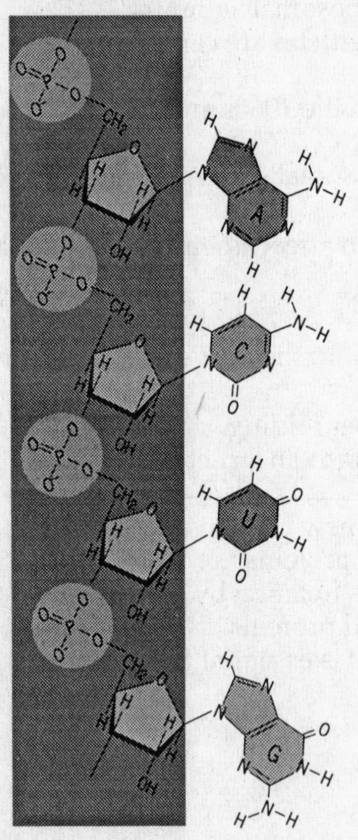

Questions for Thought:

1. When the structure of DNA was first described by Watson and Crick, they offhandedly stated, "It has not escaped our notice that the specific pairing we have postulated immediately suggests a possible copying mechanism for the genetic material" (Nature 171: 737-738. 1953). What mechanism did they have in mind?

2. Are the sugars and phosphates of the individual nucleotides of RNA interchangeable? Are they interchangeable between DNA and RNA? Why or why not?

3. It has been said that the nucleic acids of a single human cell contain as much information as several million encyclopedias, whereas the information content of glycogen or cellulose is equivalent to that of wallpaper. Explain.

4. The human genome contains over 3 billion nucleic acid base-pairs. Try to imagine some ways in which so much nucleic acid might be condensed into the tiny nucleus of each human cell. Now see how cells package DNA by looking at Chapter 12 of your textbook.

Review Problems

Short Answer

1. Give the significance (not the definition) of the following terms or phrases. Say what they do or why they are important. For example:
   water: *the fluid matrix of the cell; water is a solvent for the polar molecules of the cell, and determines the structure of biological molecules and their interactions.*

   a. electronegativity

   b. free radicals

   c. carbon

   d. functional groups

   e. ß(1 → 4) glycosidic linkages in cellulose

   f. amphipathic nature of phospholipids

   g. protein primary structure

   h. site-directed mutagenesis

   i. Anfinsen's experiments with ribonuclease

   j. molecular chaperones

2. Compare and contrast the following:
   a. K_{eq} for the dissociation of water vs. K_w

   b. glycogen vs. starch

   c. saturated vs. unsaturated hydrocarbons

   d. fat vs. oil

   e. purines vs. pyrimidines

Multiple Choice

1. The element neon (Ne) has eight electrons in its outermost electron shell. How many covalent bonds will Ne readily form?
   a. none
   b. one
   c. two
   d. four

2. Of the following elements, which is likely to form the least polar covalent bonds with hydrogen:
   a. nitrogen
   b. oxygen
   c. carbon
   d. phosphorus

3. Of the following chemical bonds, which is the strongest in water?
   a. $HOH_2C–CH_2OH$
   b. Na–Cl
   c. $CH_3COO–H$
   d. HO–H

4. Which of the following characteristics does not apply to water?
   a. The water molecule is asymmetric.
   b. The water molecule readily forms hydrophobic interactions.
   c. The covalent bonds in water are highly polarized.
   d. All three atoms in the water molecule readily form hydrogen bonds.

5. The major cellular macromolecules include:
   a. proteins, lipids, amino acids, and carbohydrates.
   b. proteins, amino acids, lipids, and nucleic acids.
   c. proteins, nucleic acids, carbohydrates, and antibodies.
   d. proteins, nucleic acids, carbohydrates, and lipids.

6. If you were to take a triacylglycerol molecule and chemically break the ester bonds, you would end up with two different kinds of molecules. What is the name of the kind of molecule that would be soluble in water?
   a. fatty acid
   b. steroid
   c. glycogen
   d. glycerol

7. Which of the following has been used as evidence that primitive life forms lacked both DNA and enzymes?
   a. RNA can both code genetic information and act as a catalyst.
   b. DNA and enzymes are only present in the most advanced cells.
   c. Advanced cells lack RNA.
   d. All of the above.

8. How many different isomers of a five-carbon hydrocarbon can be formed?
   a. one
   b. two
   c. three
   d. four

9. Cytoplasm has a pH of about 7. What is the concentration, in moles/liter, of the hydrogen ion?
   a. 7
   b. 7×10^{-7}
   c. 1×10^{7}
   d. 1×10^{-7}

10. Which of the following acid/base pairs act as natural buffers in living systems?
   a. H_2CO_3/HCO_3^-
   b. $H_2PO_4^-/HPO_4^{2-}$
   c. histidine$^+$/histidine
   d. all of the above

11. What types of isomers are these?

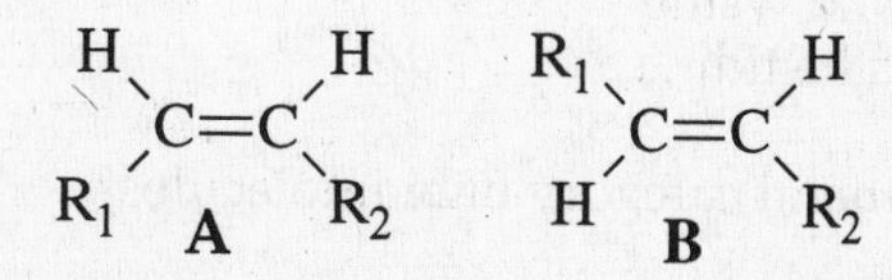

   a. A is a cis isomer and B is a trans isomer.
   b. A is a trans isomer and B is a cis isomer.
   c. A is a D isomer and B is an L isomer.
   d. A is an L isomer and B is a D isomer.

12. Which of the following pairs of functional groups might form weak bonds with one another?
    a. methyl and phosphate
    b. phosphate and carboxyl
    c. carboxyl and amino
    d. carbonyl and methyl

13. If you could grab the ends of each of the following polysaccharides, which one might you be able to stretch the most?
    a. glycogen
    b. starch
    c. cellulose
    d. none of the above

14. What two features of the structure of cholesterol make it somewhat amphipathic?
    a. The single hydroxyl along with the hydrocarbon nature of the remainder of the molecule
    b. The long, branched hydrocarbon along with the four hydrocarbon rings
    c. The five methyl groups along with the hydrocarbon chain
    d. The three six-membered rings along with the single five-membered ring

15. Proteins and macromolecular structures take on their higher order structures
    a. by self-assembly.
    b. with the help of molecular chaperones.
    c. with the help of precursor sequences that are removed from the final structure.
    d. all of the above.

16. The different orders of protein structure are determined by all of the following bond types except:
    a. peptide bonds.
    b. phosphodiester bonds.
    c. disulfide bridges.
    d. hydrogen bonds.

17. Among different antibody molecules, the most variable regions are
    a. the entire kappa chain.
    b. the entire light chain.
    c. the entire heavy chain.
    d. the antigen binding sites.

18. Which amino acids would most likely reside in the membrane-anchoring domain of a membrane-embedded protein?
    a. isoleucine, valine and phenylalanine
    b. phenylalanine, valine, and aspartate
    c. leucine, threonine and lysine
    d. lysine, arginine and histidine

19. ATP is from which general category of biomolecules?
    a. polysaccharides
    b. proteins
    c. nucleotides
    d. amino acids

20. Which of these structures would be least likely to occur in cells?

a. COO^- — C — CH_3, H, NH_3^+

b. COO^- — C — CH_3, NH_3^+, H

c. COO^- — C — OH, NH_3^+, H

d. COO^- — C — CH_2CH_3, NH_3^+, H

Problems and Essays

1. The energy of a mole of photons of light in the electromagnetic spectrum decreases logarithmically as a function of increasing wavelength. In the visible range, light has between 40 and 80 kcal/mole of photons. Below about 400 nm wavelengths, the energy goes up drastically. Using what you know about the energy of chemical bonds, make an argument for the use of sun screen with UV blocker on a sunny day.

2. Water has been described as an excellent thermal buffer for cells. What does that mean, and what accounts for this property of water?

3. Indicate whether these characteristics describe a protein motif or a protein domain, and why:
   a. Proteins that bind calcium often have a helix-loop-helix structure with hydrophilic amino acids at critical positions.

   b. Several proteins, including tissue plasminogen activator (tPA) and the digestive enzyme chymotrypsin, have a 200 to 300 amino acid region that catalyzes the breakdown of protein. This region is attached to the rest of the protein by an amino acid hinge.

   c. Proteins that reside in membranes have long sequences of hydrophobic amino acids that anchor the proteins in the lipid membrane.

   d. Proteins that bind DNA are characterized by a zinc finger: a combination of one α helix and two ß strands stabilized with a zinc atom. This structure fits into the groove of DNA.

4. For the following questions, refer to these chemical structures:

A.

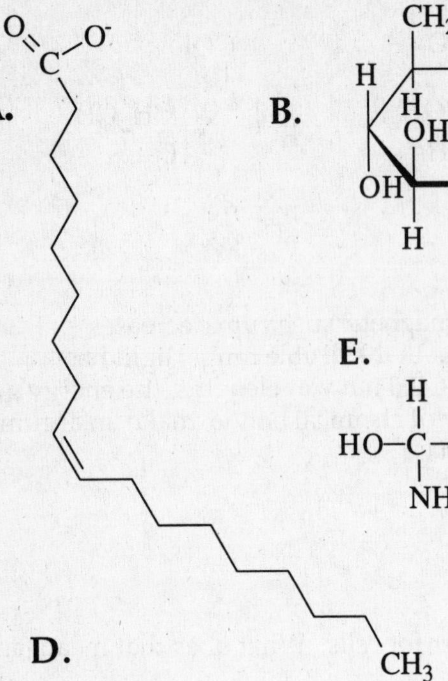

B.

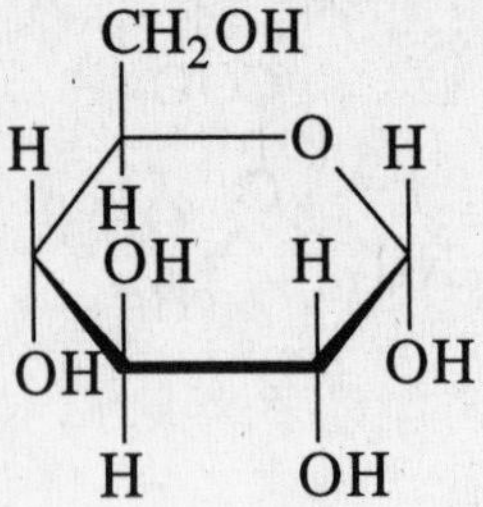

C.

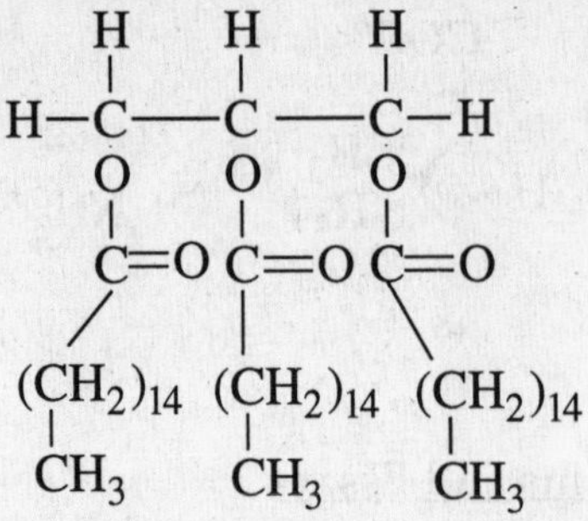

D.

E.

F.

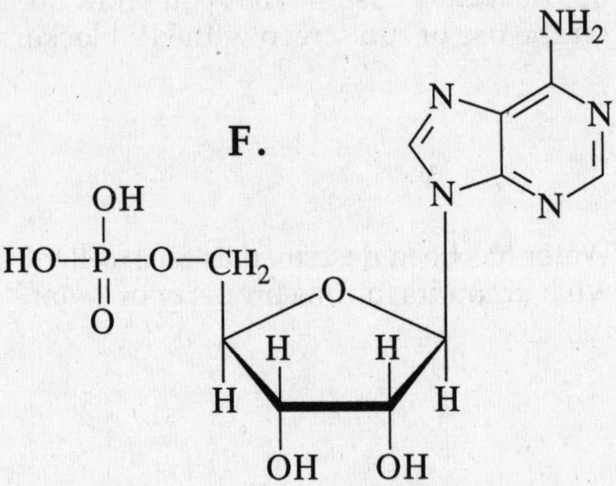

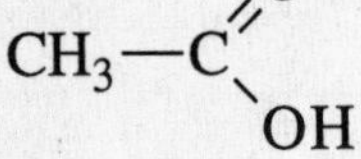

a. Which molecule is amphoteric?

b. Which molecules are hydrophobic?

c. Which molecule is amphipathic?

d. Which molecule is a building block of carbohydrates?

e. Which molecules have asymmetric carbons?

f. Which molecules have free rotation around all their carbon-to-carbon bonds?

g. Name as many of these molecules as you can, and tell what general category of compounds each one belongs to.

5. Draw the structure of a tripeptide that would partition itself in the core of the cell membrane.

6. When nucleic acids pair in forming the double helix of DNA, the guanine-cytosine pairing results from the formation of three hydrogen bonds between the purine and the pyrimidine, and the adenine-thymine pairing results from the formation of two hydrogen bonds.
   a. Estimate the bond energy present in a mole of guanine-cytosine pairs; a mole of adenine-thymine pairs.

   b. Which would be more stable: a strand of DNA rich in G–C pairs, or one rich in A–T pairs?

7. Imagine an amphipathic protein such as the one illustrated in Figure 2.33(c) of your textbook. If one surface of the helix is hydrophilic, and the other is hydrophobic, and there are 3.6 amino acids per each complete turn of the helix, what can you conclude about the occurrence of nonpolar amino acids in the primary sequence? Write a sequence of ten amino acids that might have the secondary structure of the amphipathic protein in Figure 2.33(c).

Notes:

CHAPTER THREE

Energy, Enzymes, and Metabolism

Learning Objectives

When you have finished this chapter, you should be able to:

1. Define energy and distinguish between different forms of energy.
2. Define thermodynamics, and understand the implications of the first and second laws.
3. Understand the concepts of free energy and changes in free energy.
   a. Know the relationship between free energy and thermodynamics, and between free energy and equilibrium.
   b. Compare free energy in different chemical reactions and understand standard free energy, $\Delta G°'$.
   c. Know how free energy determines which direction reactions will proceed in cells.
4. Know the properties of enzymes, including how they affect the rates of reactions in cells, and what factors influence those rates.
   a. Describe the mechanisms of enzyme catalysis.
   b. Describe features of the active site.
   c. Define the kinetic constants, K_M and V_{max}, how they are measured, and what insights they can provide about enzymatically catalyzed reactions.
   d. Give examples of how enzymes are regulated in cells.
5. Define metabolism, and distinguish between anabolism and catabolism.
6. Recognize redox reactions in cellular metabolism, and the common coenzymes that are associated with oxidation and reduction.
7. Give an overview of the anaerobic pathways for metabolizing glucose. Distinguish between glucose catabolism and glucose anabolism.

Key Terms and Phrases

energy	thermodynamics	enzyme
potential energy	first law of thermodynamics	catalyst
kinetic energy	second law of thermodynamics	ribozyme
total energy (E and ΔE)	spontaneous reaction	cofactor
free energy (ΔG)	entropy (S and ΔS)	coenzyme
standard free energy ($\Delta G°'$)	enthalpy (H and ΔH)	kinetic stability
endergonic	$T\Delta S$	substrate
exergonic	equilibrium	activation energy (E_a)
equilibrium constant (K_{eq})	steady state	transition state
coupled reactions	thermodynamic stability	ES complex
substrate orientation	induced fit	enzyme kinetics
initial velocity	maximal velocity (V_{max})	Michaelis constant (K_M)
turnover number	irreversible inhibitor	competitive inhibitor
noncompetitive inhibitor	metabolism	metabolic pathway

metabolic intermediate
anabolism
$NADP^+/NADPH$
reduced
dehydrogenase
fermentation
phosphorylase
feedback inhibition
end product
ATP
redox reactions
reducing agent
substrate level phosphorylation
reducing power
protein kinase
glycolysis
catabolism
$NAD^+/NADH$
oxidized
oxidizing agent
transfer potential
transhydrogenase
allosteric modulation
gluconeogenesis

Reviewing the Chapter

I. Introduction

A. Function is closely related to structure, in enzymes and at other levels of organization.
B. Enzymes speed up biological reactions.
C. Thermodynamic analysis reveals which processes occur spontaneously.
D. Chemical reactions in cells are linked sequentially to form metabolic pathways.

II. Energy

A. Cells require energy to survive.
B. Energy is the capacity to do work.
    1. Potential energy is the potential to do work.
    2. Kinetic energy is in the process of doing work.
C. Thermodynamics is the study of changes in energy.
    1. The first law of thermodynamics:
        a. Energy cannot be created or destroyed.
        b. Energy can be transduced or converted from one form to another.
            (1) Cells are capable of energy transformation.
            (2) Photosynthesis is the transduction of light energy into chemical energy.
            (3) All cells rely, directly or indirectly, on photosynthesis for energy.
        c. The universe can be divided into system and surroundings.
            (1) The system is a subset of the universe under study.
            (2) The surroundings are everything in the universe that is not part of the system.
        d. Systems can be closed or open.
            (1) Closed systems exchange energy with the surroundings.
            (2) Open systems exchange matter and energy with the surroundings.
        e. When there is an energy transduction (ΔE) in a system, heat content may increase or decrease, and work may be done.
        f. Total energy changes (ΔE) are independent of pathway.
        g. The first law does not predict whether an energy change will be positive or negative.
    2. The second law of thermodynamics:
        a. In any energy transformation, the entropy of the universe is increased.
            (1) Entropy is a measure of randomness or disorder.
            (2) Entropy is energy not available to do additional work.
        b. Living systems maintain a state of order, or low entropy.
            (1) Order in living systems occurs at the expense of order in the environment.
            (2) Order in living systems requires a constant input of energy.

3. Free energy:
   a. The first and second laws can be combined and expressed mathematically.
      (1) In a change of state, total energy (also called enthalpy, or ΔH) has an entropy component (ΔS) and a free energy component (ΔG).
      (2) This can be expressed as $\Delta H = \Delta G + T\Delta S$.
      (3) Free energy, ΔG, represents energy available to do work.
      (4) ΔG is a measure of spontaneity of a reaction.
         (a) $\Delta G < 0$, transformation is spontaneous (exergonic).
         (b) $\Delta G > 0$, transformation requires energy input (endergonic).
      (5) The spontaneity of a reaction depends on both the enthalpy, ΔH, and the entropy, ΔS.
   b. Free energy changes apply to chemical reactions occurring in cells.
      (1) All chemical reactions are theoretically reversible.
      (2) All chemical reactions proceed toward equilibrium.
      (3) The rates of chemical reactions are proportional to the concentration's of reactants.
         (a) For $A + B \rightleftharpoons C + D$, forward rate = $k_1[A][B]$
         (b) Backward rate = $k_2[C][D]$.
         (c) At equilibrium, forward rate = backward rate, so $k_1[A][B]=k_2[C][D]$.
         (d) Rearranged, $k_1/k_2 = [C][D]/[A][B] = K_{eq}$.
         (e) K_{eq} is called the equilibrium constant.
      (4) At equilibrium, the free energies of the products and reactants are equal, or $\Delta G=0$.
   c. Free energy changes of reactions are compared under standard conditions.
      (1) Standard free energy changes, $\Delta G^{\circ\prime}$, are described for each reaction under specific, standard, conditions.
      (2) Standard conditions do not typify cellular conditions, but are useful for making comparisons.
      (3) Standard free energy changes are related to equilibrium: $\Delta G^{\circ\prime} = -RT \ln K'_{eq}$.
   d. Free energy changes of reactions under non-standard conditions are calculated by correcting $\Delta G^{\circ\prime}$ for prevailing conditions: $\Delta G = \Delta G^{\circ\prime} + RT \ln [C][D]/[A][B]$.
      (1) Prevailing conditions may cause ΔG to be negative, even when $\Delta G^{\circ\prime}$ is positive.
      (2) Making ΔG negative may involve coupling endergonic and exergonic reactions in a sequence.
      (3) Making ΔG negative may involve coupling endergonic and exergonic reactions simultaneously.
         (a) Simultaneously coupled reactions have a common intermediate.
         (b) ATP hydrolysis is often coupled to endergonic reactions in cells.
4. Equilibrium versus steady-state metabolism.
   a. Cellular metabolism is nonequilibrium metabolism.
   b. Cells are open thermodynamic systems.
   c. Cellular metabolism exists in a steady state.
      (1) Concentrations of reactants and products remain constant, but not at equilibrium.

(2) New substrates enter; products are removed.
(3) Maintaining a steady state requires a constant input of energy, whereas maintaining equilibrium does not.

III. Enzymes as Biological Catalysts

A. Enzymes speed up chemical reactions.
1. Enzymes can function when removed from cells, as discovered in 1897 by Hans and Eduard Büchner.
2. Enzymes are almost always proteins, as discovered in 1926 by James Sumner.
3. Enzymes may be conjugated with nonprotein components.
a. Cofactors are inorganic enzyme conjugates.
b. Coenzymes are organic enzyme conjugates.

B. Properties of Enzymes:
1. Enzymes are present in cells in small amounts.
2. Enzymes are not permanently altered during the course of a reaction.
3. Enzymes cannot affect thermodynamics of reactions, only rates.
4. Enzymes may increase the rate of a reaction one trillion (10^{12}) fold.
5. Enzymes are highly specific for their particular reactants, called substrates.
6. Enzymes produce only appropriate metabolic products.
7. Enzymes can be regulated to meet the needs of a cell.

C. Overcoming the activation energy barrier:
1. A small energy input, the activation energy, is required to initiate even favorable (exergonic) reactions.
a. The activation energy barrier slows the progress of thermodynamically unstable reactants.
b. Thermodynamically unstable compounds can be kinetically stable.
c. Reactant molecules that reach the peak of the activation energy barrier are in the transition state.
2. Catalysts such as enzymes lower the activation energy.
a. Without a catalyst, only a few substrate molecules reach the transition state.
b. With a catalyst, a large proportion of substrate molecules can reach the transition state.

D. The active site and molecular specificity:
1. An enzyme interacts with its substrate to form enzyme-substrate (ES) complex.
2. The substrate binds to a portion of the enzyme called the active site.
3. The active site and the substrate have complementary shapes accounting for substrate specificity.

E. Mechanisms of enzyme catalysis:
1. Substrate orientation means enzymes hold substrates in the optimal position for the reaction.
2. Changes in the reactivity of the substrate temporarily stabilize the transition state.
a. Acidic or basic R groups on the enzyme may change the charge of the substrate.
b. Charged R groups on the enzyme may attack, bend, or link with the substrate.
3. Inducing strain in the substrate
a. Shifts in conformation after binding cause an induced fit between enzyme and substrate.
b. Covalent bonds of the substrate are strained.
c. New techniques reveal great enzyme flexibility.

F. Enzyme kinetics:
   1. Kinetics is study of rates of enzymatic reactions under various experimental conditions.
   2. The rates of enzymatic reactions increase with increasing substrate concentrations until the enzyme is saturated.
      a. At saturation every enzyme molecule is working at maximum capacity.
      b. The velocity at saturation is called maximal velocity, V_{max}.
      c. The turnover number is the number of substrate molecules converted to product per minute per enzyme molecule at V_{max}.
   3. The Michaelis constant (K_M) is the substrate concentration at one-half of V_{max}.
      a. The units of K_M are concentration units.
      b. The K_M may reflect the affinity of the enzyme for the substrate.
   4. Temperature and pH can greatly influence enzymatic reaction rates.
   5. Plots of the inverses of velocity versus substrate concentrations, such as the Lineweaver-Burk plot, facilitate estimating K_M and V_{max}.

G. Enzyme inhibitors slow the rates of enzymatic reactions
   1. Irreversible inhibitors bind tightly to the enzyme.
   2. Reversible inhibitors bind loosely to the enzyme.
      a. Competitive inhibitors:
         (1) compete with the enzyme for active sites.
         (2) usually resemble the substrate in structure.
         (3) can be overcome with high substrate/inhibitor ratios.
      b. Noncompetitive inhibitors:
         (1) bind to sites other than active sites.
         (2) usually do not resemble the substrate in structure.
         (3) cannot be overcome with high substrate/inhibitor ratios.

IV. Metabolism

A. Metabolism is the collection of biochemical reactions that occur within a cell.

B. Metabolic pathways are subsets of metabolism.
   1. A metabolic pathway is a sequence of chemical reactions.
   2. Each reaction in the sequence is catalyzed by a specific enzyme.
   3. Pathways are usually confined to specific cellular locations.
   4. Enzymes of a given pathway may be physically linked to one another.
   5. Pathways convert substrates into end products via a series of metabolic intermediates.

C. An overview of metabolism:
   1. Catabolic pathways break down complex substrates into simple end products.
      a. They provide raw materials for the cell.
      b. They provide chemical energy for the cell.
   2. Anabolic pathways synthesize complex end products from simple substrates.
      a. They require energy.
      b. They use ATP and NADPH from catabolic pathways.
   3. Anabolic and catabolic pathways are interconnected.
      a. In stage I, macromolecules are hydrolyzed into their building blocks.
      b. In stage II, building block molecules are further degraded into a few common metabolites of small molecular weight.
      c. In stage III, small molecular weight metabolites like acetyl CoA are further degraded, yielding ATP.
   4. Many metabolic pathways are common to all cells; they arose early in the course of evolution.

D. Oxidation and reduction– a matter of electrons:
   1. Oxidation-reduction (redox) reactions involve a change in the electronic state of reactants.

a. When a substrate gains electrons, it is reduced.
b. When a substrate loses electrons, it is oxidized.
c. When one substance gains (or loses) electrons, some other substance must donate (or accept) those electrons; redox substrates must occur in pairs.
   (1) In a redox pair, the substrate that gains electrons is an oxidizing agent.
   (2) The substrate that donates electrons is a reducing agent.

2. Redox reactions with organic substrates involve atoms that are bound to carbon.

E. The capture and utilization of energy:
1. Reduced atoms can be oxidized, releasing energy to do work.
2. The more a substance is reduced, the more energy can be released.
   a. Fats and carbohydrates are highly reduced.
   b. Fats and carbohydrates are oxidized in cells to release energy to do work.
   c. Glucose is a carbohydrate central to the energy-producing pathways of cells.

F. Glycolysis and ATP formation:
1. Glycolysis is the first stage in the catabolism of glucose.
   a. It occurs in the soluble portion of the cytoplasm.
   b. It begins with glucose and results in the formation of pyruvate.
2. Of the reactions of the glycolytic pathway, all but three are near equilibrium ($\Delta G = 0$) under cellular conditions.
3. The driving forces of the glycolytic pathway are the three reactions that are far from equilibrium.
4. There are 10 reactions of glycolysis:
   a. Glucose is phosphorylated to glucose 6-phosphate at the expense of ATP.
   b. Glucose-6-phosphate is isomerized to fructose-6-phosphate.
   c. Fructose-6-phosphate is phosphorylated to fructose 1,6-bisphosphate using another ATP.
   d. Fructose 1,6-bisphosphate is split into two three-carbon phosphorylated compounds.
   e. NAD^+ is reduced to NADH when glyceraldehyde 3-phosphate is converted to 1,3-bisphosphoglycerate.
      (1) Dehydrogenase enzymes oxidize and reduce cofactors.
      (2) NAD^+ is a nonprotein cofactor loosely associated with glyceraldehyde phosphate dehydrogenase.
      (3) NAD^+ can undergo oxidation and reduction at different places in the cell.
      (4) NADH donates electrons to the electron transport chain in the mitochondria.
      (5) Energy from the redox reaction of the electron transport chain is used to make ATP via oxidative phosphorylation.
   f. ATP is formed when 1,3 bisphosphoglycerate is converted to 3-phosphoglycerate by 3-phosphoglycerate kinase.
      (1) Kinase enzymes transfer phosphate groups.
      (2) Substrate-level phosphorylation occurs when ATP is formed by a kinase enzyme.
      (3) ATP formation is only moderately endergonic compared with other phosphate transfer reactions in cells.
   g. 3-phosphoglycerate is converted to pyruvate via three sequential reactions, one of them a kinase that phosphorylates ADP.
5. Glycolysis can generate a net of two ATPs for each glucose.
6. Glycolysis occurs in the absence of oxygen; i.e., it is anaerobic.

7. The end product, pyruvate, can enter anaerobic or aerobic catabolic pathways.

G. Fermentation:
   1. Fermentation restores NAD^+ from NADH.
      a. Under anaerobic conditions, glycolysis depletes the supply of NAD^+ by reducing it to NADH.
      b. In fermentation, NADH is oxidized to NAD^+ by reducing pyruvate.
         (1) In muscle cells and tumor cells, pyruvate is reduced to lactate.
         (2) In yeast and other microbes, pyruvate is reduced and converted to ethanol– the basis of the alcoholic beverage industry.
   2. Glycolysis followed by fermentation is inefficient: only about 10% of the energy of glucose is captured in ATP.

H. Reducing power:
   1. Anabolic pathways require a source of electrons to form large molecules.
   2. NADPH donates electrons to form large biomolecules.
      a. NADPH is a nonprotein cofactor similar to NADH.
      b. The supply of NADPH represents the cell's reducing power.
      c. $NADP^+$ is formed by phosphate transfer from ATP to NAD^+.
   3. NADPH and NADH are interconvertible, but have different metabolic roles.
      a. NADPH is oxidized in anabolic pathways.
      b. NAD^+ is reduced in catabolic pathways.
      c. The enzyme transhydrogenase catalyzes the transfer of hydrogen atoms from one cofactor to the other.
         (1) NADPH is favored when energy is abundant.
         (2) NADH is used to make ATP when energy is scarce.

I. Metabolic regulation:
   1. Cellular activity is regulated according to need.
   2. Regulation may involve controlling key enzymes of metabolic pathways.
   3. Enzymes are controlled by alterations in active sites.
      a. Covalent modification:
         (1) Enzymes are often regulated by phosphorylation:
            (a) Protein kinases transfer a phosphate group from ATP to the enzyme being regulated.
            (b) Phosphate groups are added to tyrosine residues by one type of protein kinase.
            (c) Phosphate groups are added to serine or threonine residues by a second type of protein kinase.
         (2) Activation of phosphorylase by phosphorylase kinase was first described in the 1950s by Fischer and Krebs.
      b. Allosteric modulation:
         (1) Enzymes can be regulated by compounds binding to allosteric sites.
         (2) In feedback inhibition, the product of the pathway allosterically inhibits one of the first enzymes of the pathway.

J. Separating catabolic and anabolic pathways:
   1. Synthesis of biomolecules does not proceed via the same reactions as breakdown of those molecules, although the two pathways may have steps in common.
      a. Some catabolic steps are essentially irreversible due to large ΔG values.
      b. Irreversible steps in catabolic pathways are accomplished by different mechanisms and different enzymes in anabolic pathways.
   2. Glycolysis and gluconeogenesis are the catabolic and anabolic pathways of glucose metabolism.

a. Synthesis of fructose 1,6-bisphosphate is coupled to hydrolysis of ATP by phosphofructokinase in glycolysis.
b. Breakdown of fructose 1,6-bisphosphate is via simple hydrolysis by fructose 1,6-bisphosphatase in gluconeogenesis.
c. Phosphofructokinase is regulated by feedback inhibition with ATP as the allosteric inhibitor.
d. Fructose 1,6-bisphosphatase is regulated by covalent modification using phosphate binding.
e. ATP levels are highly regulated.

V. Determining the Mechanism of Lysozyme Action (Experimental Pathways)

A. Lysozyme breaks glycosidic bonds of bacterial cell walls.

B. The mechanism of action was theorized in 1966 by David Phillips.
   1. The mechanism was first modeled using x-ray crystallography information.
   2. Predictions based on the model were tested and found to support the model.
   3. The enzyme and the substrate fit together best while in the transition state.
   4. The best inhibitors mimic the substrate in its transition state and are called TSAs, or transition state analogs.

C. Recent work using site-directed mutagenesis has further confirmed the mechanism proposed by Phillips.

Key Figure

Figure 3.9. Activation energy and enzymatic reactions.

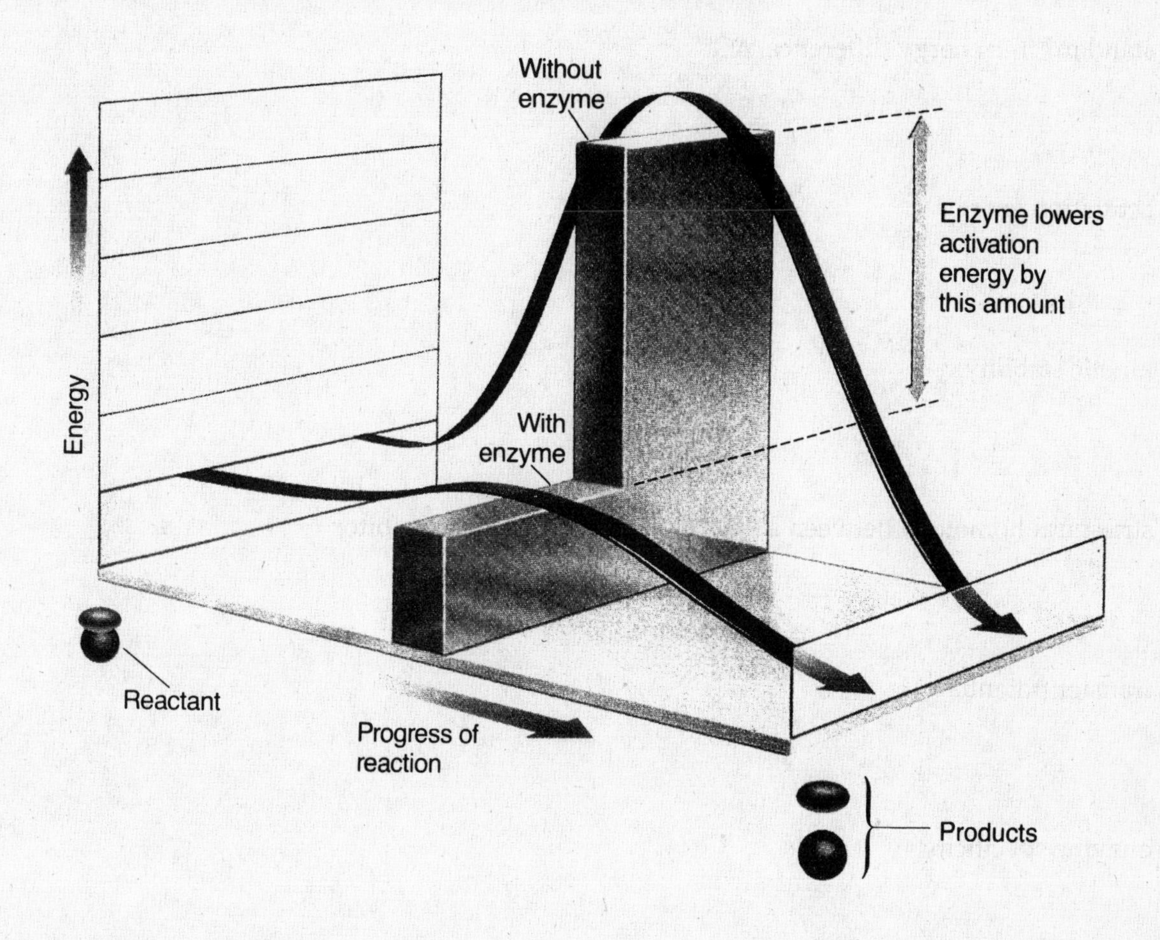

Questions for Thought:

1. How does the activation energy prevent reactants from reaching thermodynamic equilibrium?

2. Give some examples of how enzymes lower the activation energy.

3. Why would it be inappropriate for cells to overcome the energy of activation by raising the mean energy level of all the reactant molecules instead of lowering the energy of activation barrier?

4. Can lowering the activation energy make a reaction go in either direction spontaneously? What if the concentrations of products and reactants were altered making the reverse reaction a thermodynamically favored reaction. Could the enzyme function to speed up the reverse reaction in those circumstances?

Review Problems

Short Answer

1. Give the significance (not the definition) of the following terms or phrases. Say what they do or why they are important. For example:
   Second law of thermodynamics: *predicts the direction that a reaction will proceed spontaneously.*

   a. standard free energy difference, $\Delta G^{\circ\prime}$

   b. activation energy

   c. kinetic stability

   d. structural homology between a substrate and an enzyme inhibitor

   e. transfer potential

   f. enzyme specificity

   g. reciprocal plots of enzyme velocity vs. substrate concentration

   h. fermentation

   i. reducing power

   j. feedback inhibition

2. Compare and contrast the following:
   a. equilibrium vs. steady state

   b. kinetic stability vs. thermodynamic stability

   c. theoretically reversible reactions vs. essentially irreversible reactions

   d. $NAD^+/NADH$ vs. $NADP^+/NADPH$

   e. competitive vs. noncompetitive inhibition

Multiple Choice

1. Water behind a dam is an example of
   a. kinetic energy.
   b. potential energy.
   c. free energy.
   d. entropy.

2. For the following questions, answer either **I**-increased, **D**-decreased or **U**-unchanged:
   a. When sugar is dissolved in coffee, the entropy of the sugar is__________.
   b. When seawater is evaporated, leaving behind salt crystals, the entropy of the water is __________.
   c. The entropy of the salt crystals in (b) above is__________.
   d. When a fertilized egg cell develops into an embryo, the entropy of the living system is __________.
   e. The entropy of the universe in (d) above is__________.

3. For a reaction with a standard free energy change, $\Delta G'^{\circ} = +0.7$ kcal/mole, which of the following statements is true?
   a. The reaction is exergonic.
   b. The reaction would never occur spontaneously.
   c. The reaction could be made to occur by altering the concentrations of reactants and products.
   d. None of the above.

4. When a chemical reaction is at equilibrium, which of the following statements is true?
   a. The rate of the forward reaction is equal to the rate of the backward reaction.
   b. The concentrations of reactants and products are equal.
   c. The ratio of the concentrations of reactants and products equals 1.0.
   d. The reaction is proceeding in the forward direction at a rate equal to K'_{eq}.

5. Reactions that have positive standard free energy changes ($\Delta G'^{\circ}>0$) can be made to occur in cells by:
   a. coupling them with exergonic reactions via a common intermediate.
   b. manipulating the concentrations of products and reactants such that $\Delta G'<0$.
   c. coupling them to the hydrolysis of ATP.
   d. all of the above.

6. Cellular metabolism is an example of:
   a. equilibrium metabolism.
   b. steady-state metabolism.
   c. a series of reactions occurring under standard conditions.
   d. a series of reactions that defy the second law of thermodynamics.

7. Any system that is not at equilibrium:
   a. is thermodynamically unstable, although it may be kinetically stable.
   b. is kinetically unstable, although it may be thermodynamically stable.
   c. is rushing toward equilibrium at a very rapid rate.
   d. requires an enzyme to reach equilibrium.

8. Protein catalysts:
   a. increase the rates of chemical reactions in cells while remaining unchanged.
   b. are sometimes referred to as ribozymes.
   c. can change the standard free energy of a cellular reaction.
   d. all of the above.

9. Which of the following statements is not true of enzymes:
   a. Enzymes are proteins that bind to specific substrates and increase the velocity of reactions involving those substrates.
   b. Enzymes function by overcoming the activation energy barrier of a reaction.
   c. Enzymes can only make thermodynamically favorable reactions proceed; they cannot make unfavorable reactions occur.
   d. Enzymes only function when they are in intact cells; isolated enzymes loose all their catalytic ability.

10. As opposed to an uncatalyzed reaction, in an enzymatically catalyzed reaction:
    a. the activation energy of the reaction is lowered, and a larger proportion of substrate molecules has sufficient energy to overcome it.
    b. the activation energy of the reaction remains unchanged, but the substrate molecules are highly energized.
    c. the activation energy of the reaction is increased, thus decreasing the likelihood that any substrate molecules will overcome it.
    d. the activation energy of the reaction is lowered, but fewer substrate molecules can overcome it.

11. When an enzyme is functioning at V_{max}, the rate of the reaction is limited by
    a. the number of collisions between enzyme and substrate.
    b. the number of substrate molecules in the reaction.
    c. the concentration of the substrate.
    d. the rate at which the enzyme can convert substrate to product.

12. The K_M of a reaction is:
    a. a measure of the affinity of the enzyme for the substrate under some conditions.
    b. a measure of the substrate concentration at exactly one-half of V_{max}.
    c. a kinetic value that can be determined from Lineweaver-Burk plots of enzyme velocities.
    d. all of the above.

13. A certain inhibitor of an enzymatic reaction was determined to act as a competitive inhibitor. Which one of following statements about the inhibitor is most probably true?
    a. The inhibitor acts at a site other than the substrate binding site.
    b. The inhibitor increases K_M but has no effect on V_{max}.
    c. The inhibitor decreases K_M but has no effect on V_{max}.
    d. The inhibitor lowers the maximum velocity of the enzyme.

14. During catabolism, only about 40% of the energy available from oxidizing glucose is used to synthesis ATP. What becomes of the other 60%?
    a. It is lost as heat.
    b. It is used to reduce NADP.
    c. It remains in the products of metabolism.
    d. It is stored as fat.

15. During glycolysis, a net of how many ATPs are produced for each glucose that enters the pathway?
    a. 1
    b. 2
    c. 4
    d. ≥35

16. For most of the reactions of glycolysis, the ΔG under cellular conditions is close to 0. Why does the pathway continue in the direction of glucose catabolism?
    a. There are three essentially irreversible ($\Delta G << 0$) reactions that act as the "driving force" for the pathway.
    b. High levels of ATP keep the pathway going in a forward direction.
    c. The enzymes of glycolysis only function in one direction.
    d. Glycolysis occurs in either direction.

17. Which of the following statements is not true of glycolysis?
    a. ADP is phosphorylated to ATP via substrate-level phosphorylation.
    b. The pathway does not require oxygen.
    c. The pathway oxidizes two moles of NADH to NAD^+ for each mole of glucose that enters.
    d. The pathway requires two moles of ATP to get started catabolizing each mole of glucose.

18. What is the most important function of fermentation from the point of view of the cell?
    a. Fermentation is the basis of the wine, beer and liquor industry.
    b. Fermentation regenerates NAD^+ so that glycolysis can continue.
    c. Fermentation provides a supply of lactate to metabolizing muscle cells.
    d. Fermentation is an essential step in aerobic metabolism.

19. Which of the following mechanisms are used by cells to regulate metabolism?
    a. Enzymes are regulated by covalent modification.
    b. Enzymes are regulated by allosteric modification.
    c. Metabolic pathways are regulated by compartmentalization.
    d. Cells use all of the above to regulate metabolism.

20. All of the following techniques were used to study the mechanism of lysozyme action except:
    a. x-ray diffraction.
    b. the use of synthetic, chemically altered substrates.
    c. equilibrium thermodynamics.
    d. site directed mutagenesis.

Problems and Essays

1. For the reaction:

   dihydroxyacetone phosphate $\rightleftharpoons$ glyceraldehyde 3-phosphate $\Delta G^{o\prime} = 1.8$ kcal/mole

   What ratio of product/reactant must be maintained in order to keep the reaction going in a forward direction?

2. For the reaction:

   fumarate $\rightleftharpoons$ malate $K_{eq}' = 0.2818$

   What is the change in free energy when one mole of fumarate reacts to make one mole of malate under standard conditions?

3. For the reaction:

   fructose 6-phosphate $\rightleftharpoons$ glucose 6-phosphate $\Delta G^{o\prime} = -0.4$ kcal/mole

   Starting with a 0.50 M solution of fructose-6-phosphate, what is the equilibrium concentration of both fructose-6-phosphate and glucose-6-phosphate?

4. Given the reaction:

   succinate + FAD $\rightleftharpoons$ fumarate + $FADH_2$ $\Delta G^{o\prime} = 0$ kcal/mole

   a. Starting with a solution of 0.01 M succinate and 0.01 M FAD, how much fumarate will be present at equilibrium?

b. How will adding the enzyme succinic dehydrogenase change your answer to part (a) above?

5. For the complete oxidation of glucose:

$$C_6H_{12}O_6 + 6O_2 \rightleftharpoons 6CO_2 + 6H_2O \qquad \Delta G^{o\prime} = -686 \text{ kcal/mole}$$

Assume that conditions in a cell are such that only 40% of glucose oxidation is captured as ATP, and that concentrations of reactants and products are kept at standard levels.

$$ADP + P_i \rightleftharpoons ATP \qquad \Delta G^{o\prime} = +7.3 \text{ kcal/mole}$$

How many moles of ATP are formed for each mole of glucose that is oxidized?

6. Palmitic acid is a saturated fatty acid with a chain length of 16 carbons. For the complete oxidation of palmitic acid:

$$COOH(CH_2)_{14}CH_3 + 23\ O_2 \rightleftharpoons 16\ CO_2 + 16\ H_2O \qquad \Delta G^{o\prime} = -2385 \text{ kcal/mole}$$

Assume that conditions in a cell are such that only 40% of the free energy of fatty acid oxidation is captured in the form of ATP. How many moles of ATP are formed for each mole of palmitic acid oxidized?

7. Given the answers to numbers (5) and (6) above, would it be better to eat sugar or fat if you wanted to gain weight? Why? What if you wanted to lose weight?

8. For the following, put either **C** for competitive inhibition, **N** for noncompetitive inhibition, or **B** for both, depending on which type of inhibition the statement best describes.

   ________ a. The rate of product formation at subsaturating substrate concentrations is adversely affected.

   ________ b. The inhibitor binds to the active site of the enzyme.

   ________ c. At very high substrate concentrations, the initial velocity is lower in the presence of the inhibitor.

   ________ d. The V_{max} is unchanged in the presence of the inhibitor.

   ________ e. The ES complex can bind the inhibitor.

   ________ f. In the presence of the inhibitor the Lineweaver-Burk plot is rotated around the y intercept.

9. Given the following data:

	Initial Velocity (nmol product·(min·mg)$^{-1}$)			
[S] (mM)	no inhibitor	inhibitor A	inhibitor B	inhibitor C
0.100	0.400	0.258	0.200	0.333
0.125	0.476	0.316	0.238	0.400
0.167	0.588	0.400	0.294	0.500
0.250	0.769	0.546	0.385	0.667
0.500	1.111	0.857	0.556	1.000
1.000	1.429	1.200	0.714	1.333
2.000	1.667	1.500	0.833	1.600

   a. Plot the Lineweaver-Burk graph for the enzymatic reaction without an inhibitor, and with one of the three inhibitors, A, B, or C.

   b. What is the K_M and V_{max} of the reaction with and without the inhibitor?

   c. What kind of inhibitor is it, competitive or noncompetitive? Why?

CHAPTER FOUR

The Structure and Function of the Plasma Membrane

Learning Objectives

When you have finished this chapter, you should be able to:

1. Visualize the three-dimensional structure of the plasma membrane.
2. Explain the many dynamic roles of the membrane in cellular function.
3. Describe the types of lipids, proteins, and carbohydrates that make up the plasma membrane, how they are arranged in the bilayer, and the forces that hold them together.
4. Explain the relevance of membrane physical state to function. Give examples of techniques that have been used to measure membrane fluidity.
5. Describe the major integral and peripheral proteins of the erythrocyte.
6. Define the ways in which substances can cross membranes. Distinguish between the energetics of transport for charged and uncharged solutes.
7. Characterize the features of the nerve cell that enable it to produce and transmit impulses.
   a. Define the action potential in molecular terms.
   b. Describe how impulses are transmitted between cells.
   c. Give examples of some neurotransmitters and a neurotransmitter receptor.

Key Terms and Phrases

plasma membrane	selective permeability	signal transduction
receptor	lipid bilayer	fluid-mosaic model
amphipathic	phosphoglyceride	head group
sphingolipid	glycolipid	ceramide
cholesterol	leaflet	liposome
integral protein	peripheral protein	lipid-anchored protein
monotopic protein	bitopic protein	polytopic protein
freeze-fracture replication	freeze-etching	detergent
E face	P face	transition temperature
phospholipid asymmetry	flippase enzyme	cell fusion
FRAP	membrane domains	cell polarity
band 3	glycophorin A	spectrin
ankyrin	net flux	diffusion
electrochemical gradient	partition coefficient	osmosis
hypertonic	hypotonic	isotonic
turgor	plasmolysis	conductance
voltage-gated channel	chemical-gated channel	facilitated diffusion
facilitative transporter	protein isoforms	active transport
Na^+-K^+ ATPase	red blood cell ghosts	electrogenic pump

P-type pumps	V-type pumps	ABC transporters
cystic fibrosis	CFTR	bacteriorhodopsin
cotransport	symport	antiport
membrane potential	resting potential	depolarization
threshold	action potential	refractory period
all-or-none law	nerve impulse	myelin
saltatory conduction	synapse	synaptic cleft
synaptic vesicle	neurotransmitter	acetylcholine receptor

Reviewing the Chapter

I. Introduction

A. The plasma membrane is the outer boundary of the cell.
B. Plasma membranes of all cell types exhibit characteristic trilaminar appearance in electron micrographs.

II. A Summary of Membrane Functions

A. Compartmentalization: Membranes form continuous sheets that enclose intracellular compartments.
B. Selectivity barrier: Membranes allow regulated exchange of substances between compartments.
C. Transporting solutes: Membrane proteins facilitate the movement of substances between compartments.
D. Responding to external signals: Membrane receptors transduce signals from outside the cell in response to specific ligands.
E. Intercellular interaction: Membranes mediate recognition and interaction between adjacent cells.
F. Locus for biochemical activities: Membranes provide a scaffold that organizes enzymes for effective interaction.
G. Energy transduction: Membranes transduce photosynthetic energy, convert chemical energy to ATP, and store energy in ion and solute gradients.

III. An Overview of the Structure of the Plasma Membrane

A. The fluid-mosaic model is the "central dogma" of membrane biology.
   1. The core lipid bilayer exists in a fluid state, capable of dynamic movement.
   2. Membrane proteins form a mosaic of particles penetrating the lipid to varying degrees.
B. Membrane composition:
   1. The lipid and protein components of membranes are bound together by non-covalent forces.
   2. The lipid and protein compositions of different membranes vary.
   3. Protein/lipid ratios vary considerably among different membrane types.
C. Membrane lipids:
   1. Membrane lipids are amphipathic.
   2. Phosphoglycerides are diglycerides with small functional head groups linked to the glycerol backbone by phosphate ester bonds.
   3. Sphingolipids are ceramides formed by the attachment of sphingosine to fatty acids.
   4. Animal cell membranes contain cholesterol.
D. The lipid bilayer:
   1. The bilayer was proposed to account for the 2:1 ratio of lipid to cell surface area.

2. The most energetically favorable orientation for the polar head groups is facing the aqueous compartments outside of the bilayer.
3. Amphipathic lipids spontaneously form bilayered sheets or spherical vesicles, called liposomes, in artificial systems.
4. The cohesion and spontaneous assembly of bilayers make cells deformable and facilitate splitting and fusion of membranes.
5. Spontaneously formed lipid bilayers may have provided the first stable environment for the development of self-replicating molecules.

E. Membrane Ccarbohydrates:
1. Membranes contain carbohydrates covalently linked to lipids and proteins on the extracellular surface of the bilayer.
2. Glycoproteins have short, branched carbohydrates for interactions with other cells and structures outside the cell.
3. Glycolipids have larger carbohydrate chains that may be cell-to-cell recognition sites.

F. Membrane Proteins:
1. Membrane proteins attach to the bilayer asymmetrically, giving the membrane a distinct "sidedness."
2. Integral proteins are embedded in the membrane and removal requires detergents.
   a. Integral proteins are amphipathic, with hydrophobic domains anchoring them in the bilayer and hydrophilic regions forming functional domains outside of the bilayer.
   b. Integral proteins are classified as monotopic, bitopic, or polytopic according to how many times they span the membrane.
   c. Transmembrane segments normally contain nonpolar amino acids in α-helices.
   d. Channel proteins have hydrophilic cores that form aqueous channels in the membrane-spanning region.
3. Peripheral proteins are attached to the membrane by weak bonds and are easily solubilized.
4. Lipid-anchored membrane proteins are distinguished by both the types of lipid anchor and their orientation.
   a. Glycophosphatidylinositol (GPI)-linked proteins found on the outer leaflet can be released by inositol-specific phospholipases.
   b. Some inner-leaflet proteins are anchored to membrane lipids by long hydrocarbon chains.
5. The distribution of integral proteins can be analyzed by freeze-fracture and freeze-etching techniques.
   a. Freeze-fracture allows integral proteins to be visualized on the internal and external leaflet.
   b. Freeze-etching reveals more detailed information about membrane topology by sublimating a layer of water from the surface of the membranes.

IV. Membrane Lipids and Membrane Fluidity

A. Lipids exist in solid or liquid phases depending on temperature, lipid composition and saturation, and the presence of cholesterol.

B. The importance of membrane fluidity:
1. The fluidity of membranes is a compromise between structural rigidity and complete fluidity.
2. Membrane fluidity makes it possible for proteins to move in the membrane and for membranes to assemble and grow.

C. Maintaining membrane fluidity:
   1. Organisms (other than birds and mammals) maintain membrane fluidity as temperature changes by altering the composition of membrane lipids.
   2. Remodeling lipid bilayers involves saturation or desaturation of acyl chains and replacement of acyl chains by lipases or transferases.
   3. The importance of these mechanisms has been verified using mutants unable to carry out certain desaturation reactions in response to cold.

D. Asymmetry of membrane lipids:
   1. The inner and outer membrane leaflets were shown to have different lipid compositions, using lipases that can only access the outer membrane.
   2. Lipid asymmetry gives the membrane leaflets different physical and chemical properties appropriate for the different interactions occurring at the two membrane faces.
   3. Membrane lipids move easily within a leaflet but only rarely "flip-flop" between leaflets.

<u>V. The Dynamic Nature of the Plasma Membrane</u>

A. The Diffusion of Membrane Proteins after Cell Fusion
   1. Cells can be induced to fuse by certain inactivated viruses, or with polyethylene glycol.
   2. Labeled proteins have shown that membrane proteins can move between fused cells.

B. Restraints on protein mobility:
   1. Proteins can be labeled and tracked by fluorescence recovery after photobleaching (FRAP) and single particle tracking (SPT).
   2. Protein movements in membranes are slower than predicted by protein size and membrane viscosity alone.
      a. Protein movements are limited by interactions with the cytoskeleton, other proteins, and extracellular materials.

C. Membrane domains and cell polarity:
   1. Differences in protein distribution are evident in cells of organized tissues.
   2. In epithelia, the proteins of the apical membrane are distinct from those of the lateral and basal membranes.
   3. Specialized protein domains may also be present on nonpolarized cells like sperm.

D. The plasma membrane of the red blood cell:
   1. Highly homogenous preparations of membrane "ghosts" can be prepared inexpensively and simply from erythrocytes, by osmotic lysis.
   2. Membrane proteins can be purified and characterized by fractionation using SDS-PAGE electrophoresis.

E. Integral proteins of the erythrocyte membrane:
   1. Band 3 is a homodimeric, polytopic glycoprotein that exchanges Cl^- and HCO^{3-} across the red cell membrane.
   2. Glycophorin A is a monotopic protein with 16 oligosaccharide chains bearing negative charges that may prevent red cells from clumping.

F. The erythrocyte membrane skeleton:
   1. The major component of the internal membrane skeleton of erythrocytes is spectrin.
   2. Spectrin molecules are attached to the membrane surface by noncovalent bonds to ankyrin, a peripheral membrane protein which is noncovalently bonded to band 3.
   3. Spectrin is linked to other cytoplasmic proteins, including actin and tropomyosin, which maintains the integrity of the membrane.

VI. The Movement of Substances Across Cell Membranes

A. Selective permeability allows for separation and exchange of materials across the plasma membrane.
   1. Net flux is the difference between influx and efflux of materials by passive diffusion and/or active transport.

B. The energetics of solute movement:
   1. Diffusion is the spontaneous movement of material from a region of high concentration to a region of low concentration.
   2. The free-energy change during diffusion of nonelectrolytes depends on the concentration gradient.
   3. The free-energy change during diffusion of electrolytes depends on the electrochemical gradient.

C. Diffusion of substances through membranes:
   1. Diffusion requires both a concentration gradient and membrane permeability.
   2. Lipid permeability is determined by the partition coefficient, molecular size, and polarity.

D. The diffusion of water through membranes:
   1. The diffusion of water through a semipermeable membrane is called osmosis.
   2. Water diffuses from areas of lower solute concentration to areas of higher solute concentration.
   3. Cells swell in hypotonic solutions, shrink in hypertonic solutions and remain unchanged in isotonic solutions.
   4. Plant cells develop turgor in hypotonic solutions because cell walls prevent swelling. In hypertonic solutions the plant cell undergoes plasmolysis.

E. The diffusion of ions through membranes:
   1. Ions cross membranes through ion channels.
   2. Ion channels are selective and bidirectional, allowing diffusion in the direction of the electrochemical gradient.
   3. Superfamilies of ion channels have been discovered by cloning, analysis of protein sequences, site directed mutagenesis, patch clamping experiments.
   4. Ion channels can be open or closed. They are voltage- or ligand-gated.
   5. The voltage-gated potassium channel contains six membrane-spanning helices.
      a. The N-terminus forms a flexible ball that can block the channel pore.
      b. The S4 transmembrane helix is voltage sensitive.

F. Facilitated diffusion:
   1. Large or hydrophilic substances require facilitation to cross membranes.
   2. Facilitated diffusion is passive, specific, saturable, and regulated.
   3. The glucose transporter:
      a. The gradient for glucose entry into the cell is maintained by phosphorylation of glucose in the cytoplasm.
      b. Insulin stimulates glucose uptake by causing the insertion into the cell membrane of vesicles containing preformed glucose transporters.
      c. Type I diabetes is due to lack of insulin, type II diabetes is due to lack of a response to insulin.

G. Active transport:
   1. Active transport maintains the gradients for potassium, sodium, calcium, and other ions across the cell membrane.
   2. Active transport couples the movement of substances against gradients to ATP hydrolysis.

H. Coupling active transport to ATP hydrolysis: The properties of the Na^+-K^+ ATPase.
   1. The Na^+-K^+ ATPase requires K^+ outside, Na^+ and ATP inside, and is inhibited by ouabain.
   2. The ratio of Na^+:K^+ pumped is 3:2.

3. The ATPase is a P-type pump. Phosphorylation causes changes in conformation and ion affinity that allow transport against gradients.
4. The Na^+-K^+ ATPase is found only in animals and evolved early as a means to regulate volume and create large Na^+ and K^+ gradients.

I. Other ion transport systems:
1. Other P-type pumps include H^+ and Ca^{2+} ATPases.
2. Vacuolar (V-type) pumps use ATP, but are not phosphorylated during pumping.
3. ATP-binding cassette (ABC) transporters have regulatory ATP-binding sites.

J. Cystic fibrosis:
1. CF is a genetic disease characterized by abnormal fluid secretions from tissues and caused by a defective chloride channel.
3. Genetic analysis revealed an ABC transporter (the CFTR polypeptide) with two nucleotide-dependent regulatory sites.
4. A defect prevents normal insertion of the CFTR polypeptide into the membrane.
5. Gene therapy involving delivery of a corrected gene to airway tissues using viruses or liposomes has had some temporary success.

K. Bacteriorhodopsin utilizes light energy for active transport of H^+.

L. Gradients created by active ion pumping store energy that can be coupled to other transport processes.

VII. Membrane Potentials and Nerve Impulses

A. Potential differences exist when charges are separated.
1. Membrane potentials have been measured in all types of cells.
2. K^+ gradients maintained by the Na^+-K^+ ATPase are responsible for the resting membrane potential.
3. The Nernst equation is used to calculate the voltage equivalent of the concentration gradient for particular ions.
4. The negative resting membrane potential is near the negative Nernst potential for K^+ and far from the positive Nernst potential for Na^+.

B. The action potential:
1. When cells are stimulated, Na^+ channels open causing membrane depolarization.
2. When cells are stimulated to threshold, voltage-gated Na^+ channels open triggering an action potential.
3. Sodium channels cannot reopen immediately following an action potential, producing a short refractory period when the membrane cannot be stimulated.
4. Excitable membranes exhibit "all-or-none" behavior.

C. Propagation of action potentials as an impulse:
1. Action potentials produce local membrane currents depolarizing adjacent regions of the membrane that propagate the action potential.
2. The speed of a neural impulse depends on the axon diameter and whether the axon is myelinated.
   a. Resistance to local current flow decreases as diameter increases.
   b. Myelin sheaths cause saltatory conduction.

E. Neurotransmission– jumping the synaptic cleft:
1. Presynaptic neurons communicate with postsynaptic neurons (synapses) or muscles (neuromuscular junctions) across a gap (synaptic cleft).
2. Chemical transmitters released from the presynaptic cell diffuse to receptors on the postsynaptic cell.
3. The bound transmitter can depolarize (excite) or hyperpolarize (inhibit) the postsynaptic cell.

4. Transmitter action is terminated by reuptake or enzymatic breakdown.
5. Synapses control the directional flow of information in the nervous system.

VIII. The Acetylcholine Receptor (Experimental Pathways)

A. Claude Bernard discovered that curare paralyzed muscle function without blocking either nerve or muscle impulses.
B. Langley postulated a "chemical transmitter" and "receptive substance" that bound both curare and nicotine.
C. Loewi used two hearts to show that "vagusstoff" (acetylcholine) formed in one heart could stop contraction in the second.
D. The electric fish Torpedo is an excellent source of nicotinic acetylcholine (nACh) receptors and acetylcholinesterase.
E. The snake venom α-bungarotoxin and the nonionic detergent Triton X-100 allowed affinity purification of the nACh receptor.
F. Reconstituting purified receptors into artificial lipids proved that the nACh receptor was a cation channel.
G. The structure of the receptor has been studied by both electron microscopy and genetic methods.
H. A 43K protein is shown to anchor the receptor to the postsynaptic region.

Key Figure

Figure 4.4*b*. The structure of the plasma membrane.

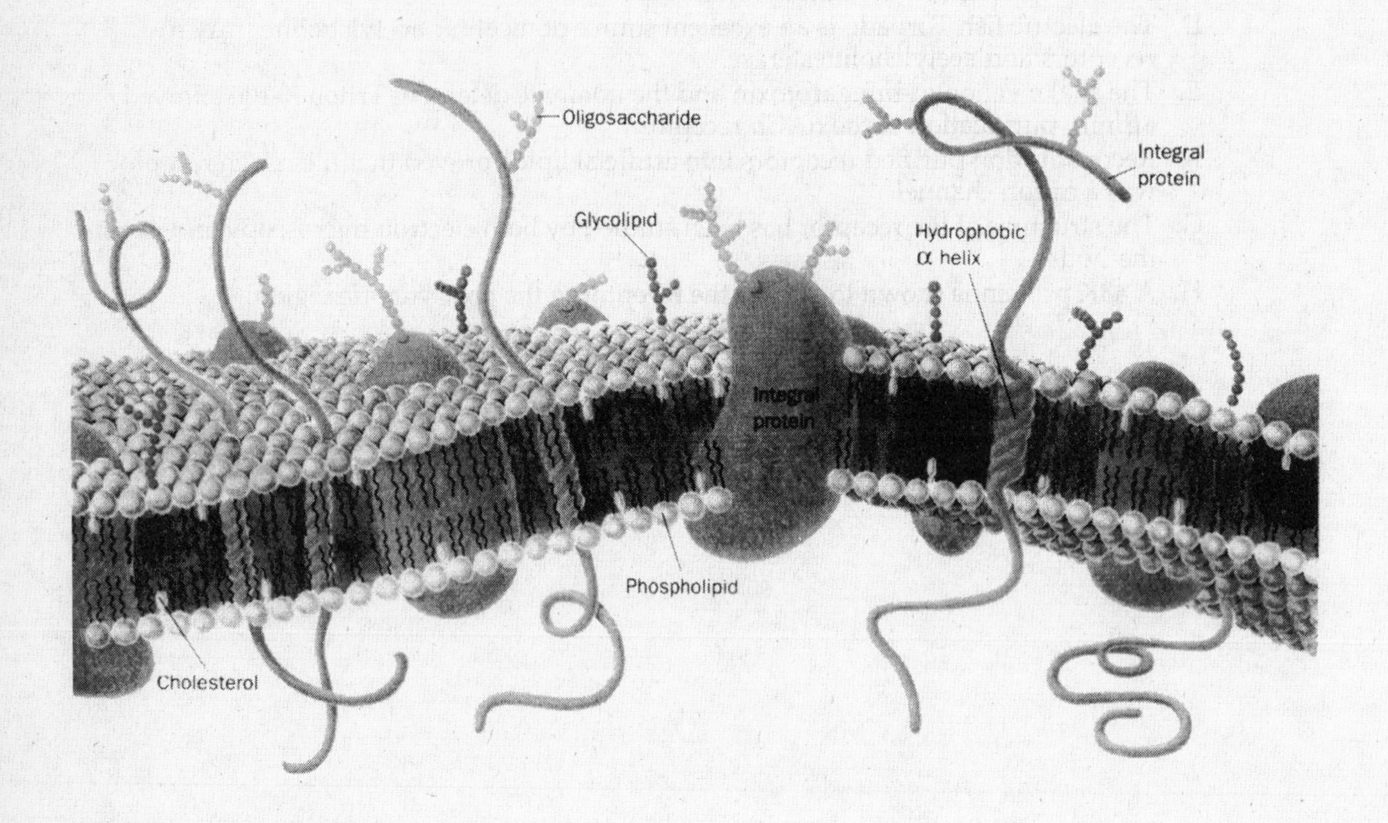

Questions for Thought:

1. The inner and outer leaflets of the cell membrane are different in both composition and function. What leaflet asymmetries are apparent from this cartoon drawing of a membrane? What asymmetries exist that are not apparent in this drawing?

2. In this drawing, the membrane proteins are evenly distributed throughout the lipid matrix. Can you think of a membrane, or portion of a membrane, in which proteins might be clumped together? What functional role might clumps of proteins serve in membranes?

3. Can you identify which proteins in this diagram, if any, are peripheral, and which are integral?

4. What factors influence the frequency with which membrane-bound proteins collide with one another?

Review Problems

Short Answer

1. Give the significance (not the definition) of the following terms or phrases. Say what they do or why they are important. For example:
amphipathic nature of phospholipids: *results in the spontaneous formation of the bilayer structure characteristic of cell membranes.*

   a. membrane leaflet asymmetry

   b. partition coefficient

   c. flippase enzymes

   d. freeze-fracture and freeze etching

   e. detergents

   f. turgor pressure in plant cells

   g. chemical-gated channels

   h. Na^+-K^+ ATPase

   i. membrane potentials

   j. myelin

2. Compare and contrast the following:
   a. the membrane surface facing inside the cell vs. that facing outside

   b. membrane lipid mobility vs. membrane protein mobility

   c. diffusion vs. facilitated diffusion

   d. membrane potential vs. resting potential

   e. myelinated nerve cells vs. large-diameter nerve cells

Multiple Choice

1. Of the four classes of cellular macromolecules, which one is not a component of cell membranes?
   a. proteins
   b. lipids
   c. nucleic acids
   d. carbohydrates

2. How do membranes function as a locus of biochemical reactions?
   a. Membranes separate the reactants from the products of enzymatic reactions.
   b. Membranes hold enzymes of sequential reactions in close proximity.
   c. Membrane lipids catalyze important cellular reactions.
   d. Membranes increase the activation energy for catalysis.

3. Features of the fluid mosaic model of membrane structure include:
   a. a lipid bilayer.
   b. dynamic motion of both membrane lipids and proteins.
   c. proteins that may rest on the surface, or penetrate the membrane.
   d. all of the above.

4. The relative proportion (by weight) of lipid and protein in a membrane:
   a. is always about equal.
   b. varies depending upon the source of the membrane.
   c. is always characterized by more lipid than protein.
   d. is always characterized by more protein than lipid.

5. Which of the following is not true of membrane phosphoglycerides?
   a. They include phosphatidylcholine, phosphatidylethanolamine and cholesterol, among others.
   b. They are amphipathic.
   c. They have two fatty acids attached to a glycerol molecule, and a polar head group attached via a phosphate.
   d. They can have both saturated and unsaturated fatty acids.

6. Gorter and Grendel's earliest evidence that membranes were formed from a lipid bilayer included:
   a. microscopic examination of red blood cell membranes.
   b. measuring the surface area of lipids extracted from red blood cells, and comparing that to the surface area of the cells.
   c. measuring the mobility rates of membrane proteins.
   d. all of the above.

7. Membrane carbohydrates occur as:
   a. either glycoproteins or glycolipids, but are not free in the membrane.
   b. either glycoproteins, or free in the membrane, but are not attached to lipids.
   c. either glycolipids or free in the membrane, but are not attached to proteins.
   d. none of the above.

8. A membrane-affiliated protein is isolated by detergent extraction, and found to be amphipathic and lacking any nonprotein components. To which class of membrane proteins does it belong?
   a. integral proteins
   b. peripheral proteins
   c. lipid-anchored proteins
   d. none of the above

9. An integral membrane protein is isolated and its amino acids are sequenced. It is found to have four segments of 22 hydrophobic amino acids. To which class of integral proteins does it belong?
   a. monotopic proteins
   b. bitopic proteins
   c. polytopic proteins
   d. none of the above

10. A membrane is examined using freeze-fracture and freeze-etch analysis. The surfaces are shown to have almost no pits or bumps. Which of the following is the most likely source of the membrane?
    a. tmitochondria, inner membrane
    b. myelin
    c. red blood cell
    d. all of the above

11. Which one of the following organisms has the highest proportion of unsaturated fatty acids in their membranes?
    a. Antarctic fish
    b. cactus
    c. bacteria from thermal hot springs
    d. humans

12. Which of the following statements about the mobility of membrane components is true?
    a. Proteins and lipids move equally rapidly within the plane of the membrane.
    b. Lipids generally move more rapidly than proteins within the plane of the membrane.
    c. Proteins generally move more rapidly than lipids within the plane of the membrane.
    d. Lipids and proteins "flip-flop" between membrane leaflets more readily than they move within the plane of the membrane.

13. Substance X is a charged molecule of small molecular weight. It is found at higher concentrations outside of cells than inside. How might substance X enter cells?
    a. diffusion through a channel
    b. diffusion through the lipid bilayer
    c. active transport
    d. none of the above

14. The distribution of K^+ across an artificial membrane was measured, and the concentrations were found to be equal on both sides. Which of the following statements is true about the distribution of K^+?
    a. K^+ must be at equilibrium across the cell membrane.
    b. K^+ cannot be at equilibrium across the membrane.
    c. There cannot be a membrane potential under these conditions.
    d. More information is needed to determine whether K^+ is at equilibrium.

15. If a freshwater plant were put in seawater, its cells would:
    a. undergo plasmolysis.
    b. lyse.
    c. swell with turgor pressure.
    d. all of the above.

16. The cell in number (15) above would be in an environment that was:
    a. hypertonic.
    b. hypotonic.
    c. isotonic.
    d. hypo-osmotic.

17. Which of the following is an example of an ion transporter?
    a. Na^+-K^+ ATPase
    b. ABC transporter
    c. CFTR
    d. all of the above

18. Arrange the statements below in the proper sequence of events that occur in an action potential:

    I. The resting potential is reestablished.
    II. A stimulus depolarizes the membrane to threshold.
    III. Na^+ gates open and Na^+ rushes into the cell.
    IV. K^+ gates open and K^+ rushes out of the cell.

    a. II, III, IV, I
    b. III, IV, II, I
    c. II, IV, III, I
    d. I, II, IV, III

19. A neurotransmitter that binds to a postsynaptic cell and opens K^+ channels:
    a. excites the postsynaptic cell.
    b. inhibits the postsynaptic cell.
    c. depolarizes the postsynaptic cell.
    d. must be acetylcholine.

20. Much of what we know about the nicotinic acetylcholine receptor has come from studies of:
    a. the Torpedo marmarota electric organ.
    b. the human heart.
    c. simple, single-celled organisms.
    d. E. coli bacteria.

Problems and Essays

1. In discussing the compartmentalizing role of cell membranes, your textbook states that "intermixing of [the contents of the various cellular spaces] would be disastrous." Based on what you have learned in this and earlier chapters, give two examples of how intermixing of intracellular contents would be harmful.

2. Some integral membrane enzymes depend upon the lipids in their microenvironment not only for scaffolding, but for enzymatic activity. The Na^+-K^+ ATPase is one example. Using techniques described in chapter 4, design an experiment to test the influence of membrane fluidity on the velocity of the Na^+-K^+ ATPase.

3. Some of these "experimental results" are real, and others have been concocted to test your knowledge. Indicate which results are consistent with our current understanding of the structure of cell membranes (**C**), and which are not consistent (**N**).
    a.__________Freeze-fracture studies of artificial bilayers show no membrane particles, whereas those of natural membranes show many pits and bumps.
    b.__________When lipids are extracted from red blood cell membranes and spread in a monolayer on the surface of water, they occupy an area equal to about twice the surface area of the cells from which they were extracted.
    c.__________Treatment of natural membranes with high salt solutions results in complete separation of all the lipid from the protein and carbohydrate components.
    d.__________If the membrane of a cell is punctured using a micropipet, the membrane will reseal itself after the pipet is removed.
    e. __________Stains that specifically bind to carbohydrates will stick to the inner surface of a cell membrane, but not the extracellular surface.

4. Indicate whether each of teh following proteins is integral (**I**), peripheral (**P**), or lipid-anchored (**LA**).

   a. ___________ Cytochrome c is affiliated with the inner mitochondrial membrane where it participates in the transfer of electrons in aerobic respiration. It can be dissociated from the membrane by treatment with 3 M KCl, resulting in a fully active, soluble protein.

   b. ___________ Ca^{2+} ATPase is a protein affiliated with the internal membrane system of muscle cells. It can be partially isolated from the membrane by treatment with detergents, but usually retains a few molecules of phospholipid. When the detergent is dialyzed away, the protein forms a cloudy suspension.

   c. ___________ Acetylcholinesterase is a protein associated with the synaptic membranes of nerve cells. The enzyme can be separated from the membrane in a fully active form by treatment with phospholipase, but not with strong salt solutions. Analysis of its primary structure reveals no long stretches of hydrophobic amino acids.

   d. ___________ Porin is present in the membranes of E. coli. Its function is to permit the flow of small molecules across the membrane.

   e. ___________ The N-terminus of the bacterial protein bacteriorhodopsin is located on the extracellular surface of the cell, and the C-terminus is on the cytoplasmic side.

   f. ___________ Both the N-terminus and the C-terminus of cytochrome b_5 are restricted to the same side of the membrane. The cytochrome remains firmly anchored to the membrane in the presence of strong salt solutions, but can be extracted with detergents. The primary structure reveals one stretch of hydrophobic amino acids.

5. Urea is an uncharged molecule that is produced as waste during the digestion of protein. The kidney has the task of ridding the body of excess urea. The concentration of the urea in the plasma (i.e., on the "inside") is about 0.04 M. The typical urine concentration is 3.00 M. What is the cost, in kcal, of moving a mole of urea out of the body and into the urine?

6. It is possible for the distribution of some substances across a cell membrane to be at equilibrium, even if the concentrations inside and outside the cell are not equal? Explain this.

7. On the planet Loligo, where all living creatures have body temperatures of 298 °K, elements have different symbols than on Earth. The major ions in the nerve cells of the Loligotians and their concentrations (in mM) are as follows:

ion:	Ψ^+	Ξ^+	Θ^-
inside cells	180	6	34
outside cells	21	244	51

Unlike Earthling nerve cells, the resting potential of Loligotian nerve cells is +100 mV, with the inside of the cells positive relative to the outside.

a. Which of the ions is closest to equilibrium in the resting nerve cell?

b. To which of the ions is the Loligotian cell membrane most permeable; In other words, for which ion would you expect to find the most leak channels?

c. Earthling action potential begins by opening membrane channels specific for the ion that is furthest from equilibrium, i.e., voltage-gated Na^+ channels, and allowing such ions to rush toward equilibrium, thus temporarily reversing the resting potential. If the same holds true for the Loligotians, what kind of channels might instigate the action potential on that planet?

Notes:

CHAPTER FIVE

Aerobic Respiration and the Mitochondrion

Learning Objectives

When you have finished this chapter, you should be able to:

1. Explain the relationship between the structure and function of mitochondria. Relate this to mitochondrial evolution.
2. Summarize the major metabolic pathways that occur in mitochondria.
   a. Understand the central role that the TCA cycle plays in cellular metabolism.
   b. Describe the four types of electron carriers in the electron transport chain, and the four complexes into which they are organized.
3. Describe the transfer of energy in oxidative metabolism, including the forms in which it is captured, stored and transduced.
4. Define the proton-motive force, how it arises, and its electrical and chemical components.
5. Explain how ATP is synthesized, including the machinery, energetics, and regulation.
6. Describe how proteins are targeted for mitochondria.

Key Terms and Phrases

aerobic	anaerobic	fermentation
outer and inner membranes	cristae	matrix
intermembrane space	porins	"high energy" electrons
coenzyme A	tricarboxylic acid (TCA) cycle	oxidative phosphorylation
chemiosmotic mechanism	fast-twitch fibers	slow twitch fibers
electron-transfer potential	oxidation-reduction potential	electron-transport chain
flavoprotein	cytochrome	prosthetic group
ubiquinone (coenzyme Q)	electron transport complex	iron-sulfur proteins
NADH dehydrogenase	succinate dehydrogenase	cytochrome bc_1
cytochrome c oxidase	proton-motive force (Δp)	electrochemical gradient
2,4-dinitrophenol (DNP)	F_1 ATPase	F_o ATPase
ATP synthase	submitochondrial particles	ADP-ATP translocase
microbodies	peroxisomes	glyoxysomes

Reviewing the Chapter

I. Introduction

A. Oxygen accumulated in the primitive atmosphere after the cyanobacteria appeared.
B. Aerobes evolved to use oxygen to extract more energy from organic foodstuffs.
C. In eukaryotic cells, aerobic respiration takes place in the mitochondria.

II. Mitochondrial Structure and Function

A. Mitochondria have characteristic morphologies despite variability in appearance.
   1. Typical mitochondria are bacteria-sized, sausage-shaped organelles but may be round or threadlike.
   2. The size and number of mitochondria reflect the energy requirements of the cell.
   3. Inner and outer mitochondrial membranes enclose two spaces: the matrix and intermembrane space.

B. Mitochondrial membranes. Dofferemces between the inner and outer membranes:
   1. The outer membrane is less than 50% protein; the inner membrane is more than 75% protein.
   2. The inner membrane contains cardiolipin but not cholesterol, both of which are components of bacterial membranes.
   3. The outer membrane contains a large pore-forming protein called porin.
   4. The inner membrane is impermeable to even small molecules and ions. The outer membrane is permeable to even some proteins.

C. The mitochondrial matrix:
   1. The matrix contains a circular DNA molecule, ribosomes, and enzymes
   2. RNA and proteins can be synthesized in the matrix.

D. Mitochondria self-replicate by binary fission.

III. The Role of Mitochondria in Metabolism

A. The first steps in oxidative metabolism are carried out in glycolysis.
   1. Glycolysis produces pyruvate, NADH, and two molecules of ATP.
   2. Aerobic organisms use molecular oxygen to extract more than 30 additional ATPs from pyruvate and NADH.
   3. Pyruvate is transported across the inner membrane and decarboxylated to form acetyl CoA, which enters the tricarboxylic acid (TCA) cycle.

B. The tricarboxylic acid cycle:
   1. The TCA cycle is a stepwise cycle where substrate is oxidized and its energy conserved.
   2. The two-carbon acetyl group from acetyl CoA is condensed with the four-carbon oxaloacetate to form a six-carbon citrate.
   3. During the cycle, two carbons are oxidized to CO_2, regenerating the four-carbon oxaloacetate needed to continue the cycle.
   4. Four reactions in the cycle transfer a pair of electrons to NAD^+ to form NADH, or to FAD^+ to form $FADH_2$.
   5. Reaction intermediates in the TCA cycle are common compounds generated in other catabolic reaction,s making the TCA cycle the central metabolic pathway of the cell.

C. The importance of reduced coenzymes in the formation of ATP:
   1. The reduced coenzymes $FADH_2$ and NADH are the primary products of the TCA cycle.
   2. NADH formed during glycolysis enters the mitochondria via malate-aspartate or glycerol phosphate shuttles.
   3. Electrons pass from $FADH_2$ or NADH to O_2, the terminal electron acceptor, through a chain of carriers in the inner membrane.
   4. As electrons move through the electron-transport chain, H^+ are pumped out across the inner membrane.
   5. ATP is formed by the controlled movement of H^+ back across the membrane through an ATP-synthesizing enzyme.
   6. The coupling of H^+ translocation to ATP synthesis is called chemiosmosis.

7. Three molecules of ATP are formed from each pair of electrons donated by NADH. Two molecules of ATP are formed for each pair of electrons donated by $FADH_2$.

D. The role of anaerobic and aerobic metabolism in exercise (The Human Perspective):
   1. ATP hydrolysis increases 100-fold during exercise, rapidly exhausting the ATP "on hand."
   2. Muscles use stored creatine phosphate (CrP) to rapidly generate ATP but must rely on aerobic or anaerobic synthesis of new ATP for sustained activity.
   3. Fast-twitch muscle fibers contract rapidly, have few mitochondria and produce ATP anaerobically.
      a. Anaerobic metabolism produces fewer ATPs per glucose but produces them very fast.
      b. Anaerobic metabolism rapidly depletes available glucose and builds up lactic acid which reduces cellular pH.
   4. Slow-twitch fibers contract slowly, have many mitochondria and produce most of their ATP by aerobic metabolism.
      a. Aerobic metabolism initially uses glucose as a substrate.
      b. Free fatty acids are oxidized during prolonged exercise.
   5. The ratio of fast- to slow-twitch fibers is variable and depends on the normal function of the muscle.

IV. The Role of Mitochondria in the Formation of ATP.

A. ATP can be formed by substrate level phosphorylation or oxidative phosphorylation.

B. Oxidation-reduction potentials:
   1. Strong oxidizing agents have a high affinity for electrons; strong reducing agents have a weak affinity for electrons.
   2. Redox reactions $A_{(ox)} + B_{(red)} \rightleftharpoons A_{(red)} + B_{(ox)}$ are accompanied by a decrease in free energy: $\Delta G°' = -nF \Delta E'_o$.
   3. The transfer of electrons causes charge separation that can be measured as an oxidation-reduction (redox) potential.
   4. The voltage change ($\Delta E'_o$) for the oxidation of NADH to NAD^+ by O_2 is +1.14 V, equal to a $\Delta G°'$ of -52.6 kcal/mol, enough to synthesize several ATPs ($\Delta G°'$ = +7.3 kcal/mol).
   5. Oxidation of isocitrate, α-ketoglutarate and malate are coupled to the reduction of NAD^+. Oxidation of succinate is coupled to the reduction of FAD.

G. Electron transport:
   1. Electrons move through the inner membrane via a series of carriers of decreasing redox potential.
   2. There are four types of electron carriers:
      a. Flavoproteins are polypeptides bound to either FAD or FMN.
      b. Cytochromes contain heme groups bearing Fe or Cu metal ions.
      c. Ubiquinone (coenzyme Q) is a lipid-soluble molecule made of five-carbon isoprenoid units.
      d. Iron-sulfur proteins contain Fe in association with inorganic sulfur rather than heme.
   3. Electron-transport complexes;
      a. Complex I (NADH dehydrogenase) catalyzes transfer of electrons from NADH to ubiquinone and transports three to four H^+ per pair.
      b. Complex II (succinate dehydrogenase) catalyzes transfer of electrons from succinate to FAD to ubiquinone without transport of H^+.
      c Complex III (cytochrome bc_1) catalyzes the transfer of electrons from ubiquinone to cytochrome c and transports four H^+ per pair.

d. Complex IV (cytochrome c oxidase) catalyzes transfer of electrons to O_2 and transports H^+ across the inner membrane.
   (1) Cytochrome c oxidase is a large well studied complex that adds four electrons to O_2 to form two molecules of H_2O.
   (2) The metabolic poisons CO, N_3^- and CN^- bind catalytic sites in Complex IV.

4. Electrons donated by NADH enter through complex I. When $FADH_2$ is the donor, electrons enter through complex II.
5. Free energy released at sites in complexes I, III and IV is sufficient to move protons across the inner membrane.
6. Ubiquinone and cytochrome c are independent of the four complexes.

V. Translocation of Protons and the Establishment of a Proton-Motive Force

A. There are two components to the proton gradient:
   1. The concentration difference between the matrix and intermembrane space creates a pH gradient (ΔpH).
   2. The voltage difference resulting from the separation of charge across the membrane creates an electric potential gradient (Ψ).
   3. Approximately 80% of the proton-motive force (Ψ + ΔpH) is due to voltage, 20% to the proton gradient.

B. Dinitrophenol (DNP) uncouples glucose oxidation and ATP formation by increasing the permeability of the inner membrane to H^+, dissipating the proton gradient.

VI. The Machinery for ATP Formation

A. Coupling factor 1, the spheres on the matrix side of the inner membrane, is a reversible ATPase (ATP synthase).

B. The structure of the ATP synthase:
   1. The F_1 particle is the catalytic subunit represented by the beads seen on the inner membrane in electron micrographs.
   2. The F_o particle attaches to F_1 and is embedded in the inner membrane.

C. The mechanism of ATP formation:
   1. ATP is formed when protons move back across the inner membrane through the ATP synthase.
   2. In the "binding-change" hypothesis, movement of protons through the F_o and F_1 subunits alters the affinity of the synthase for ADP, P_i and ATP.
   3. Binding sites on the catalytic subunit exhibit three states: tight, loose, and open.
   4. Binding of H^+ causes allosteric changes in the enzyme, shifting affinity of the binding sites between the three states.
   5. Energy is required to bind the substrates to the enzyme.
   6. In the bound state, the addition of P_i to ADP occurs spontaneously.

D. Additional roles for the proton-motive force:
   1. The H^+ gradient drives transport of ADP into and ATP out of the mitochondria.
   2. Transport of Ca^{2+} and some proteins into the mitochondria is energized by the proton-motive force.

VII. Control of Respiration

A. ADP level is the most important factor controlling the respiration rate.

VIII. Importing Mitochondrial Proteins

A. Mitochondrial proteins are encoded by nuclear and mitochondrial genes.

B. A sequence in the N-terminus directs transport of polypeptides from the cytoplasm into the matrix.

C. Proteins enter the mitochondria at specialized sites where the outer and inner membranes make contact.
D. Translocation is powered by the voltage gradient.
E. A cytosolic chaperone (Hsp70) unfolds the polypeptide for transport through the membrane.
F. Some proteins enter the matrix and are then threaded back either into the inner membrane or completely through it into the intermembrane space.
G. Proteins from mitochondrial genes are synthesized on ribosomes attached to the inner membrane.

IX. Peroxisomes

A. Peroxisomes and glyoxysomes are microbodies that also perform oxidative metabolism and import proteins posttransationally from the cytoplasm.
B. Peroxisomes oxidize a number of organic compounds and synthesize cholesterol, bile acids and plasmalogens.
C. Hydrogen peroxide (H_2O_2) formed in peroxisomes is a reactive and toxic compound broken down by the enzyme catalase.
D. Glyoxysomes convert stored fatty acids to glucose for use by newly developing plants.

X. Diseases That Result from Abnormal Mitochondrial or Peroxisomal Function (The Human Perspective)

F. Patients with Zellweger syndrome lack peroxisomal enzymes due to defects in translocation of proteins from the cytoplasm into the peroxisome.
   1. Defects include an inability to metabolize substrates, reduced ATP formation, and failure to stop oxidizing substrate when ADP levels are low.
   2. Muscle and nerve tissues are most severely affected due to their high metabolic rates.
G. Adrenoleukodystrophy is caused by the lack of a single peroxisomal enzyme, allowing fatty acids to accumulate in the brain and damage the myelin sheaths of nerves.

XI. Coupling of Oxidation to Phosphorylation (Experimental Pathways)

A. Mitchell, in 1961, first proposed the chemiosmotic theory to explain the coupling of oxidation and phosphorylation.
   1. The pH and electrical gradient resulting from transport of protons links oxidation to phosphorylation.
   2. Carriers accepting both H^+ and electrons should be located on the inner side of the membrane to take up protons.
   3 When electrons are passed to carriers only able to accept electrons, the H^+ is translocated across the inner membrane.
B. A number of testable predictions followed from Mitchell's chemiosmotic hypothesis.
   1. Addition of O_2 to oxygen-starved mitochondria led to the predicted release of H^+ in the medium.
   2. The contribution of pH and electric potential to the proton-motive force was measured.
   3. The number and mechanism of H^+ translocations at each site have been vigorously debated and are still not completely understood.
   4. The generation of artificial proton gradients in chloroplasts proved that a pH gradient alone was sufficient to power phosphorylation.
   5. Reconstitution of bacteriorhodopsin (a light-driven H^+ pump) and ATP synthase into vesicles produced ATP upon illumination.
   6. These last experiments proved that electrons transport and ATP synthesis are not directly coupled.

Key Figure

Figure 5.13. The use of inhibitors to determine the sequence of carriers in the electron-transport chain.

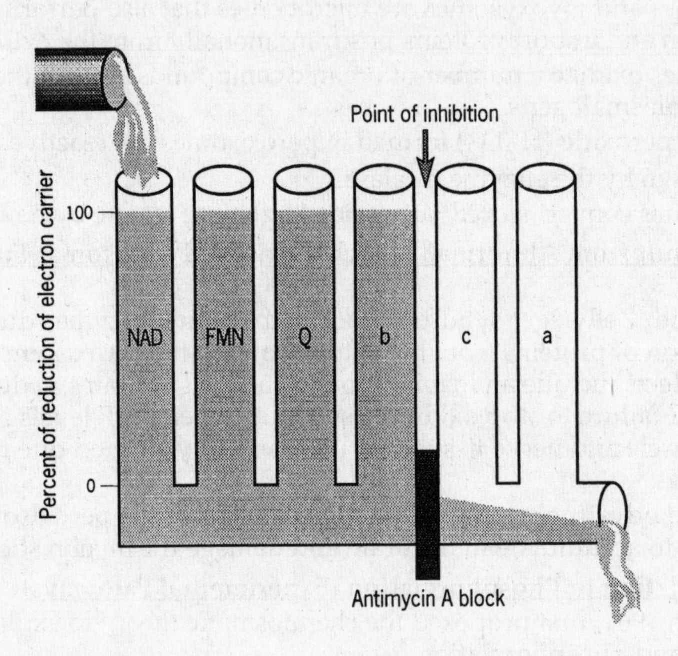

Questions for Thought:

1. In this hydraulic analogy, what does the water represent in the electron transport chain?

2. When inhibitor studies are done, what techniques are used to determine which carriers are "full" and which are "empty"? Give the appropriate term for a carrier that is "full" of electrons.

3. In addition to the electron-transport chain, can you think of another type of cellular process whose sequence might be elucidated by the use of inhibitors?

4. How would this diagram differ if the inhibitor were azide or carbon monoxide?

Review Problems

Short Answer

1. Give the significance (not the definition) of the following terms or phrases. Say what they do or why they are important. For example:
   proton-motive force: *provides the energy for the synthesis of ATP.*

   a. evolutionary success of the cyanobacteria

   b. cristae

   c. porins

   d. spatial arrangement of carriers in the electron-transport chain

   e. only 20% of Δp is ΔpH

   f. inner mitochondrial membrane is impermeable to H^+

   g. levels of ADP in the mitochondria

   h. fermentation

   i. catalase in the peroxisomes

   j. Na^+, K^+ ATPase from red blood cells can be made to synthesize ATP

2. Compare and contrast the following:
   a. inner vs. outer mitochondrial membranes

   b. "high energy" vs. "low energy" electrons

   c. substrate level vs. oxidative phosphorylation

   d. chemical vs. electrical gradient

   e. redox loop vs. proton pump

Multiple Choice

1. The fact that mitochondria were known in the ninteenth century to be osmotically active was evidence that:
   a. the mitochondrial matrix is hypotonic to the cellular cytoplasm.
   b. the mitochondrial matrix is hypertonic to the cellular cytoplasm.
   c. mitochondria were derived from a symbiotic bacterium.
   d. mitochondria are surrounded by a semipermeable membrane.

2. Which cell type is likely to have the most mitochondria?
   a. fast-twitch muscle
   b. slow-twitch muscle
   c. liver cells
   d. adipose cells

3. The composition of the inner mitochondrial membrane is most like that of:
   a. the outer mitochondrial membrane.
   b. the cell's plasma membrane.
   c. some bacterial plasma membranes.
   d. the nuclear membrane.

4. Which of the following statements about the outer mitochondrial membrane is not true?
   a. The outer membrane is impermeable to hydrogen ions.
   b. The outer membrane is highly permeable to substances of small molecular weight.
   c. The outer membrane is about 50% lipid and 50% protein by weight.
   d. The outer membrane contains porins.

5. Which of the following would not be found as part of the mitochondrial matrix?
   a. ribosomes
   b. DNA
   c. enzymes
   d. All of the above can be found in the mitchondrial matrix.

6. In what form does the product of glycolysis enter the TCA cycle?
   a. acetyl CoA
   b. pyruvate
   c. NADH
   d. glucose

7. Why is the TCA cycle the central pathway of the cell?
   a. It occurs in the center of the cell.
   b. Its metabolites are the same compounds generated by most of the cell's catabolic pathways.
   c. All other metabolic pathways depend upon it.
   d. None of the above.

8. Which of the following are reduced coenzymes?
   a. NADH and $FADH_2$
   b. NAD^+ and FAD
   c. ATP and GTP
   d. coenzyme A and ubiquinone

9. Which of the following is not a source of energy in active muscle cells?
   a. creatine phosphate
   b. ATP
   c. lactic acid
   d. glucose

10. Which of the following is not a feature of oxidative phosphorylation?
   a. direct transfer of phosphate from a substrate molecule to ADP
   b. an electrochemical gradient across the inner mitochondrial membrane
   c. a membrane-bound ATP synthase
   d. a proton-motive force

11. Given the following redox couples and standard oxidation-reduction potentials:

	E'_0 (V)
2 cytochrome c_{ox} + 2 e- $\rightleftharpoons$ 2 cytochrome c_{red}	+0.254 V
2 cytochrome $a3_{ox}$ + 2 e- $\rightleftharpoons$ 2 cytochrome $a3_{red}$	+0.385 V

which way would electrons move between these couples under standard conditions?
   a. From cytochrome c_{red} to cytochrome $a3_{ox}$
   b. From cytochrome $a3_{red}$ to cytochrome c_{ox}
   c. From cytochrome c_{red} to cytochrome c_{ox}
   d. From cytochrome $a3_{red}$ to cytochrome $a3_{ox}$

12. Why do some redox reactions in cells result in the transfer of electrons from a reductant with a higher standard redox potential to an oxidant with a lower standard redox potential?
   a. Redox potentials are defined under standard conditions, and cellular conditions are typically not standard conditions.
   b. The rules governing oxidation and reduction apply only in vitro, and not in vivo.
   c. Electrons always go from reductants with high redox potentials to oxidants with low redox potentials. It's the rule, not the exception.
   d. None of the above.

13. What is the source of free energy (ΔG) for moving protons out of mitochondria against their concentration and electrical gradients?
    a. glycolysis
    b. creatine phosphate
    c. the redox reactions of electron transport
    d. ATP

14. Which of the following is not anelectron carrier class in the electron transport chain?
    a. flavoproteins
    b. cytochromes
    c. iron-sulfur proteins
    d. cytochrome c oxidase

15. If complex III were incorporated into an artificial lipid vesicle in order to demonstrate it's proton translocating ability in isolation, which of the following would be an appropriate electron donor?
    a. cytochrome a_{red}
    b. cytochrome c_{red}
    c. UQH_2
    d. cytochrome c oxidase

16. In glycolysis, ATP is formed by the transfer of a high energy phosphate from 1,3-bisphosphoglycerate to ADP. No such high energy phosphate donor has ever been isolated in mitochondria. Why not?
    a. The techniques for isolating the phosphate donor are not refined enough .
    b. No such phosphate donor exists.
    c. The high energy phosphate donor is very short-lived and difficult to isolate.
    d. None of the above.

17. Which of the following could act as an "uncoupler" of electron transport and ATP synthesis?
    a. the F_o base piece of the ATP synthase (without the F_1 subunit)
    b. dinitrophenol
    c. neither a nor b
    d. both a and b

18. Which of the following statements about mitochondrial proteins is true?
    a. All mitochondrial proteins are coded for and synthesized in the mitochondria.
    b. Some mitochondrial proteins are coded for and synthesized in the mitochondria, and others are imported from the cytoplasm.
    c. All mitochondrial proteins are imported from the cytoplasm.
    d. Mitochondrial proteins arise from other mitochondrial proteins.

19. Using differential centrifugation, you are trying to isolate peroxisomes and glyoxysomes from a mixture of cellular organelles. After a few centrifugation steps, you think you may have a relatively pure suspension. How might you determine that your suspension does, indeed, have these organelles?
    a. Do an assay for the enzyme catalase.
    b. Do an assay for the enzyme succinate dehydrogenase.
    c. Look in the microscope for a double-membrane bound organelle.
    d. Assay your suspension for nucleic acid.

20. Which of the following experimental results does not support Mitchell's chemiosmosis theory?
    a. Electron transport in isolated mitochondria was shown to result in acidification of the medium.
    b. Addition of dinitrophenol to isolated mitochondria was shown to inhibit electron transport, but had no effect on ATP synthesis.
    c. Addition of dinitrophenol to isolated mitochondria during electron transport was shown to inhibit acidification of the medium.
    d. An artificial proton gradient across the inner mitochondrial membrane drives ATP synthesis in the absence of electron transport.

Problems and Essays

1. List three mitochondrial features consistent with the view that mitochondria first arose as bacterial symbionts residing within the cytoplasm of other prokaryotic host cells.

2. a. Write the overall, simplified, balanced reaction for the TCA cycle.

   b. In the first reaction of the TCA cycle, oxaloacetate combines with acetyl CoA to form citrate, yet there is no net disappearance of oxaloacetate. Explain.

Use this table to answer questions (3) through (5):

Oxidants		Reductants	E_o' (V)
$NAD^+ + 2H^+ + 2e^-$	$\rightleftharpoons$	$NADH + H^+$	-0.320
2 cytochrome b_{ox} + 2e	$\rightleftharpoons$	2 cytochrome b_{red}	+0.030
ubiquinone$_{ox}$ + $2H^+ + 2e^-$	$\rightleftharpoons$	ubiquinone$_{red}$	+0.100
2 cytochrome c_{ox} + $2e^-$	$\rightleftharpoons$	2 cytochrome c_{red}	+0.254
$1/2\ O_2 + 2H^+\ 2e^-$	$\rightleftharpoons$	H_2O	+0.816

3. a. Write the simplified, balanced reaction that would occur between ubiquinone and cytochrome c under standard conditions.

b. Calculate the standard free-energy change, $\Delta G^{\circ\prime}$, of the reaction in part (a).

c. Would this reaction yield enough free energy to synthesize a mole of ATP under standard conditions?

4. a. In the electron-transport chain, electrons move between ubiquinone and cytochrome b. Which molecule would act as the oxidant and which would be the reductant under standard conditions?

b. Could the roles you've assigned in part (a) be reversed under cellular conditions? If so, what conditions would have to be maintained?

5. a. Calculate the free energy released when one mole of NADH reduces one mole of oxygen under standard conditions.

b. This reaction occurs as part of oxidative metabolism, albeit quite indirectly. (Recall from Chapter 3 that the free-energy change of a reaction is independent of the pathway the reaction takes.) Based on your answer to part (a), propose an explanation for why cells have evolved the electron-transport chain to carry electrons from NADH to oxygen rather than a direct redox reaction between the pair.

6. Mitchell and Moyle determined that the total proton-motive force across the inner mitochondrial membrane was 230 mV, of which 80% was due to the electrical potential. What was the pH gradient across the membrane?

7. Based on the chemiosmotic mechanism of oxidative phosphorylation, briefly predict the results of the following experiments:

   a. When inside out submitochondrial particles are engaged in electron transport, the interior of the vesicles____.

   b. When electron transport is blocked in intact mitochondria, and an artificial pH gradient is imposed by lowering the pH of the medium, the synthesis of ATP__________________.

   c. FCCP is a drug that partitions itself into membranes and opens a channel specific for the passage of protons. When FCCP is added to a suspension of mitochondria, electron transport______________________, and ATP synthesis______________________.

   d. Under anaerobic conditions, the pH gradient________________________(decreases, increases or stays the same) relative to aerobic conditions.

Notes:

CHAPTER SIX

Photosynthesis and the Chloroplast

Learning Objectives

When you have finished this chapter, you should be able to:

1. List the major evolutionary steps leading from the earliest autotrophs to higher plants.
2. Describe the chloroplast, paying particular attention to membranes and how chloroplast membranes function in photosynthesis.
3. Distinguish between the light-dependent reactions and the light-independent reactions of photosynthesis. Summarize the major steps in each of these pathways, including their substrates and ultimate products.
4. List the major electron carriers in photosynthesis. Describe how the oxidation and reduction energy of these carriers is stored, and then used to make ATP, NADPH, and ultimately carbohydrate.
5. Distinguish between productive and nonproductive reactions that occur in the chloroplast, and describe some of the adaptations that have evolved to maximize the productive pathways and minimize the nonproductive ones.

Key Terms and Phrases

heterotrophs	autotrophs	chemoautotrophs
photoautotrophs	photosynthesis	chloroplast
thylakoid	grana	lumen
stroma	stroma thylakoids (stroma lamellae)	light-dependent reactions
light-independent reactions	photon	ground state
excited state	chlorophyll	pigment
chromophore	absorption spectrum	action spectrum
accessory pigments	carotenoids	photosynthetic unit
reaction center	antenna	photosystem I (PSI)
photosystem II (PSII)	P680	P700
primary electron acceptor	light-harvesting complex II (LHCII)	Z scheme
photolysis	pheophytin	plastoquinone
plastocyanin	appressed	nonappressed
phylloquinone (A_1)	noncyclic photophosphorylation	CO_2 fixation
cyclic photophosphorylation	C_3 pathway	C_4 (Hatch-Slack) pathway
Rubisco	ribulose bisphosphate carboxylase	photorespiration
transpiration	stomata	mesophyll cells
bundle sheath cells	CAM plants	photoinhibition

Reviewing the Chapter

I. Introduction

A. The earliest living organisms were heterotrophs that obtained organic substrates from the environment.

B. Autotrophs manufacture organic nutrients from CO_2, H_2S ,and H_2O.

C. Synthesis of complex molecules from CO_2 requires energy.

1. Chemoautotrophs use energy stored in inorganic molecules to synthesize organic compounds.
2. Photoautotrophs use radiant energy to carry out photosynthesis, the transformation of sunlight into chemical energy.

D. Photosynthesis converts low energy electrons from a donor molecule into high energy electrons that can be used in anabolic reactions.

1. The first photoautotrophs used hydrogen sulfide (H_2S) as an electron source.
2. Water (H_2O) is a more abundant source of electrons; photosynthesis using water produces molecular oxygen (O_2) as a by-product.

II. Chloroplast Structure and Function

A. Photosynthesis, like aerobic respiration, requires a membrane and an electron-transport system to generate a proton electrochemical gradient.

B. In eukaryotic cells, photosynthesis takes place in large, membranous cytoplasmic organelles called chloroplasts.

C. Chloroplasts have a double membrane, like the mitochondria.

1. The outer membrane contains porin and is permeable to large molecules.
2. The inner membrane contains light-absorbing pigment, electron carriers and ATP-synthesizing enzymes.

D. The inner membrane is folded into flattened sacs called thylakoids, arranged into flattened stacks called grana.

E. Chloroplasts are semiautonomous, self-replicating organelles containing their own small, circular DNA molecules.

F. Thylakoid membranes contain a high percentage of unsaturated neutral glycolipids.

III. An Overview of Photosynthetic Metabolism

A. In the 1930s it was shown that photosynthesis is an oxidation-reduction reaction transferring an electron from water to CO_2.

B. Experiments using ^{18}O demonstrated that the O_2 released during photosynthesis came from two molecules of H_2O, not CO_2, as previously believed.

C. During photosynthesis, water is oxidized to form oxygen; in respiration, oxygen is reduced to form water.

1. Respiration removes high energy electrons from reduced organic substrates to form ATP and NADH.
2. Photosynthesis boosts low energy electrons to form ATP and NADPH, which are then used to reduce CO_2 to carbohydrate.

D. Photosynthesis occurs in two stages:

1. Light-dependent (light) reactions absorb light, converting it into chemical energy in the forms of ATP and NADPH.
2. Light-independent (dark) reactions use the energy of ATP to form carbohydrate.

IV. The Absorption of Light

A. Absorption of photons lifts electrons from inner to outer orbitals, elevating molecules from the ground state to an excited state.

1. The energy in a mole of photons depends on the wavelength of the light.

2. The energy required to shift electrons varies for different molecules.
3. Therefore, molecules absorb specific wavelengths of light.

B. Energy from excited molecules can be reradiated at a lower wavelength (fluorescence) or, in the case of chloroplasts, transferred to electrons carriers.

C. Pigmented molecules contain chromophores, groups capable of absorbing light.
1. Chlorophyll contains a porphyrin ring that absorbs light and a hydrophobic tail anchoring it to the chloroplast membranes.
2. A conjugated system of single and double bonds in the porphyrin ring creates a cloud of delocalized electrons that strongly absorb light and broaden the absorption spectrum.

D. Different photosynthetic organisms have different classes of chlorophyll.
1. Chlorophyll a is present in all oxygen-producing photosynthetic organisms.
2. Chlorophyll b is present in higher plants and green algae.
3. Chlorophyll c is found in brown algae, diatoms and photosynthetic protozoa.

E. Accessory pigments broaden the range of absorbed light able to stimulate photosynthesis and protect photosynthetic molecules from reactive forms of oxygen.

V. Photosynthetic Units and Reaction Centers

A. Each photosynthetic unit contains 300 chlorophyll molecules that harvest light.

B. Polypeptides facilitate energy transfer by holding pigment molecules in fixed orientations.

C. Excitation energy is transferred to a reaction-center chlorophyll that passes electrons to electron acceptors.

D. Oxygen evolution:
1. Two photosystems act in series to boost electrons from H_2O to NADPH.
   a. Photosystem II (PSII) boosts electrons from below the energy level of water to a point midway between water and $NADP^+$.
   b. Photosystem I (PSI) boosts electrons to a level above $NADP^+$.
2. The reaction center of PSII is referred to as P680, and that of PSI as P700.
3. The Z scheme describes the flow of electrons from water to $NADP^+$.

E. PSII operations– splitting water:
1. Two peptides, D1 and D2, bind the P680 chlorophyll molecule and perform reactions required to oxidize water.
2. Light is harvested by a pigment-protein complex called the light-harvesting complex II (LHCII).
3. Harvested energy is passed through a core antenna complex to P680.
   a. Transfer of energy from P680 to a primary electron acceptor (Pheo) generates a pair of oppositely charged species $P680^+$ and $Pheo^-$.
   b. This redox pair is capable of photolysis, the splitting of water.
   c. The electrons from water pass through the D1 polypeptide to $P680^+$, while $Pheo^-$ passes its electrons through plastoquinone (PQ) intermediates to the opposite side of the chloroplast membrane.
   d. Electrons are passed to H^+ in the stroma forming a pH gradient in the thylakoid lumen.
   e. The four electrons required to form 1 molecule of O_2 are transferred successive cycles through $P680^+$ to four Mn^{2+} ions, allowing the oxygen-evolving complex of PSII to catalyze the removal of four electrons from two molecules of H_2O.
4. Electrons are transferred from PQ to P700 via cytochrome b_6f and plastocyanin.

F. PSI and the reduction of $NADP^+$:
   1. Photons harvested by antenna pigments in PSI (LHCI) oxidize chlorophyll a in P700, forming $P700^+$.
   2. The electrons formed by oxidation of P700 pass to another molecule of chlorophyll a acting as a primary electron acceptor (A_o^-).
   3. The redox potential of the $P700^+/A_o^-$ pair reduces $NADP^+$.
   4. Transfer of electrons from $P700^+$, on the luminal side, through intermediates to $NADP^+$, on the stromal side, adds to the proton gradient.
   5. Electrons from PSI can be diverted to reduce nitrate, ammonia or sulfate forming compounds necessary for life.

VI. Photophosphorylation

A. The ATP synthase in chloroplasts is constructed of polypeptides homologous to mitochondrial enzymes and believed to function by a similar mechanism.

B. The synthase headpieces (CF_1) are aggregated in exposed regions of the grana stacks and project into the stroma so H^+ move through the synthase down the H^+ concentration gradient from lumen to stroma.

C. Isolated chloroplasts engaged in photosynthesis alkalinize the incubation medium creating a pH gradient of 3 to 4 pH units.

D. The electromotive force created in chloroplasts byelectron transport does not have a large voltage component because other ions are transported simultaneously.

E. The movement of electrons during photosynthesis, a process resulting in the formation of oxygen, is linear and called noncyclic photophosphorylation.

F. Cyclic photophosphorylation is carried out independently by PSI and involves the recycling of the electrons from ferredoxin back to the electron-deficient reaction center.

G. The degree of cyclic vs. non-cyclic photophosphorylation depends on the energy needs of the cell at the time.

VII. Carbon Dioxide and the Formation of Carbohydrate

A. The movement of carbon in the cell can be followed during photosynthesis using $[^{14}C]O_2$ as a tracer.

B. The incorporation of labeled carbon into reduced organic compounds is very fast.
   1. The first compound to be identified was 3-phosphoglycerate (PGA), a three-carbon intermediate in glycolysis. The pathway was called the C_3 pathway.
   2. CO_2 is condensed with a five-carbon compound, ribulose 1,5-bisphosphate (RuBP) to form a six-carbon compound that splits into two molecules of PGA.
   3. The condensation and splitting are two activities of a single enzyme, ribulose bisphosphate carboxylase (Rubisco).

C. The C_3 pathway (called the Calvin cycle) includes three basic steps:
   1. Carboxylation of RuBP and splitting to form PGA.
   2. Reduction of PGA to glyceraldehyde 3-phosphate (GAP) using NADPH and ATP formed by electron transport.
   3. Regeneration of RuBP.

D. The GAP molecules can be used in the cytosol as metabolic substrates, or converted to sucrose and carried to nonphotosynthetic tissues.

E. Alternatively, GAP can be converted to starch in the chloroplast where it is stored for use during periods when photosynthesis has stopped.

F. The formation of carbohydrate is expensive (12 NADPH and 18 ATP per six-carbon sugar) reflecting the fact that CO_2 is the most highly oxidized form of carbon.

G. Enzymes of the Calvin cycle must be inhibited in the dark to prevent a "futile" cycle– catabolism of stored energy to fuel synthesis of new carbohydrate.

H. During carbon fixation, the active site on Rubisco binds RuBP, making RuBP susceptible to attack by either CO_2 or O_2.
   2. Photorespiration is the utilization of O_2 to form glycolate and the subsequent release of CO_2 in the mitochondria.
   3. Whether photorespiration or photosynthesis occurs depends on the CO_2/O_2 ratio.

I. C_4 plants combine CO_2 with the three-carbon compound phosphoenolpyruvate (PEP) to form four-carbon skeletons, oxaloacetate or malate.
   1. The first step in the C_4 (or Hatch-Slack) pathway is the attachment of CO_2 to PEP by phophoenolpyruvate carboxylase.
   2. The C_4 plants are primarily tropical grasses and can utilize CO_2 at levels far below the 50 ppm threshold where photosynthesis ceases in C_3 plants.
   3. In hot, dry environment,s C_4 plants can acquire enough CO_2 for photosynthesis while keeping their stomata partially closed to prevent excessive water loss.
   4. C_4 products are transported to bundle sheath cells, where they are protected from the atmosphere. The CO_2 is cleaved off and concentrations of CO_2 high enough for utilization by Rubisco are attained.

J. CAM (Crassulacean acid metabolism) plants fix CO_2 at night using PEP carboxylase.

K. Peroxisomes, chloroplasts, and mitochondria shuttle intermediates among themselves.

L. Photoinhibition protects the plant from toxic oxygen radicals formed by overstimulation of PSII.

M. Genetic engineering may be useful in conferring herbicide resistance to some plants thereby increasing the effectiveness of herbicides in control of weeds.
   1. Some common herbicides act by binding the D1 protein of PSII.
   2. Strains of purple bacteria contain a photosynthetic reaction center similar to PSII but which is resistant to certain herbicides.
   3. Geneticists would like to modify the D1 gene by inserting DNA from resistant mutants of purple bacteria.
   4. Rubisco would be less susceptible to photorespiration if its oxygenase activity could be reduced without effecting its carboxylase activity.

VIII. Organization of the Thylakoid Membrane (Experimental Pathways)

A. Stacked membranes of grana thylakoids are called appressed membranes, while stroma thylakoid membranes are called nonappressed.

B. Early biochemical and freeze-fracture studeies showed PSII complexes located only in appressed membranes and PSI complexes in both appressed and nonappressed region.

C. Using vesicle preparations it was later shown that PSI and ATP synthase are only found in nonappressed membranes and PSII in appressed membranes.
   1. Preparation of inside-out and right-side-out vesicles showed that the activities of PSII and PSI/ATP synthase are localized in different membrane subpopulations.
   2. Only the b_6f complexes connecting PSII and PSI is distributed between the two membrane types.

D. Gold-labeled antibodies have provided further evidence for the localization of the component systems.

E. The ratio of monogalactosyldiacylglycerol (MGDG) and digalactosyldiacylglycerol (DGDG) is also higher in appressed vs. nonappressed membranes.

F. The spatial separation of components requires mobile electron carriers.
   1. Measurements of membrane fluidity indicate that thylakoids are highly fluid.
   2. Diffusion rates for PQ in the membrane correspond to reduction half-times for b_6f following exposure to flashes of light.

G. The activities of PSII and PSI are coordinated.
   1. Phosphorylation of LHCII by a protein kinase was revealed by $^{32}P_i$ labeling.
   2. LHCII does not become labeled in the dark but the kinase can be activated in the presence of a strong reducing agent.
   3. The terminal electron receptor in PSII is PQ; it activates the kinase if the activity of PSII and PSI are out of balance.
   4. Phosphorylation of LHCII increases energy transfer to PSI by allowing the movement of LHCII nearer to the PSI center.
   5. Phosphorylation of light-harvesting proteins increases negative charge repulsion, perhaps providing the force driving LHCII migration.
   6. Migration of LHCII may also help protect PSII, which is more sensitive to high light intensity.

Key Figure

Figure 6.8. Action spectrum for photosynthesis.

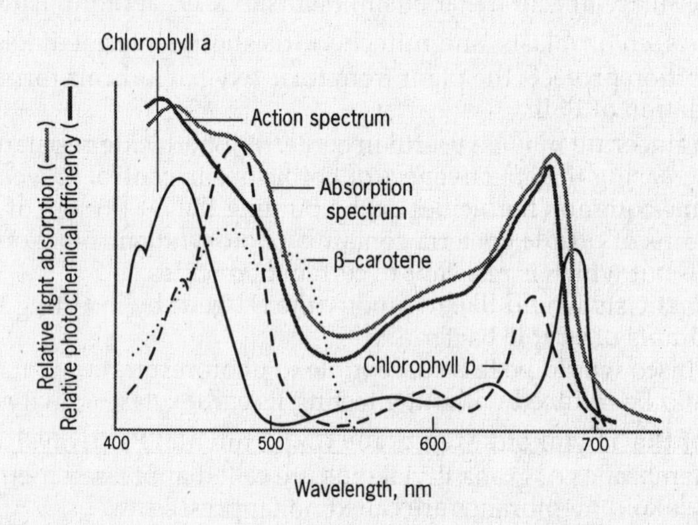

Questions for Thought:

1. Assuming the area under the action-spectrum curve represents the total amount of photosynthesis at all wavelengths, which pigments are responsible for the bulk of photosynthesis?

2. Design an experiment that would give this kind of data. How might you measure the rate of photosynthesis (the y axis)?

3. Based on their absorption spectra, do you think you could distinguish between chlorophylls a and b just by looking at pure preparations of each?

4. How might the action spectrum from a prokaryote that lacks chlorophyll b differ from the one illustrated here? What advantage do plants with both chlorophylls a and b have over organisms with only chlorophyll a?

Review Problems

Short Answer

1. Give the significance (not the definition) of the following terms or phrases. Say what they do or why they are important. For example:
   photoautotrophs: *capture light energy from the sun. The rest of the biotic world depend upon them for energy.*

   a. the switch from H_2S to H_2O as a source of electrons for photosynthesis

   b. antenna

   c. fluidity of the thylakoid membranes

   d. carotenoids

   e. two photosystems, PS I and PS II

   f. alternate routes that electrons can take from ferredoxin (other than to NADPH)

   g. thylakoid membranes are impermeable to protons

   h. Rubisco binds to RuBP, not CO_2 or O_2

   i. peroxisomes are closely associated with chloroplasts

   j. C_4 and CAM pathways

2. Compare and contrast the following:
   a. chemoautotrophs vs. photoautotrophs

   b. grana thylakoids vs. stroma thylakoids

   c. light-dependent vs. light-independent reactions

   d. cyclic vs. noncyclic photophosphorylation

   e. mesophyll cells vs. bundle sheath cells

Multiple Choice

1. Of the following, the most decisive event in biological evolution was:
   a. the evolution of heterotrophy.
   b. the evolution of chemoautotrophy.
   c. the evolution of photoautotrophy.
   d. the evolution of carbon dioxide.

2. The products of the light reactions of photosynthesis are:
   a. ATP and NAD^+.
   b. ATP and NADPH.
   c. glucose.
   d. sucrose.

3. The energy content of a photon of light depends upon:
   a. wavelength.
   b. temperature.
   c. the excited state.
   d. whether or not it is emitted in a vacuum.

4. The energy of an excited electron of a chlorophyll molecule:
   a. is released as heat.
   b. is released as fluorescence.
   c. is passed on to an electron acceptor.
   d. All of the above may be true, depending on circumstances.

5. If the reaction-center chlorophyll absorbs light of wavelength 700 nm, then the antenna pigment molecules of that photosynthetic unit must absorb:
   a. light with wavelengths less than 700 nm.
   b. light with wavelengths greater than 700 nm.
   c. light with a wavelength of exactly 700 nm.
   d. light of all wavelengths.

6. In photosystem II, P680 is oxidized to $P680^+$ and pheophytin is reduced to $Pheo^-$. What prevents $Pheo^-$ from reducing $P680^+$ back to P680?
   a. They are not a compatible redox pair.
   b. The opposite charges are moved to different sides of the thylakoid membrane.
   c. The redox potentials of $P680^+$ and $Pheo^-$ make it thermodynamically unfavorable.
   d. None of the above.

7. The driving force for photolysis is:
   a. an electron gradient across the thylakoid membrane.
   b. ATP.
   c. the redox potential of $P680^+$.
   d. NADPH.

8. Which of the following must be true of the chloroplast membranes?
   a. Thylakoid membranes are impermeable to protons.
   b. The ATP synthase is located in the outer chloroplast membrane.
   c. The ATP synthase is located in the inner chloroplast envelope.
   d. All of the above.

9. Plastoquinone is to ubiquinone as the thylakoid membrane is to
   a. the inner mitochondrial membrane.
   b. the outer mitochondrial membrane.
   c. the stromal membranes.
   d. the inner chloroplast membrane.

10. Which of the following does not directly contribute to the proton gradient across the thylakoid membrane?
    a. photolysis
    b. ferredoxin-$NADP^+$ reductase
    c. cytochrome b_6f complex
    d. plastocyanin

11. Which of the following is not part of cyclic photophosphorylation?
    a. PSII
    b. PSI
    c. ATP synthase
    d. cytochrome b_6f complex

12. Which of the following is not a characteristic of the enzyme ribulose bisphosphate carboxylase?
    a. It links CO_2 to a two-carbon acceptor, resulting in the synthesis of 3-phosphoglyceric acid.
    b. It is the most abundant protein on Earth.
    c. It catalyzes both carbon fixation and photorespiration.
    d. It is soluble in the stroma of the chloroplast.

13. The functional role of glucose in animals is to sucrose in plants as glycogen in animals is to:
    a. glucose in plants.
    b. ATP in plants.
    c. starch in plants.
    d. plants.

14. The most highly oxidized form of carbon is:
    a. carbohydrate.
    b. carbon in protein.
    c. carbon dioxide.
    d. hydrocarbon.

15. What are the most likely products of the reaction catalyzed by Rubisco under normal atmospheric conditions, where O_2 is abundant and CO_2 is low?
    a. 3-phosphoglycerate
    b. RuBP
    c. glycine and serine
    d. 2-phosphoglycolate

16. What products of photorespiration can be salvaged even though it is generally inefficient to do so?
    a. NADPH
    b. Amino acids
    c. ATP
    d. Important components of cell membranes

17. If you wanted to isolate the enzyme phosphoenolpyruvate carboxylase, where would you look?
    a. bundle sheath cells of C_4 plants
    b. mesophyll cells of C_4 plants
    c. chloroplasts of C_3 plants
    d. mitochondria of CAM plants

18. If you wanted to isolate large quantities of malate from CAM plants, when would you look?
    a. early morning
    b. midday on a sunny day
    c. midnight
    d. time of day wouldn't matter

19. Why might ATP synthase molecules be excluded from appressed membranes?
    a. They need to be near PSI complexes.
    b. They require large amounts of MGDG to function.
    c. They have large CF_1 subunits that interfere with membrane-membrane contact.
    d. They are not excluded from appressed membranes.

20. Which of the following are not objectives of plant genetic engineers?
    a. reducing the level of photorespiration in crop plants
    b. decreasing the carboxylase activity of Rubisco from crop plants
    c. increasing the resistance of crop plants to herbicides
    d. increasing photosynthetic efficiency of crop plants

Problems and Essays

1. Given the following oxidation potentials:

Oxidants		Reductants	E_o' (V)
$NADP^+ + H^+ + 2e^-$	$\rightleftharpoons$	NADPH	-0.320
$S^\circ + 2H^+ + 2e^-$	$\rightleftharpoons$	H_2S	+0.030
$1/2\ O_2 + 2H^+\ 2e^-$	$\rightleftharpoons$	H_2O	+0.816

a. How much energy is required for H_2O to reduce one mole of $NADP^+$ under standard conditions? With H_2S as the electron donor?

b. Given the following equations and constants:

$$E = h\nu \ \ and \ \ \nu = \frac{c}{\lambda}$$

$$h = 1.58 \times 10^{-34} \ cal \cdot sec \ (Planck's \ constant)$$

$$c = 3 \times 10^{8} \ m / sec \ (the \ speed \ of \ light \ in \ a \ vacuum)$$

$$1 \ mole = 6.02 \times 10^{23}$$

$$1 \ m = 10^{9} \ nm$$

How much energy does a mole of photons have at a wavelength of 680 nm? At a wavelength of 485 nm?

$\mathcal{E} = h\, c/680$

c. Assuming that the reactions are taking place under standard conditions and that all of the energy of a mole of photons is captured, can one mole of photons at 680 nm provide enough energy for H_2O to reduce a mole of $NADP^+$? What about a mole of photons at 485 nm?

d. Making the same assumptions as in part (c), how many moles of photons of wavelength 680 nm are required to provide the energy for H_2O to reduce one mole of $NADP^+$?

2. Refer to the Calvin cycle illustrated in figure 6.22. What ratio of ATP to NADPH is required to synthesize one mole of six-carbon sugar? Assuming that noncyclic photosynthesis produces equimolar amounts of ATP and NADPH, how must plants supplement their ATP levels to maintain the appropriate ratio of ATP to NADPH?

3. The proton gradient that drives ATP synthesis is generated at four different points in noncyclic photophosphorylation, as listed below. For each step, indicate whether a gradient is formed by proton translocation across the thylakoid (**PT**), proton consumption on the stromal side of the membrane (**SC**), or proton release on the lumen side (**RL**):

______________ photolysis

______________ QH_2

______________ cytochrome b_6f complex

______________ ferredoxin-$NADP^+$ reductase

4. Indicate whether the following experimental results apply to C_3, C_4, and/or CAM plants:

___C_4___ a. The plant was incubated in the presence of $^{14}CO_2$, and the first organic compounds in which radioactivity appeared were malate and oxaloacetate.

___C_3___ b. The plant was incubated in the presence of $^{14}CO_2$, and the radioactivity appeared both in 3-phosphoglycerate and glycolate.

___CAM___ c. The plant was incubated in the presence of $^{14}CO_2$, and newly incorporated radioactivity appeared in organic compounds during the night, but not during the day.

___CAM___ d. The leaves of the plant were examined under the microscope and were found to have two concentric cylinders of cells.

5 a. The following are lists of pigments (and their absorption maxima) that form the photosynthetic units of photosystems I and II. For each list, indicate the pigment molecule that is acting as the reaction center. Assuming all the pigments participate, indicate the pathway that energy takes in going to the reaction center.

PSI	PSII
Chl a (660)	Chl a (660)
Chl b (500)	Chl a (670)
Chl a (678)	Chl a (678)
carotenes (490)	Chl b (500)
Chl a (685)	Chl a (680)
Chl a (700)	xanthophylls (450)
Chl a (670)	Chl a (685)
Chl a (690)	

b. Plants with the pigments listed above were used in a series of experiments that measured the effect of wavelength on photosynthesis. When plants were given light of 700 nm, oxygen production ceased, even though the absorption spectrum indicated that some chlorophyll could absorb light of that wavelength. If the 700 nm light was supplemented with a beam of light at 650 nm, oxygen production resumed. Explain.

6. The membranes of the chloroplast play an essential role in photosynthesis. List three ways in which photosynthesis is dependent upon the membranes of the chloroplast.

Notes:

CHAPTER SEVEN

Interactions Between Cells and Their Environment

Learning Objectives

When you have finished this chapter, you should be able to:

1. Describe the extracellular matrix. Give some examples of tissues whose characteristic properties are determined by their respective extracellular matrices.
2. List the major components of the extracellular matrix, and include the functional role of each.
3. Discuss the families of proteins that mediate cell-to-cell and cell-to-substrate adhesion. Explain the roles of some of these proteins in inflammation and cancer.
4. Describe the cell surface specializations that enhance attachment of cells to substrates and to other cells.
   a. Give examples of cells that express each of these types of connections, respectively.
   b. Distinguish between the functional features of each type of cell-to-cell connection.
5. Describe the origin and role of plant cell walls.

Key Terms and Phrases

glycocalyx
collagen
fibronectin
tenascin
integrin
hemidesmosomes
NCAM and VCAM
desmosomes
metastasis
connexon
plasmodesmata
hemicellulose
cell plate
extracellular matrix (ECM)
proteoglycans
heparan sulfate proteoglycans (HSPG)
entactin
RGD sequence
immunoglobulin superfamily (IgSF)
cadherins
intercellular junctional complex
tight junctions
gap junction intercellular communication (GJIC)
cell walls
pectins
primary cell wall
basement membrane
glycosaminoglycans
laminin
thrombospondin
focal adhesions
selectins
adherins junctions
transmembrane signaling
connexin
microfibrils
lignins
secondary cell wall

Reviewing the Chapter

I. Introduction

A. Cells interact with extracellular material and other cells to form tissues.
B. These interactions mediate diverse activities: cell migration, growth, differentiation and tissue organization during embryogenesis.

II. The Extracellular Space

A. The glycocalyx (cell coat) is formed by carbohydrates projecting from membrane lipids and proteins.

B. The extracellular matrix (ECM) is an organized network beyond the plasma membrane.
   1. A basement membrane (basal lamina) underlies epithelia and surrounds muscle and fat cells.
      a. The basement membrane maintains epithelial cell polarity.
      b. The basement membrane assists in cell migration.
      c. The basement membrane forms a barrier to macromolecules.
   2. The ECM gives connective tissues their identifiable properties.

C. Components of the ECM:
   1. Collagens are a family of glycoproteins found only in the ECM.
      a. Collagen is a trimer containing three polypeptide chains.
      b. The structure of fibrillar collagens provides high tensile strength.
      c. The arrangement of collagen fibers determines the properties of tissues such as tendons and the cornea.
   2. Proteoglycans (PGs) are protein-polysaccharide complexes with a core protein attached to glycosaminoglycan (GAG) chains.
      a. GAGs have repeating disaccharide structure (A-B-A-B-A-B)
      b. Negatively charged GAGs attract small cations and water, forming a porous, hydrated gel.
      c. Heparan sulfate proteoglycans (HSPGs) are integral membrane proteins and interact with the cytoskeleton.
   3. Some PGs, e.g., fibronectin and laminin, bind cells to the ECM; others, e.g., tenascin, inhibit adhesion.

III. Adhesion of Cells to Noncellular Substrates

A. Integrins are membrane-spanning heterodimers with noncovalently linked chains.
   1. Large extracellular domains have Mg^{2+}- or Ca^{2+}-dependent ligand-binding sites specific for arginine-glycine-aspartic acid (RGD) sequences.
   2. Small cytoplasmic domains bind to cytoskeletal proteins.
   3. RGD-containing drugs or toxins inhibit normal integrin function.

B. Focal adhesions:
   1. In vitro cells form stable attachments on contact with substrates and flatten over time.
   2. In vivo, the tightest cell-substrate are mediated by hemidesmosomes, basal attachments of epithelial cells to the basement membrane.
   3. Contact between intracellular and extracellular structures occurs via integrins.

IV. Adhesion of Cells to Other Cells

A. Cells have surface-recognition sites that maintain organization within tissues.

B. Selectins bind to specific carbohydrates on surfaces of cells.
   1. Extracellular domains contain multiple ligand-binding sites.
   2. Selectins mediate interactions between circulating leukocytes and vessel walls.

C. Immunoglobulins and integrins:
   1. Most immunoglobulin superfamily (IgSF) proteins are involved in immunity.
   2. NCAM and VCAM IgSF proteins mediate adhesion between nonimmune cells directly or through integrins.

D. Cadherins are glycoproteins mediating Ca^{2+}-dependent cell adhesion.
   1. Cadherins join cells of similar type.
   2. Expression of different cadherins during embryogenesis allows tissues to rearrange during development.

E. Adherens junctions and desmosomes mechanically bind adjacent cells.
   1. The zonulae adherens in epithelia forms a belt around the cell's apical surface.
      a. Cell-to-cell contact in adherens junctions is cemented by extracellular domains of cadherin molecules.

b. Intracellular domains of cadherin molecules are linked by catenins to cytoskeletal proteins.

2. Maculae adherens are disc-shaped junctions between cells.
   a. Intracellular domains are linked to intermediate filaments.
   b. Intermediate filaments span the cell.

F. Adhesion receptors in transmembrane signaling:
   1. Binding of integrins and cadherins to extracellular ligands can transfer information into the cell.
   2. "Inside-out" signaling occurs when cellular changes alter the binding of extracellular domains of integral proteins.

V. The Role of Cell Adhesion in Inflammation and Metastasis (The Human Perspective)

A. Inflammation is a normal response to injury or infection but can be triggered inappropriately.

B. Inflammation begins with the recruitment of leukocytes to the site of injury.
   1. Neutrophils attach to P-selectins on the walls of the venules.
   2. Platelet activating factor is released by the venules.
      a. PAF alters the binding of neutrophilic integrins to IgSF (ICAMs).
      b. Neutrophils slip through the vessel walls (extravasation) into tissues.

C. Abnormal cell adhesion may be responsible for the spread (metastasis) of some cancers.

VI. Tight Junctions: Sealing the Extracellular Space

A. Tight junctions restrict the movement of material between epithelial cells.
B. Integral proteins (occludins) form intermittent cell contacts near junctional complexes.
C. Continuous parallel strands of integral proteins encircle the cell.
D. Tight junctions form the blood-brain barrier.
E. Cells of the immune system can cross tight junctions.

VII. Gap Junctions and Plasmodesmata: Mediating Cell Communication

A. Gap junctions connect cells via connexons, complexes of six connexin molecules.
B. Connexons have pores allowing diffusion of molecules less than 1000 daltons.
C. Gap junction intercellular communication (GJIC) integrates the activity of adjacent cells.
D. Plasmodesmata are channels through cell walls, connecting adjacent cells in plants.

VIII. Cell Walls

A. Plant cell walls provide protection against abrasion, osmotic stress, and pathogens.
B. Microfibrils of cellulose form the fibrous component of the cell wall.
C. The matrix of the cell wall contains hemicellulose, pectins, and hydroxyproline-rich, proline-rich, and glycine-rich structural proteins.
D. Primary cell walls are flexible to allow growth after division.
E. Mature secondary cell walls are thicker and more rigid.

IX. The Role of Gap Junctions in Intercellular Communication (Experimental Pathways)

A. An electronic synapse was first proposed to explain the rapid spread of depolarization between adjacent cells.
B. Electrical coupling of epithelial cells was observed when ions injected into an epithelial cell spread to adjacent cells.
C. Fluorescent dyes injected into cells spread between, but not out of, cells.
D. Electron micrographs established the structure of gap junctions.
E. Many types of cancer cell lack normal GJIC.
   1. GJIC decreases progressively as some cancers grow.
   2. Defective genes encoding gap junctions may be involved in some types ofcancer.

Key Figure

Figure 7.28. An overview of the interactions at the cell surface.

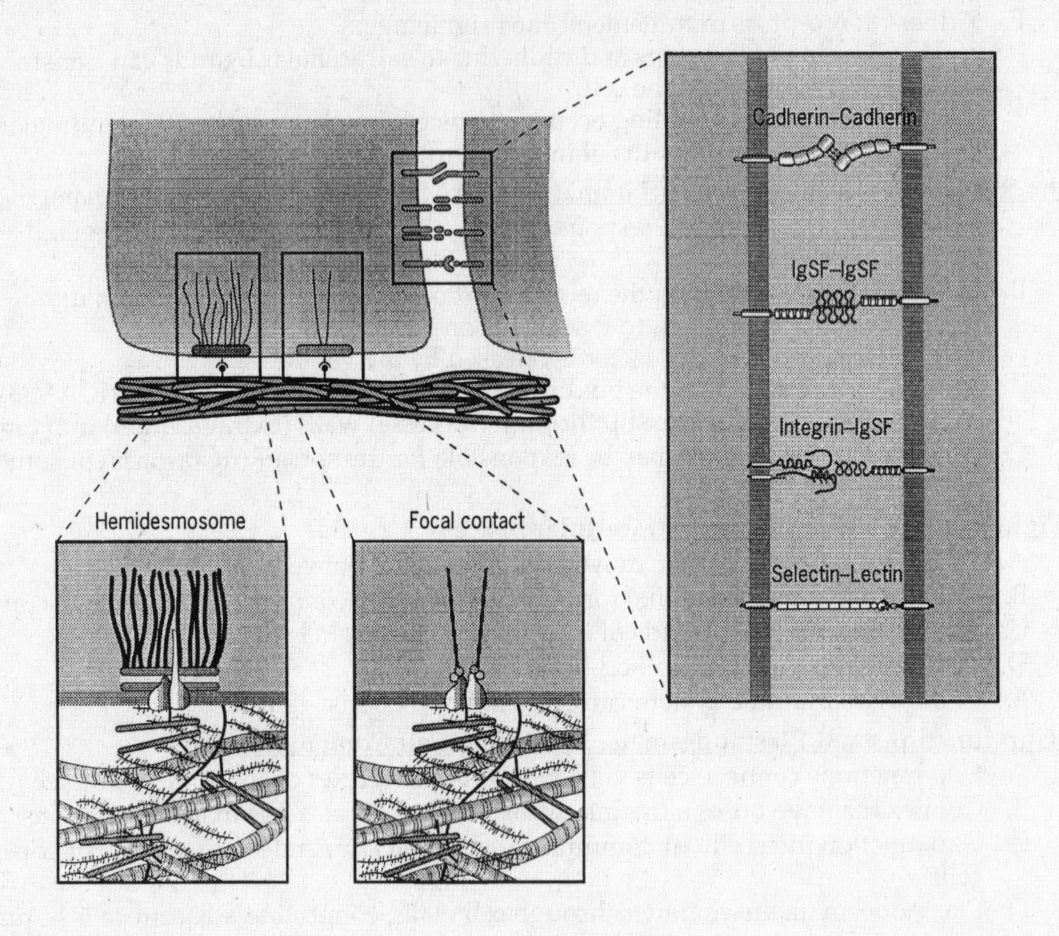

Questions for Thought:

1. How do the proteins that participate in these different cell-to-cell contacts differ? What features do they have in common?

2. What do each of these kinds of adhesive structures do for the cell and the tissue in which they are found? What kinds of tissues might have an abundance of each respective kind of adhesive unit?

3. Which proteins could be involved in transmembrane signaling? Which proteins probably are not? We usually think of transmembrane signaling as a transfer of information from the outside of the cell to the inside. How might information go from inside the cell to the outside compartment?

4. Some metastatic cancers are characterized by nonfunctioning cell-adhesion proteins. Which proteins illustrated here might contribute to metastasis in mutant form?

Review Problems

Short Answer

1. Give the significance (not the definition) of the following terms or phrases. Say what they do or why they are important. For example:
   extracellular matrix: *plays a key role in determining the shape and activities of the cell and tissue.*

   a. glycocalyx

   b. basement membrane

   c. integrins

   d. RGD sequence

   e. hemidesmosomes

   f. cadherins

   g. tight junctions

   h. gap junctions

   i. plasmodesmata

   j. pectins

2. Compare and contrast the following:
   a. glycocalyx vs. extracellular matrix

   b. fibronectin vs. integrin

   c. focal contacts vs. hemidesmosomes

   d. tight junctions vs. gap junctions

   e. primary cell walls vs. secondary cell walls

Multiple Choice

1. Which of the following is not an example of a type of extracellular matrix?
   a. basement membrane
   b. the cornea of the eye
   c. intermediate filaments
   d. bone

2. Which of the following proteins of the ECM has a membrane bilayer-spanning domain?
   a. fibrillar collagen
   b. heparan sulfate proteoglycans
   c. fibronectin
   d. laminin

3. Embryos injected with antibodies to fibronectin show inhibited movement of neural crest cells during development of the nervous system. These experiments show that:
   a. development of the neural crest involves the expression of antibody genes.
   b. developing neurons must synthesize fibronectin.
   c. fibronectin-antibody complexes form pathways for neural migration.
   d. neurons in embryos must transiently bind to fibronectin during migration.

4. Although the protein families that comprise the ECM are very different, they have one of the following characteristics in common:
   a. They all have two or more distinct domains with specific binding sites.
   b. They are all derived from proteins of the immune system.
   c. They all have membrane-spanning domains.
   d. None of them are attached to the cell.

5. Which of the following is not a role of integrins?
   a. Integrins form a loose, hydrated gel in the ECM.
   b. Integrins anchor cells to the substrate.
   c. Integrins transmit signals to the intracellular compartment.
   d. Integrins participate in specialized cell-to-cell adhesive structures

6. Integrin is to a focal adhesion as __?__ is to an adherens junction.
   a. integrin
   b. selectin
   c. fibronectin
   d. cadherin

7. The addition of a peptide with an RGD sequence would probably__?__ the binding of cultured cells to a fibronectin-coated dish.
   a. enhance
   b. inhibit
   c. have no effect on
   d. none of the above

8. The addition of a peptide with an RGD sequence would probably__?__ the binding of cultured cells to a collagen-coated or laminin-coated dish.
   a. enhance
   b. inhibit
   c. have no effect on
   d. none of the above

9. Focal adhesions are to cultured cells as __?__ are to cells in vivo.
   a. plasmodesmata
   b. gap junctions
   c. cell walls
   d. hemidesmosomes

10. Actin filaments are to focal adhesions as__?__are to hemidesmosomes.
    a. myosin filaments
    b. keratin filaments
    c. microfilaments
    d. cadherins

11. If a laboratory mouse had a mutation rendering the gene that codes for the L-selectin protein nonfunctional, which of the following symptoms would the animal exhibit?
    a. blistering of the skin
    b. clotting deficiencies
    c. metastasis
    d. inability to fight infection in the tissues

12. Calcium-dependent adhesion is to the cadherins as calcium-independent adhesion is to the_?.
    a. IgSF proteins
    b. fibronectins
    c. integrins
    d. glycocalyx

13. To which category of proteins listed below are NCAM and VCAM most closely related?
    a. proteoglycans
    b. antibodies
    c. integrins
    d. selectins

14. Which of the following cell types would you expect to have the highest number of desmosomes?
   a. smooth muscle cells
   b. red blood cells
   c. epithelial cells of the skin
   d. nerve cells

15. How might anti-selectin antibodies act as anti-inflammatory drugs?
   a. Anti-selectin antibodies prevent activated endothelial cells from expressing selectin.
   b. Anti-selectin antibodies compete with carbohydrate ligands on neutrophil surfaces for selectin binding sites.
   c. Anti-selectin antibodies prevent neutrophils from transiently binding blood vessel walls.
   d. Anti-selectin antibodies prevent activated endothelial cells from producing platelet activating factor.

16. In going from the apical to the basal end of an epithelial cell, in what order do the various cell surface junctions appear?
   a. tight junction → adherens junction → desmosomes → hemidesmosomes
   b. desmosomes → hemidesmosomes → adherens junction → tight junction
   c. adherens junction → tight junction → desmosomes → hemidesmosomes
   d. tight junction → adherens junction → hemidesmosomes → desmosomes

17. Cells of the heart must contract in synchrony to produce an effective heart beat. The electrical signal telling each cell to contract must reach every cell of the heart at the same time. What kind of cell-to-cell junctions would you expect to find in heart tissue?
   a. gap junctions
   b. tight junctions
   c. desmosome
   d. adherens junctions

18. Cells of the pancreatic acini produce and secrete powerful digestive enzymes. The enzymes are secreted into the acinar lumen where they flow into the pancreatic duct and ultimately into the intestine. Backleakage of these enzymes could cause self-digestion. What types of junctions would you expect to find between pancreatic acinar cells?
   a. gap junctions
   b. tight junctions
   c. desmosome
   d. adherens junctions

19. Fluorscein that is injected into one plant cell__?__ show up in adjacent cells; DNA injected into one plant cell __?__ show up in adjacent cells.
   a. would not; would not
   b. would not; would
   c. would; would not
   d. would; would

20. That cancer cells often lack the gap junctions common to non-malignant cells of the same type is evidence that
   a. cancer is a disease of gap junctions.
   b. GJIC is an element of controlled cell growth.
   c. cancers all involve mutations in connexin genes.
   d. connexin is responsible for cell adhesion.

Problems and Essays

1. Your studies of the development of the insect nervous system have lead you to suspect that nerve cells that grow out of the central nervous system follow paths of extracellular laminin. These pathways, however, are rich in both laminin and collagen, and you cannot be sure that its the laminin and not the collagen that these cells are following. Design an experiment in vitro to distinguish between these two possibilities.

2. Indicate whether the following are extracellular (EX), intracellular (IN), or integral, membrane-spanning (MS) proteins:

   ________ laminin

   ________ collagen

   ________ heparan sulfate proteoglycans (HSPG)

   ________ connexin

   ________ fibronectin

   ________ keratins affiliated with hemidesmosomes

   ________ integrin

   ________ selectin

   ________ NCAM

   ________ catenins

   ________ actin

3. The textbook suggests that unwanted blood clots may be treated by injecting synthetic RGD-containing peptides into the circulatory system, but then warns that such treatments might interfere with a "wide variety of other processes". Name three other processes that might be influenced by this treatment.

4. In your lab you have four murine (mouse) cell lines growing in culture. Culture 1 is an epithelial line originated from embryonic glial cells. Culture 2 is a line of neural ectoderm, also started from embryonic cells. Culture 3 is derived from culture 1, but has been transfected with a gene for N-cadherin. Culture 4 is derived from culture 2, but has been transfected with a gene for E-cadherin. When you grow cells from each of the different cultures as a monolayer, then add radiolabeled, suspended cells from each of the cultures to each respective monolayer, you note that some labeled cells stick to some monolayers and not to others, and other labeled cells stick to different monolayers.

a. Indicate in the table which combinations of suspended and monolayered cells aggregate (+) and which do not (-).

b. A fifth culture of murine fibroblast cells, called L cells, express no cadherins. Add a line to your table indicating which combinations that involve culture 5 will form aggregates.

Cells in suspension

Cells in monolayer	1	2	3	4	5
1					
2					
3					
4					
5					

5. Metastatic cancer cells often exhibit changes in characteristics of cell surface proteins. Assume you are comparing the cell surface proteins of a particularly invasive cancer with those of normal cells. Describe how the level of each of these proteins differs in the cancer cells. Include a sentence describing how the change contributes to the malignant character of the cell.

cell surface protein	expression increased or decreased in cancer cell	contribution to cell's malignant character
laminin receptors		
fibronectin receptors		
integrins		
cadherins		
gap junctions		

6. The following are descriptions of some tissues with specialized functions as described. Indicate what types of cell-to-cell junctions probably hold these cells together, and why.

   a. Smooth muscle surrounds the digestive tract. The muscle must contract in synchrony to create the peristaltic movements that propel the contents through the intestine.

   b. The acini of the pancreas are clumps of cell that surround a lumen and secrete powerful digestive enzymes into the lumen destined for the intestine. Back-leakage of the secreted enzymes could cause severe cell damage to the tissues surrounding the acini.

   c. The smooth muscle of the uterus must withstand extreme stretching during pregnancy without tearing.

   d. The skin of a frog actively transports salts from the environment into the body, creating an osmotic and electrical gradient between the inside of the animal and the pond water.

   e. The longitudinal muscles and the circular muscles that surround the cavity of a sea anemone are complementary to one another. One of the sets of muscles contracts, the other relaxes. Each group acts as a single unit.

Notes:

CHAPTER EIGHT

Cytoplasmic Membrane Systems: Structure, Function and Membrane Trafficking

Learning Objectives

When you have finished this chapter, you should be able to:

1. Explain the structural and functional relationship between the endoplasmic reticulum, Golgi complexes, lysosomes, and plasma membranes of eukaryotic cells.
   a. Describe the pathway of proteins destined for secretion, starting with synthesis and ending with exocytosis.
   b. Describe the pathway of materials taken up by the cell (including the various mechanisms of uptake), from the extracellular compartment to the final destination.
   c. Describe the trafficking of membranes in both secretory and cellular uptake processes.
2. Summarize the mechanisms whereby newly synthesized proteins are targeted and sorted for their appropriate destinations.
3. List the ways in which proteins can be modified and elaborated after they are synthesized; include the intracellular sites of these modifications.
4. Describe lysosomes and plant vacuoles. What do they have in common? Describe some human disorders associated with lysosomal malfunction.

Key Terms and Phrases

endomembrane system
secretory pathway
endocytic pathway
pulse-chase experiment
endoplasmic reticulum (ER)
cytoplasmic space
signal hypothesis
translocation
protein quality control
dolichol phosphate
trans Golgi network (TGN)
non-clathrin coated vesicles
anterograde movement
adaptin
secretory granules
turnover
vacuole
endocytosis
opsonin
low-density lipoprotein (LDL)
transport vesicle
constitutive secretion
sorting signal
cell fractionation
rough endoplasmic reticulum (RER)
smooth endoplasmic reticulum (SER)
signal recognition particle (SRP)
signal peptidase
phospholipid exchange protein
transitional elements
dictyosome
adenosylation ribose factor (ARF)
retrograde movement
N-linked and O-linked carbohydrates
exocytosis
lysosomal storage disorder
tonoplast
receptor-mediated endocytosis (RME)
endosomes
high-density lipoprotein (HDL)
biosynthetic pathway
regulated secretion
autoradiography
microsomes
cisternal space (lumen)
signal sequence
SRP receptor
flippase
glycosyltransferase
Golgi complex
coat proteins (COPs)
bulk flow
t-SNAREs and v-SNAREs
clathrin
lysosome
autophagy
phagocytosis
bulk-phase endocytosis
coated pits
triskelion

Reviewing the Chapter

I. Introduction

A. Membranes divide the cytoplasm of eukaryotic cells into distinct compartments.
B. Membrane-bound structures (organelles) are found in all cell types.

II. The Dynamic Nature of the Endomembrane System

A. Most organelles are part of a dynamic system in which vesicles move between compartments.
B. Biosynthetic pathways move proteins and lipids within the cell.
C. Secretory pathways discharge proteins from cells.
   1. Constitutive secretion is the continual, unregulated discharge of material from the cell.
   2. Regulated secretion is the discharge of products stored in cytoplasmic granules, in response to appropriate stimuli.
D. Endocytic pathways move materials into cells.
E. Sorting signals are recognized by receptors and target proteins to specific sites.

III. A Few Approaches to the Study of Cytomembranes

A. Autoradiography visualizes radiolabeled materials exposed to a photographic film.
   1. The site of protein synthesis in the cell can be visualized by incubating tissue with labeled amino acids.
   2. Brief incubation with labeled amino acids (pulse) and subsequent incubation with unlabeled amino acids (chase) allow the movements of proteins in the cell to be followed.
B. Subcellular fractionation is used to study the biochemistry of purified cell structures.
   1. Vesicles formed in cell homogenates have different properties allowing membranes from different organelles to be separated.
   2. Microsomes are heterogenous mixtures of similar-sized vesicles, formed from membranes of the endoplasmic reticulum and Golgi complex.
   3. Microsomes retain activity during purification, allowing studies of function and composition.
C. Genetic mutants provide insights into the function of normal gene products.
D. The dynamic activities of endomembrane systems are highly conserved despite the structural diversity of different cell types.

IV. The Endoplasmic Reticulum

A. Membranes of the endoplasmic reticulum (ER) enclose a luminal (cisternal) space.
B. Smooth ER (SER) is an interconnecting network of tubular membrane elements.
C. Functions of the SER include:
   1. synthesis of steroids in endocrine cells.
   2. detoxification of organic compounds in liver cells.
   3. release of glucose 6-phosphate in liver cells.
   4. sequestration of Ca^{2+}.
D. Rough ER (RER) has ribosomes on the cytosolic side of continuous, flattened sacs (cisternae).
   1. The polarity of organelles in some secretory cells reflects the flow from the site of synthesis to the site of discharge.
   2. Proteins synthesized on ribosomes of RER include secretory proteins, integral membrane proteins, and proteins of organelles.
   3. Proteins synthesized on "free" ribosomes include cytosolic proteins, peripheral membrane proteins, and proteins found in mitochondria and chloroplasts.
E. A signal sequence on the N-terminus attaches nascent secretory proteins to the ER.
   1. Messenger RNA binds to free ribosomes in the cytosol.

2. The signal sequence binds a signal recognition particle (SRP) which then attaches to an SRP receptor on the ER.
3. The release of the SRP requires a GTP-binding protein.
4. Proteins undergoing translocation pass through channels in the ER.
5. The signal sequence is cleaved, the protein is folded, and carbohydrates are attached in the cisternae of the ER.

F. Asymmetry is maintained during membrane trafficking.

G. Integral membrane proteins have hydrophobic stop-transfer sequences.

H. Lipids inserted into the outer leaflet of the ER are modified by:
1. flippases that translocate some lipids to the inner leaflet.
2. enzymes that convert one phospholipid to another.
3. selective budding of vesicles.
4. phospholipid exchange proteins that move lipids between membranes.

I. N-linked and O-linked carbohydrates are added by glycosyltransferases.
1. Sugars are attached in an ordered sequence.
1. The core carbohydrate chain is assembled on a lipid carrier, dolichol phosphate.
2. The core carbohydrate is transferred to the polypeptide by oligosaccharyltransferase and is modified en route to the Golgi complex.

J. Transport vesicles targeted for the Golgi complex arise from the smooth, apical cisternae of the RER.

V. The Golgi Complex

A. Golgi are stacks of flattened cisternae lacking ribosomes.

B. The cis face of the Golgi faces the ER; the trans face is on the opposite end of the stack.

C. Proteins are processed first in the cis, then medial, then trans Golgi cisternae.

D. In the maturation model, cisternae formed on the cis face are transient and physically move to the trans face.

E. A more recent model holds that material moves through permanent Golgi stacks in non-clathrin-coated vesicles.
1. Adenosylation ribose factor (ARF), a GTP-binding protein, is required for vesicle transfer between cisternae.
2. Proteins are directed to different sites in the ER and Golgi by specific sequences.
3. Specific v-SNARE proteins in the membranes of vesicles and t-SNARE proteins in receptor membranes direct vesicle movements between compartments.
4. Anterograde and retrograde movements through the Golgi are possible.

F. Sorting of proteins by different destinations occurs in the trans-Golgi network (TGN).
1. Nonselective, non-clathrin coated vesicles carry constitutively secreted material.
2. Selective, clathrin-coated vesicles are targeted for specific destinations.
3. Clathrin-coated vesicles contain adaptins– proteins involved in the binding of vesicles to specific receptors.

G. Lysosomal proteins are tagged with phosphate groups in the Golgi cisternae.
1. These proteins are captured by mannose 6-phosphate receptors in the TGN.
2. The cytosolic domains of the receptors bind adaptin molecules and direct the vesicles to lysosomes.

H. Oligosaccharides attach to glycoproteins and glycolipids in the Golgi, where the synthesis of complex polysaccharaides also takes place.

I. Exocytosis is the process of vesicle fusion and content discharge.
1. Cells engaged in regulated secretion have numerous clathrin-coated vesicles.
2. The process of exocytosis is triggered by local increases in Ca^{2+} concentration.
3. Fusion proteins found on the vesicle and membrane mediate the formation of fusion pores.

4. Vesicles may or may not become absorbed into the plasma membrane.

VI. Lysosomes

A. Lysosomes contain acid hydrolases that can digest every kind of biological molecule.
B. Internal proton concentration is kept high by H^+-ATPases.
C. Glycolsylated proteins, lgp-A and lgp-B, may protect the lysosome from self-digestion.
D. Lysosomes are identified by the presence of the enzyme acid phosphatase.
E. Lysosomes are involved in two major cell functions– phagocytosis and autophagy.
F. Primary lysosomes fuse with either phagocytic or autophagic vesicles, forming residual bodies that either undergo exocytosis or are retained in the cell as lipofuscin granules.
G. Disorders resulting from defects in lysosomal function (The Human Perspective):
    1. Lysosomal malfunction can have profound effects on human health.
    2. Silicosis and asbestosis result from the lysosomal uptake of undigestible fibers.
    3. Lysosomal storage disorders result from the absence of specific lysosomal enzymes allowing undigested material to accumulate.
    4. Most hereditary lysosomal disorders can be diagnosed prenatally.

VII. Plant Cell Vacuoles

A. Plant cells contain large vacuoles for the storage of compounds.
B. Transport systems in the tonoplast (vacuole membrane) accumulate material in the vacuole, attracting water by osmosis; this in turn generates turgor pressure, providing support for the cell.
C. Plant vacuoles, like lysosomes of animal cells, are an end point for biosynthetic pathways and may contain acid hydrolases.

IIX. Cellular Uptake of Particles and Macromolecules

A. Phagocytosis is the cellular uptake of particulate matter from the environment.
    1. Phagocytosis is initiated by cellular contact with an appropriate target.
    2. Phagocytosis may be stimulated by the opsonins that coat the foreign objects.
B. Endocytosis is the cellular uptake of material dissolved or suspended in fluid.
    1. Bulk-phase endocytosis does not require surface membrane recognition.
    2. Receptor-mediated endocytosis (RME) follows the binding of substances to membrane receptors.
        a. Ligand-bound receptors collect in specialized regions of the membrane forming coated pits.
        b. Clathrin-coated regions invaginate into the cytoplasm, forming coated vesicles.
        c. Clathrin contains three chains that form a triskelion capable of changing from a flat sheet to a cage like scaffold surrounding the vesicle.
        d. Adaptor complexes between the clathrin lattice and the membrane interact with specific signals in the receptors being internalized.
        e. The vesicle loses its clathrin coat and may fuse with an early endosome. The receptors may be recycled to the plasma membrane.
C. LDLs and cholesterol metabolism:
    1. Cholesterol is transported in the blood in lipoprotein complexes.
    2. LDLs are taken up by RME and delivered to lysosomes, releasing the cholesterol for use by the cells.
    3. HDLs transports cholesterol from tissues to the liver, for elimination.
    4. HDL is associated with lowering cholesterol levels, whereas LDL is associated with high blood cholesterol and the formation of atherosclerosis.

IX. Receptor-Mediated Endocytosis (Experimental Pathways)

A. The mechanisms of endocytosis at specialized coated pits was first proposed by Roth and Porter to explain the uptake of yolk protein by oocytes.
B. The detailed structure of the coated vesicles was revealed by electron microscopy.

C. Barbara Pearse identified clathrin as the predominant protein by SDS-PAGE electrophoresis of proteins from purified fractions of coated vesicles.
D. Cells from individuals with familial hypercholesterolemia (FH) are unable to regulate cholesterol synthesis in response to the presence of LDL.
E. Brown and Goldstein showed that in affected individuals, the primary defect is the inability of the LDL receptor to initiate RME.
F. A key tyrosine residue may be essential for the localization of ligand bound receptors in coated pits.

Key Figure

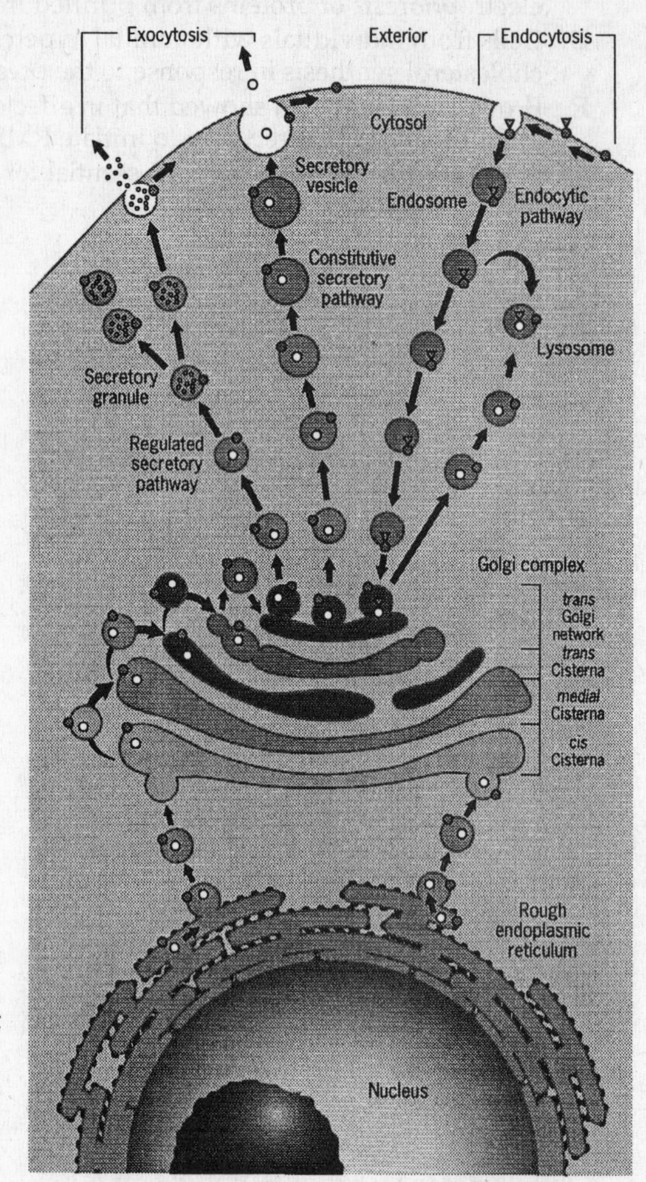

(a)

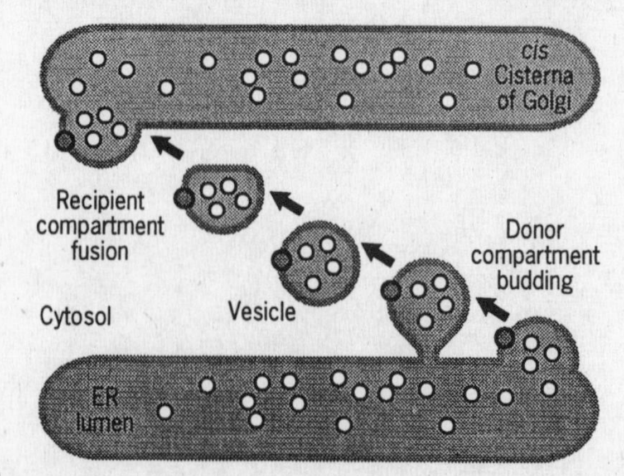

(b)

Figure 8.2. Biosynthetic and endocytic pathways unite the endomembranes in a dynamic, interconnected network.

Questions for Thought:

1. Starting with the rough endoplasmic reticulum, trace a patch of membrane out to the cell's outer boundary, then back again to the Golgi complex. What lipid modifications may take place during this migration?

2. If you were doing histochemistry on the cell illustrated here, which membrane compartments would test positive for carbohydrates?

3. Which vesicles illustrated here might be surrounded by clathrin?

4. The exact mechanism of regulated secretion is only partially understood. We know that calcium and membrane proteins on both the secretory vesicle and the plasma membrane are involved. Can you propose a mechanism that includes these components? How might you test your model?

Review Problems

Short Answer

1. Give the significance (not the definition) of the following terms or phrases. Say what they do or why they are important. For example:
   endoplasmic reticulum: *delineates an intracellular compartment that functions in the sorting of newly synthesized proteins and provides a surface for the synthesis of many proteins.*

   a. regulated secretion

   b. cell fractionation

   c. signal sequence

   d. trans Golgi network

   e. Rothman and Orci experiments on VSV-infected CHO cells

   f. phosphorylated mannose residues on TGN glycoproteins

   g. lysosomes

   h. acid hydrolases

   i. receptor-mediated endocytosis (RME)

   j. coated pits

2. Compare and contrast the following:
   a. pulse vs. chase

   b. protein synthesis on membrane-bound vs. free ribosomes

   c. synthesis of N-linked vs. O-linked oligosaccharaides

   d. autophagy vs. phagocytosis

   e. LDL vs. HDL

Multiple Choice

1. Cytoplasmic membrane systems are characteristic of what types of cells?
   a. prokaryotic cells
   b. eukaryotic cells
   c. bacterial cells
   d. all cells

2. Pancreatic cells that produce and release digestive enzymes engage in what type of secretion?
   a. regulated secretion
   b. constitutive secretion
   c. transport secretion
   d. none of the above

3. Which of the following cell types would be the best model system for studying secretory granules?
   a. muscle cells
   b. yeast cells
   c. epithelial cells
   d. pancreatic cells

4. You are interested in studying the asymmetric (coming from just one cell side) release of digestive enzymes from epthelial cells that line the digestive tract. Which of the following techniques would be best for these studies?
   a. cell fractionation
   b. homogenization
   c. autoradiography
   d. all of the above

5. Which of the following explains why microsomes can't be seen in cells viewed with the electron microscope?
   a. They are far too small.
   b. They are artifacts of homogenization and centrifugation.
   c. They are transparent to electrons.
   d. They actually can be seen in electron micrographs of cells.

6. Which of the following proteins would not be found in the smooth endoplasmic reticulum?
   a. Ca^{2+}-pumping enzymes
   b. cytochrome P450
   c. glucose 6-phosphatase
   d. signal peptidase

7. Which of the following groups of proteins probably lack a signal sequence?
   a. acid hydrolase enzymes synthesized in macrophage cells
   b. glycolytic enzymes synthesized in liver cells
   c. polypeptide hormones synthesized in endocrine cells
   d. antibody hormones synthesized in plasma cells

8. Which of the following is not an essential component of the complex that directs a nascent protein into the lumen of the RER?
   a. protein disulfide isomerase
   b. SRP
   c. SRP receptor
   d. GTP-binding protein

9. If you compared the proteins in a cis Golgi compartment with those in a trans Golgi compartment, you would find:
   a. the proteins in the two compartments are identical.
   b. the proteins in the cis compartment are glycosylated and contain modified amino acids, whereas those in the trans compartment are not modified.
   c. the proteins in the cis compartment are glycosylated, whereas those in the trans compartment are glycosylated and contain modified amino acids.
   d. the proteins of the cis compartment are shorter than those of the trans compartment.

10. You are tracing the path of a secretory protein from its synthesis to its export from a cell. You have added radioactive amino acids to a culture of cells, and then measured the amount of radioactivity that shows up in the proteins of each of the following cell fractions at different times after the addition. List the order in which the proteins of these fractions first exhibit radioactivity:

    I secretory vesicles
    II Golgi complex
    III rough ER
    IV smooth ER
    V nucleus

   a. III → II → I → out of the cell
   b. III → II → IV → V → out of the cell
   c. V → III → II → I → out of the cell
   d. IV → III → II → I → out of the cell

11. Of the following processes that occur during protein trafficking, which involve GTP?
   a. protein translocation across the ER membrane
   b. disassembly of the COP coat on a transport vesicle
   c. fusion of non-clathrin coated vesicles with target membrane
   d. all of the above

12. You have labeled the lipids on a patch of rough ER membrane with a fluorescent probe. After a few minutes, the probe shows up in the membranes of the cis Golgi. Now you treat the cells with the drug brefeldin A. Where might the fluorescent probe show up next?
    a. trans Golgi network
    b. endoplasmic reticulum
    c. plasma membrane
    d. secretory vesicles

13. VAMPs are to v-SNARE proteins as syntaxin like proteins are to ___?___.
    a. t-SNARE proteins
    b. v-SNARE proteins
    c. vesicular proteins
    d. coat proteins

14. Which type of vesicle of the trans Golgi network would be most likely to carry hormones destined for regulated secretion?
    a. lysosomal vesicles
    b. clathrin-coated vesicles
    c. non-clathrin-coated vesicles
    d. all of the above

15. The KDEL sequence is to proteins destined for the ER lumen as ___?___ is to proteins destined for lysosomes.
    a. KKXX
    b. adaptins
    c. t-SNARE
    d. phosphorylated mannose residues

16. Of the following, which would be least likely to be found in lysosomes?
    a. acid hydrolase enzymes
    b. a half-digested mitochondrion
    c. nucleic acids
    d. a high concentration of protons

17. The structures most analogous to lysosomes in plant cells are called ___?___.
    a. vacuoles
    b. peroxisomes
    c. digestive vesicles
    d. lysosomes

18. Uptake of low-density lipoproteins (LDL) occurs by ___?___, whereas retrieval of plasma membrane after extensive secretory activity occurs by ___?___.
    a. receptor-mediated endocytosis, bulk-phase endocytosis
    b. bulk-phase endocytosis, receptor-mediated endocytosis
    c. receptor-mediated endocytosis, receptor-mediated endocytosis
    d. bulk-phase endocytosis, bulk-phase endocytosis

19. Which of the following are not associated with coated pits?
    a. TGN
    b. triskelion
    c. adaptor complexes
    d. integral membrane receptors

20. Which cells of the body would probably be the richest in HDL receptors?
    a. stomach cells
    b. liver cells
    c. muscle cells
    d. epithelial cells

Problems and Essays

1. For each of the following proteins, indicate whether it is synthesized on "free" ribosomes (FR) or on the rough endoplasmic reticulum (RER).

    __RER__ fibronectin
    __Free__ cytoskeletal proteins such as actin
    __RER__ acid hydrolases
    __RER__ collagen
    __free__ triskelion
    __free__ the enzymes of glycolysis
    __R__ adaptin
    __R__ LDL receptor
    __R__ signal peptidase
    __R__ protein disulfide isomerase
    __free__ mitochondrial proteins coded for by nuclear DNA

2. Jamieson and Palade used the pulse-chase method to follow the fate of radiolabeled amino acids, from protein synthesis to secretion. Using what you know about endocytosis, use the same method (just on paper) to follow the fate of radiolabeled amino acids that are incorporated into the proteins of a bacterial cell that is engulfed by a macrophage.

3. Insulin is synthesized in pancreatic islet cells first as a precursor called proinsulin; then it is transformed into mature insulin by a proteolytic cleavage of a short peptide, via a protease enzyme. Your laboratory has antibodies specific for both proinsulin (anti-P) and insulin (anti-I). Both antibodies can be complexed with electron-dense gold particles and viewed in electron micrographs as dark spots. You have treated one specimen of thinly sliced pancreatic tissue with anti-P and another with anti-I, and viewed both in the electron microscope. The + signs on the following table indicate where the black dots appear in your micrographs:

	anti-P	anti-I	anti-clathrin
mature secretory vesicles	-	+	
trans Golgi network	+	-	
cis and medial Golgi cisternae	+	-	

a. Where does proteolytic conversion of proinsulin take place?

b. Assume you have treated a third specimen with anti-clathrin antibodies. On the table, indicate which, if any, vesicles would stain positive for clathrin?

4. A certain integral protein of the plasma membrane has two transmembrane domains in which both the C-terminus and the N-terminus are directed toward the cell exterior in the mature protein. Where are the C- and N-termini located when the protein is newly translocated into the RER membrane? Into the Golgi? Into secretory vesicles?

5. Protein synthesis can be studied in cell-free systems using free amino acids, mRNA and an extract of wheat germ, which contains ribosomes and all of the cofactors necessary for translation. You are using such a system in your lab to study translocation of nascent proteins. For each of the experimental conditions described below, indicate whether the protein is completely synthesized, whether the mature protein has a signal sequence, and whether it is translocated across a membrane.

Experimental Conditions	completely synthesized?	signal sequence?	translocated across a membrane?
a. mRNA codes for a cytosolic protein; no SRP present; no SRP receptor present:: no microsomes present			
b. mRNA codes for a secretory protein; no SRP present; no SRP receptor present; no microsomes present			
c. mRNA codes for a secretory protein; exogenous SRP added; no SRP receptor present; no microsomes present			
d. mRNA codes for a secretory protein; exogenous SRP added; free (not membrane bound) SPR receptor added; no microsomes			
e. mRNA codes for a secretory protein; exogenous SRP added; exogenous SRP receptor added; microsomes added			

6. When an endosome containing LDL bound to LDL receptors fuses with a lysosome, the drop in pH decreases the affinity of the particle for the receptor, and both the lipid and the carrier proteins are degraded in the fused vesicle. Iron, like cholesterol, is transported in the blood as a complex with the transport protein transferrin. It is the iron-transferrin complex, called ferrotransferrin, that serves as the ligand for the transferrin receptor on cell membranes. Unlike the response of LDL, a decrease in pH does not lower the affinity of the ligand for the receptor. It does, however, lower the affinity of iron for the carrier protein transferrin. Propose a sequence of events in cells that take up ferrotransferrin, and follow the fates of both transferrin and the transferrin receptor following iron uptake.

Notes:

CHAPTER NINE

The Cytoskeleton and Cell Motility

Learning Objectives

When you have finished this chapter, you should be able to:

1. Describe the three types of cytoskeletal structures.
   a. For each structure, list the features of the component proteins and explain how the proteins interact.
   b. Explain how each structure is assembled and disassembled, and how these processes are regulated.
2. Define "molecular motor." Name the three families of motor proteins.
   a. State where each type of motor protein is likely to occur and with which cytoskeletal structures each is affiliated.
   b. Explain how motor proteins convert chemical energy into cellular movement.
3. Name the cellular activities that involve the cytoskeleton, including those that affect movement and those that do not.
4. Describe the organization of cytoskeletal and accessory proteins in skeletal muscle. How does this arrangement relate to muscle contraction? How is contraction regulated?
5. Understand the techniques used to study the cytoskeleton and cellular motility. Appreciate why these techniques are well suited to these types of studies.

Key Terms and Phrases

cytoskeleton
microtubules
actin
kinesin
plus and minus ends
centriole
basal body
axoneme
sarcomere
transverse (T) tubules
actin-binding proteins
lamellipodia
intermediate filaments (IFs)
fluorescence microscopy
video microscopy
ciliary and cytoplasmic dynein
microtubule-associated protein (MAP)
microtubule-organizing center (MTOC)
cilia
muscle fiber
thin and thick filaments
neuromuscular junction
exctitation-contraction coupling
contact inhibition of movement
microfilaments
tubulin
myosin
protofilament
centrosome
pericentriolar material
flagella
myofibrils
striations
sarcoplasmic reticulum (SR)
pseudopodia
growth cone

Reviewing the Chapter

I. Introduction

A. The cytoskeleton is a dynamic structure with many roles.
   1. It serves as a scaffold, providing structural support and maintaining cell shape.
   2. It serves as an internal framework, organizing organelles within the cell.
   3. It assists in movement of materials within the cell and cellular locomotion.

4. It provides anchoring sites for mRNA.
5. It serves as a signal transducer.

B. The cytoskeleton is a network of three filamentous structures: microtubules, microfilaments, and intermediate filaments.

II. The Study of the Cytoskeleton

A. Fluorescent microscopy of live cells locates fluorescently-labeled proteins.
B. Video microscopy and in vitro motility assays enable observation of intracellular movements.
C. Molecular biology and genetic mutants identify genes for essential cytoskeletal proteins.
D. The electron microscope can be used to visualize the cytoskeleton of a treated cell.

III. Molecular Motors

A. Molecular motors convert chemical energy of ATP into mechanical energy.
B. Molecular motors move by stepwise, conformational changes corresponding to a mechanical cycle.
C. Three types of molecular motors have been described.
   1. Kinesins and dyneins move along tracks of microtubules.
   2. Myosins move along microfilament tracks.

IV. Microtubules

A. Structure and composition:
   1. Microtubules are hollow cylindrical structures found in the cytoskeleton, mitotic spindle, cilia, and flagella.
   2. The microtubule is a polymer made up of globular tubulin subunits arranged in longitudinal rows called protofilaments.
   3. Most microtubules contain 13 protofilaments.
   4. Heterodimers of α- and ß-tubulin subunits are assembled into tubules with plus and minus ends.

B. Microtubule-associated proteins (MAPs):
   1. A MAP contains a globular "head" that attaches to a microtubule and a filamentous "tail" that extends away from the microtubule's surface.
   2. MAPs not only form crossbridges, organizing microtubules into bundles, they also stabilize, alter the rigidity of, or influence the assembly of microtubules.
   3. MAPs are regulated by phosphorylation of specific amino acid residues.

C. Microtubules as structural supports:
   1. The distribution of microtubules corresponds to the shape of the cell.
   2. Nerve axons have microtubules oriented parallel to the long axis of the axon.
      a. In mature axons, microtubules function in axonal transport of vesicles.
      b. During embryonic development, microtubules play a key role in axon growth.
   3. Microtubules in plant cells organize cellulose-synthesizing enzymes that produce the cell wall.
   4. Microtubules play a role in the location of organelles.

D. Microtubules as agents of intracellular motility:
   1. Microtubules facilitate the movement of vesicles traveling between compartments.
   2. Axonal transport:
      a. Neurotransmitters are synthesized in the cell body and released from synapses at the ends of axons.
      b. Both anterograde transport (away from the cell body) and retrograde transport (toward the cell body) involve microtubules.

E. Kinesins:
   1. Kinesin is a large protein with a pair of force-generating globular heads and a fan-shaped, cargo-binding tail.
      a. The motor domains of kinesin and kinesinlike proteins are similar.
      b. Tails sequences differ reflecting the variety of cargo these proteins can haul.
   2. Kinesin is a plus end-directed microtubular motor responsible for anterograde transport.
   3. Kinesin functions via an ATP-dependent cross-bridge cycle.
   4. Kinesinlike proteins are found in all eukaryotic cells.

F. Cytoplasmic Dyneins:
   1. Dyneins are associated with cilia and flagella but are also found in the cytoplasm.
   2. Cytoplasmic dyneins are large proteins with globular, force-generating heads.
   3. Dyneins move toward the minus ends of microtubules and are involved in chromosomal movement during mitosis, minus end-directed movement of vesicles, and retrograde axonal transport.

G. Microtubule organizing centers (MTOCs):
   1. Centrosomes have two centrioles surrounded by pericentriolar material.
      a. Centrioles have nine fibrils, each with three microtubules.
      b. Centrioles are found in pairs oriented at right angles to each other.
      c. Microtubules grow from the pericentriolar material of a centrosome.
      d. Centrosomes form the poles from which the mitotic spindle emerges.
   2. Basal bodies and other MTOCs:
      a. Basal bodies are the MTOCs for cilia and flagella.
      b. Spindle pole bodies serve as MTOCs in fungi.
      c. The protein γ–tubulin occurs in all MTOCs and is essential for microtubule assembly.

H. The dynamic properties of microtubules:
   1. Microtubules assemble and disassemble readily.
   2. Tubulin dimers are added or removed at the plus end, away from the MTOC.

I. Factors that influence assembly and disassembly:
   1. GTP on the ß-subunit of the tubulin dimer is hydrolyzed during polymerization.
      a. Dimers bound to GTP have a higher affinity for one another than those bound to GDP.
      b. GTP hydrolysis provides a mechanism for regulation of assembly and disassembly.
   2. Temperature, Ca^{2+} concentration, and MAPs influence microtubule stability.

J. Cilia and flagella:
   1 Cilia and flagella have similar structures but different motions.
   2. Cilia are relatively short and occur in large numbers on cell surfaces.
   3. Flagella, longer and fewer in number, are used primarily for motility.

K. The structure of cilia and flagella:
   1. The central core (axoneme) of both cilia and flagella contain microtubules in a 9 + 2 arrangement.
      a. The peripheral doublets contain both an A and a B tubule.
      b. Radial spokes connect the A tubules to the central tubules.
   2. Cilia and flagella emerge from basal bodies.
   3. Basal bodies contain nine triplet microtubules and no central microtubules.

L. The dynein arms:
   1. The machinery for ciliary and flagellar function resides in the axoneme.
   2. Ciliary dynein is needed for ATP hydrolysis.

M. The mechanism of ciliary and flagellar locomotion:
   1. Swinging cross-bridges generate the forces for ciliary or flagellar movement.
   2. The dynein arm of an A tubule binds to a neighboring B tubule and forces a conformational change that slides the tubules past each other.
   3. Sliding alternates from one side of the axoneme to the other.
   4. Radial spokes and interdoublet links provide resistance.

N. Ciliary and flagellar locomotion is regulated by Ca^{2+} and cAMP.

V. Intermediate Filaments

A. Intermediate filaments (IFs) are a heterogeneous group of proteins, each with a central helix flanked by globular domains.

B. IFs assemble by several mechanisms that do not include nucleotide hydrolysis.
   1. The basic unit of assembly is a tetramer formed by two antiparallel dimers.
   2. Both the tetramer and the IF lack polarity.

C. IFs resist tensile forces and IFs containing keratin form the protective barrier of skin.

D. Assembly and disassembly of IFs are controlled by phosphorylation and dephosphorylation.

E. IFs include neurofilaments, the major component of the structural framework supporting neurons.

F. IFs may provide specialized mechanical support in certain cell types.
   1. Defective IFs may lead to extreme fragility.
   2. Overexpression of IFs in neurons can lead to degenerative diseases.

VI. Microfilaments

A. Microfilaments are made of actin and are involved in cell motility.

B. Using ATP, globular actin (G actin) polymerizes to form actin filaments (F actin).

C. Actin is a major contractile protein in muscle but is present in all cell types.

D. G actin is added at the plus end and removed from the minus end, causing treadmilling.

E. Most motile processes involving actin require myosin.

F. Myosin II-class proteins generate forces in muscle and some non-muscle processes.
   1. Myosin II has a tail composed of two heavy chains and two globular heads.
   2. Nonhelical sections of the heavy chain form "hinge" regions.
   3. Each head region has actin- and ATP-binding sites on the heavy chain.
   4. Myosin forms bipolar filaments with tails all oriented toward the center.

G. Myosin I is found near cell surfaces and has only a single head.

VII. Muscle Contractility

A. Skeletal muscle fibers are large, multinucleate cells packed with myofibrils that contain repeating arrays of sarcomeres.

B. Thick and thin filaments in the sarcomere overlap.
   1. Sarcomeres extend between Z lines.
   2. I bands contain only thin filaments, the H zone contains only thick filaments, and the A band contains both thick and thin filaments.

C. Skeletal muscle functions by shortening fiber length.

D. Three classes of filamentS are found in muscle cells: actin, myosin, and titin.
   1. Titin filaments provide tension in resting muscles and support myosin filaments.
   2. Tropomyosin molecules occupy the groove between the two actin molecules.
   3. Heterotrimeric troponin molecules are spaced evenly along thin filaments.
   4. Myosin head project laterally from the thick filaments.

E. During contraction, the myosin heads form cross-bridges with the actin filaments.
   1. The myosin heads bend, sliding the thin actin filaments over the thick filament.
   2. Full contraction of a sarcomere requires 50 to 100 cycles of cross-bridging.

F. Energy is provided by ATPase activity in the myosin head.

1. ATP hydrolysis provides energy for "cocking" the myosin head.
2. The binding of the head to the thin filament initiates the power stroke.
3. The binding of a new ATP releases the head from the thin filament.
4. The absence of ATP prevents dissociation of the thick and thin filaments and is the basis for rigor mortis, the condition following death.

G. Muscle fibers innervated by branches of a single motor neuron form motor units.
1. The contact between nerve and muscle occurs at the neuromuscular junction.
2. Linking the nerve impulse to the shortening of the sarcomere is called excitation-contraction coupling.
3. Action potentials are carried into the cell interior by transverse (T) tubules.
4. T tubules terminate near the sarcoplasmic reticulum (SR), which sequesters Ca^{2+}.
5. In the relaxed state, cytoplasmic Ca^{2+} levels are low. Action potentials open calcium channels in the SR, releasing Ca^{2+}.
6. The binding of Ca^{2+} to troponin C causes a conformational change, shifting the tropomyosin and exposing the myosin binding site.
7. Ca^{2+} is removed from the cytoplasm by the Ca^{2+}-ATPase activity in the SR, causing tropomyosin to hide myosin-binding sites.

VIII. Nonmuscle Motility

A. Actin-binding proteins organize actin filaments into functional assemblies.
1. Monomer-sequestering proteins bind to G actin and prevent polymerization.
2. End-capping proteins regulate the length of actin filaments.
3. Cross-linking proteins link together two or more separate actin filaments.
   a. Rod-shaped cross-linking proteins promote the formation of networks that have the properties of elastic gels.
   b. Globular actin-binding proteins bundle actin filaments into tight parallel arrays which are found in microvilli and stereocilia.
4. Filament-severing proteins shorten filaments, decreasing cytoplasmic viscosity.
5. Membrane-binding proteins link contractile proteins to the plasma membrane.

B. Actin filaments and myosin motors are responsible for cytokinesis, phagocytosis, cytoplasmic streaming, vesicle trafficking, movement of integral proteins in membranes, cell locomotion and more.

C. Stress fibers attach to the substrate at focal contacts and contain bundles of actin filaments.
1. Stress fibers contain other proteins found in muscle cells.
2. Stress fibers contract isometrically, creating tension between cell and substrate.

D. Cells lacking cilia or flagella move by "crawling" over the substrate.
1. Cultured fibroblasts crawl by forming lamellipodia.
2. The protrusion of lamellipodia involves the organization of actin filaments.
3. Force generation in lamellipodia occurs by the addition of actin monomers to cortical filaments and/or the movement of myosin motors along cortical filaments.
4. Cell movement requires both the formation and breakdown of microfilaments and adhesions, occuring simultaneously in different regions of the cells.

E. Contact inhibition of movement occurs when motile cells in culture make contact.

F. Axonal outgrowth occurs in embryonic nerve cells.
1. Highly motile growth cones contain several structures filled with actin.
2. Growth cones follow defined markers during embryonic development.

G. The cytoskeleton is responsible for cell shape changes that occur during development.

IX. The Molecular Motor That Drives Fast Axonal Transport (Experimental Pathways)

A. A techniques called video-enhanced contrast, differential interference microscopy, or AVEC-DIC, allowed high-resolution images of living cells, facilitating the study of axonal transport.

B. Allen showed that vesicular movements along linear elements are both anterograde and retrograde.

C. In 1985, Vale, Schnapp, Reese and Sheetz showed that vesicles of different sizes move along linear elements at equal rates, indicating that a single motor drives fast axonal transport.

    1. Organelles attached to single filaments move in both directions, even passing one another, suggesting that each filament has more than one track.
    2. The filaments were the diameter of microtubules and bound immunofluorescent antibodies to tubulin.

D. Vale's group reconstituted the axonal transport system from separate components.

    1. The addition of a supernatant fraction (S2) to the mixture of tubulin and organelles promoted transport in only the anterograde direction.
        a. The motor protein kinesin was isolated from S2.
        b. Kinesin was a identified as a plus end-directed motor.
    2. A retrograde motor was identified using affinity chromatography to remove kinesin from the extracts.
    3. Drug sensitivity suggests that the retrograde motor may be a dyneinlike protein.

E. A kinesin receptor molecule on transported vesicles, called kinectin, is part of a larger family of vesicle-transport proteins.

Key Figure

Figure 9.59*a*. The shortening of a sarcomere during muscle contraction.

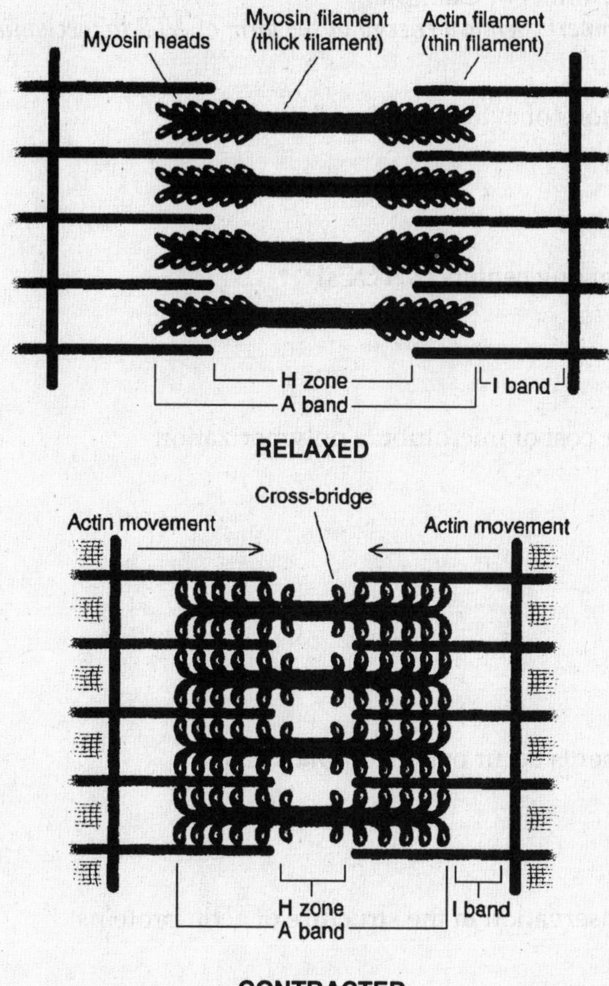

Questions for Thought:

1. Can you infer from these diagrams why it is good to stretch your muscles before exercising? Is it possible to stretch your muscles too much? Why?

2. Under normal conditions, what limits your muscles from stretching so much that the thick and thin filaments no longer overlap? Cardiac muscle is striated like skeletal muscle but is not connected to bone. What prevents cardiac muscle from overstretching?

3. Visualize a cross section of a sarcomere taken at the level of the H band. Now compare that with cross sections taken at the A band and the H zone. How do they differ?

4. Sarcomeres are force generators serially arranged in a myofibril. Myofibrils are force generators with a parallel arrangement within a muscle fiber. When a body builder builds muscle, are force generators increasing in parallel or in series?

Review Problems

Short Answer

1. Give the significance (not the definition) of the following terms or phrases. Say what they do or why they are important. For example:
   molecular motors: *convert chemical energy in the form of ATP to mechanical energy in the form of movement.*

   a. overlapping functions of cytoskeletal proteins

   b. microtubule-organizing centers (MTOCs)

   c. the high metabolic cost of microtubule polymerization

   d. γ–tubulin

   e. intermediate filaments occur only in animal cells

   f. high degree of conservation in the structure of actin proteins

   g. experiments of A. Huxley, Niedergerke, H. Huxley, and Hanson

   h. actin-binding proteins

   i. contractile stress fibers

   j. contact inhibition of movement

2. Compare and contrast the following:
   a. microtubules vs. microfilaments

   b. cytoplasmic dynein vs. kinesin

   d. basal bodies vs. centrioles

   e. muscle fibers vs. myofibrils

   f. pseudopodia vs. lamellipodia

Multiple Choice

1. The vertebrate skeleton is to bones as the cytoskeleton is to __?__.
   a. calcium
   b. tubulin, actin and dynein
   c. microtubules, microfilaments, and intermediate filaments
   d. scaffolding

2. The movement of a motor protein along a cytoskeletal filament is most analogous to:
   a. a locomotive rolling along a railroad track.
   b. a truck driving along the highway.
   c. a person on crutches walking down a sidewalk.
   d. a jet plane in flight.

3. Which of the following is not typical of the "cargo" moved by molecular motors?
   a. ATP
   b. secretory vesicles
   c. cytoskeletal filaments
   d. organelles

4. Differences in microtubular stability and function between different cells (or different locations within a cell) may be due to which of the following?
   a. the expression of different tubulin isoforms
   b. the number of protofilaments composing a single microtubule
   c. the presence of different MAPs
   d. all of the above

5. If you treated cells with a drug that interferes with microtubules, such as colchicine, which of the following would result?
   a. Cell shapewould be disrupted.
   b. Mitosis and meiosis would not occur.
   c. The intracellular location of organelles would be disrupted.
   d. All of the above would result.

6. Which of the molecular motor proteins is associated with microtubules?
   a. kinesins
   b. dyneins
   c. myosins
   d. both (a) and (b)

7. Which of the following cytoskeletal components is not found in all eukaryotic cells?
   a. dynein
   b. intermediate filaments
   c. microtubules
   d. kinesinlike proteins

8. Dynein is to retrograde movement as __?__ is to anterograde movement.
   a. kinesin
   b. myosin
   c. ATP
   d. tubulin

9. Which of the following is not an example of a mictotubule-organizing center (MTOC)?
   a. a centrosome
   b. a basal body
   c. a tubulin-GTP dimer
   d. a spindle pole body

10. First you dissolve the membrane from an intact flagellum, using the detergent Triton X-100. Next you soak the axoneme in a solution containing EDTA, which removes the Mg^{2+}. What remains of the axoneme after these treatments?
   a. peripheral tubules only
   b. peripheral tubules and central tubules, but no side arms or ATPase activity
   c. peripheral tubules, central tubules, side arms, and ATPase activity
   d. peripheral tubules, central tubules, side arms, ATPase activity, and a membrane

11. In what way are intermediate filaments (IF) similar to both microtubules and microfilaments?
   a. IFs are a chemically heterogeneous group of structures.
   b. IFs assemble and disassemble under changing cellular conditions.
   c. IFs have been identified in animal cells but not in other eukaryotes.
   d. IFs assemble into structures of varying thickness via several different pathways.

12. What kind of information can be obtained by injecting cells with fragments of heavy meromyosin (HMM) or S1 subfragments of HMM?
   a. identification of actin-containing filaments and their respective polarities.
   b. identification of actin-containing filaments but not their respective polarities.
   c. identification of myosin-containing filaments and their respective polarities.
   d. identification of myosin-containing filaments but not their respective polarities.

13. Which of the following cell types would be the best source of actin?
   a. vertebrate skeletal muscle cells
   b. asymmetric cells, such as those lining the digestive tract
   c. plant cells in the process of active cell division
   d. skin cells

14. Which of the following is essential for treadmilling to occur down the length of an actin filament?
    a. incubation with HMM
    b. asymmetry in the rates of polymerization and depolymerization occurring at the two respective ends of the filament
    c. stability of the actin filament
    d. all of the above

15. Which types of cell motility can occur in the presence of actin alone (without myosin)?
    a. skeletal muscle contraction
    b. cytokinesis
    c. the acrosomal reaction just prior to the fertilization of an egg by a sperm
    d. none (All types of motiliy using actin also require myosin.)

16. In an analogy between skeletal muscle motility and a railroad, if actin is the railroad tracks and ATP is the diesel fuel, what is the locomotive?
    a. myosin
    b. titin
    c. troponin
    d. tropomyosin

17. If you compared under the microscope a contracted sarcomere to a relaxed sarcomere, which of the following regions would not differ in width?
    a. the A band
    b. the I band
    c. the H zone
    d. the entire sarcomere

18. During the course of a muscle twitch (a contraction due to a single action potential), when is the free calcium concentration of the muscle cytoplasm highest?
    a. milliseconds before the T tubule undergoes an action potential
    b. after the T tubule undergoes an action potential but before the muscle generates tension
    c. at the height of the tension generated by the muscle
    d. at the end of the twitch, after the muscle has relaxed

19. The sarcoplasmic reticulum must have integral membrane proteins that can:
    a. release and pump Ca^{2+}.
    b. bind to tropomyosin and troponin.
    c. undergo action potentials.
    d. contract.

20. The diversity of roles played by microtubules largely results from the diversity of tubulin isoforms that make up the filaments. What accounts for the diversity of roles played by microfilaments?
    a. actin isoforms
    b. variations in the manner in which G actin aggregatse into F actin
    c. differences in the myosins that interact with actin
    d. the variety of actin-binding proteins

Problems and Essays

1. You are observing a microtubule using video-enhanced microscopy. At various time intervals, you measure the length of both ends (A and B) of the microtubule using a fixed reference point. You have plotted your data below:

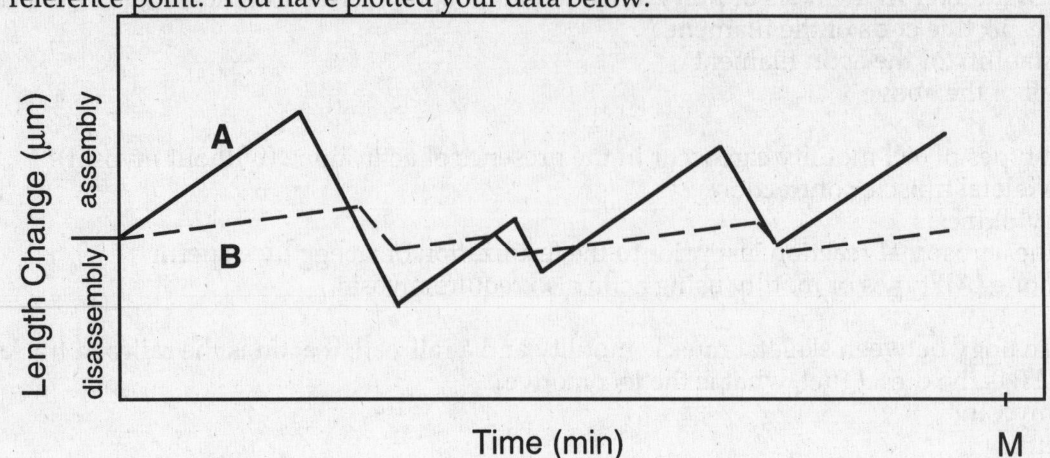

   a. Which is faster, assembly or disassembly? Hows can you tell?

   b. Which line represents the changes occurring at the plus end? Why?

   c. What is the name of the phenomenon illustrated here?

   d. What are some naturally occurring intracellular events that may correspond to the peaks and valleys of these lines?

   e. At the point labeled M, draw what would happen to this microtubule if the cell in which it resides entered mitosis.

2. Indicate which of the following processes are directly affected by colchicine (C), taxol (T), cytochalasin (CHL), and/or nonhydrolyzable ATP analogs, such as AMP-PNP :

   ______________ axonal transport

   ______________ mitotic spindle formation

   ______________ mitotic spindle disassembly

   ______________ acrosomal reaction

   ______________ phagocytosis

   ______________ cytokinesis

3. Panel A is a schematic diagram of a cross section of a sarcomere from skeletal muscle.

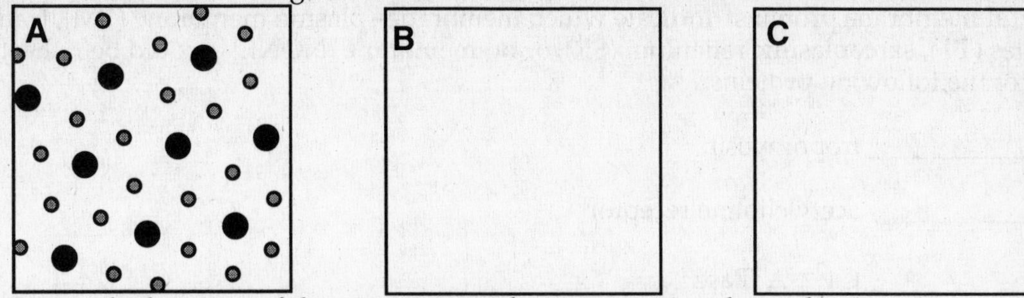

a. From which region of the sarcomere is this cross section derived?

b. In panel B, depict an H zone cross section, and in panel C, an I band cross section.

c. How might the cross section in panel A differ between contracted and relaxed muscle?

4. Colchicine, an alkaloid from the saffron plant, binds to tubulin dimers and prevents their polymerization. If applied to actively dividing cells, colchicine causes the disappearance of the mitotic spindle, thus blocking cell division. A different drug, taxol, has just the opposite effect. In the presence of taxol, microtubules polymerize or are stabilized. This, it turns out, is is just as deleterious to the dividing cell as is colchicine. Can you explain this based on what you know about the roles of microtubules in cells?

5. The isolated axoneme of a cilium can be gently treated with proteases that disrupt the nexin links between microtubules and the radial spokes between peripheral and central microtubules. When these axonemes are exposed to ATP, the fiber appears to rapidly elongate, up to nine times its original length. Explain.

6. A skeletal muscle cell has three distinct membrane systems, each with its own complement of integral membrane proteins. Indicate which membrane– plasma membrane (PM), transverse tubules (TT), sarcoplasmic reticulum (SR), or no membrane (NONE)– would be richest in each of the following proteins.

_______________ tropomyosin

_______________ acetylcholine receptor

_______________ Ca^{2+}-ATPase

_______________ titin

_______________ Ca^{2+}-release channels

7. Fill in the blanks in the following table:

	Ciliary Movement	Skeletal Muscle Movement
Filament Type	microtubule	microfilament
Motor Protein	dynein	myosin
Fuel Molecule	ATP	
Name of the Functional Unit of Movement		

CHAPTER TEN

The Nature of the Gene and the Genome

Learning Objectives

When you have finished this chapter, you should be able to:

1. Understand Mendel's rules of inheritance and their significance for genetics.
2. Define the linkage group, and explain the reasons for incomplete linkage.
3. Give an overview of the discoveries that led Watson and Crick to formulate their structure for DNA.. Describe the structure of DNA.
4. Describe the different conformations that the DNA double helix can assume.
5. Understand genome complexity and the methods used to determine it. Define three classes of DNA sequences.
6. Know the definition of transposons, their effect on gene expression, and their role in evolution.
7. Define a restriction enzyme. Give examples of restriction enzymes, and describe their role in the development of modern molecular biology.
8. Give examples of how the techniques of modern molecular biology are being used in medicine and technology.

Key Terms and Phrases

gene	allele	law of segregation
law of independent assortment	homologous chromosomes	bivalent
linkage group	T. H. Morgan	mutation
crossing over	genetic recombination	locus
polytene chromosomes	nucleotide	pyrimidine
purine	base composition analysis	Chargaff's rules
Rosalind Franklin	Maurice Wilkins	Watson and Crick
double helix	major groove	minor groove
complementarity	template	B form of DNA
A form of DNA	Z-DNA	DNA supercoiling
negative supercoiling	positive supercoiling	topoisomerase
genome	complexity of the genome	DNA denaturation
melting temperature (T_m)	DNA renaturation	DNA reannealing
nucleic acid hybridization	highly repeated sequences	tandem sequence
satellite DNA	microsatellite DNA	minisatellite DNA
in situ hybridization	moderately repeated sequences	SINEs
LINEs	nonrepeated sequences	duplication
deletion	pseudogenes	mobile genetic element
Barbara McClintock	transposition	transposable element
transposon	transposase	Alu sequence
polymorphic genes	genetic map	physical map
restriction endonuclease	restriction fragment length polymorphisms (RFLPs)	
restriction map	gene therapy	Friedrich Miescher
Frederick Griffith	Avery, Macleod, and McCarty	Hershey and Chase

Reviewing the Chapter

I. Introduction

A. Gregor Mendel's work became the foundation for the science of genetics

B. Mendel established the laws of inheritance based on his studies of pea plants.

1. Characteristics of organisms are governed by units of inheritance called genes.
   a. Each trait is controlled by two genes called alleles.
   b. Alleles can be identical or nonidentical.
   c. When they are nonidentical, the dominant allele masks the recessive allele.
2. A reproductive cell called a gamete contains one gene for each trait.
   a. Somatic cells arise by the union of male and female gametes.
   b. In somatic cells, two alleles controlling each trait are inherited; one from each parent.
3. According to the law of segregation, the pairs of genes become separated during gamete formation.
4. According to the law of independent assortment, genes controlling different traits segregate independently of each other.

II. Chromosomes: The Physical Carriers of Genes

A. The discovery of chromosomes:

1. Chromosomes were first observed in dividing cells, using the light microscope.
2. During cell division, chromosomes are divided equally between the two daughter cells.
3. Chromosomes are doubled prior to cell division.

B. Fertilization and meiosis– changing chromosome numbers:

1. There is a qualitative difference among chromosomes.
2. Meiosis prevents the doubling of the number of chromosomes between generations.

C. The chromosomes as the carriers of genetic information:

1. Chromosomes are present as pairs of homologues.
2. During meiosis, homologous chromosomes first associate and form a bivalent; they then separate into different cells.
3. Behavior of pairs of chromosomes in meiosis correlates with Mendel's laws of inheritance.

D. The chromosome as a linkage group:

1. Genes that are on the same chromosome do not assort independently.
2. All of the genes on a chromosome form a linkage group.

E. Genetic analysis in Drosophila:

1. T. H. Morgan was the first to use fruit flies in genetic research.
2. Mutant fruit flies were isolated and became the primary tool for geneticists.
3. Mutation was recognized as the raw material of evolution.
4. Correlation between the four linkage groups and the four pairs of homologous chromosomes in Drosophila cells supported the idea of genes residing directly on chromosomes.

F. Crossing over and recombination:

1. Linkage between alleles on the same chromosome is incomplete.
2. Crossing over (or genetic recombination) produces offspring with new combinations of genetic traits.
   a. The percentage of recombination between two genes is constant.
   b. The percentage of recombination between different pairs of genes can be different.
3. The position of genes along the chromosome (their locus) is fixed, and can be mapped.

4. Recombination frequency increases with distance between genes.

F. Mutagenesis and giant chromosomes:

1. Exposure to a sublethal dose of x-rays increases spontaneous mutation frequency.
2. Cells in the larval salivary gland of Drosophila have giant polytene chromosomes.
3. Polytene chromosomes display a specific pattern when they have been stained; individual genes can be correlated with specific bands.
4. Gene expression can be directly visualized by observing the "puffs" in polytene chromosomes.

III. The Chemical Nature of the Gene

A. DNA is the genetic material in all organisms

B. The structure of DNA:

1. The earliest information on DNA structure was derived from biochemical studies and x-ray diffraction analyses.
2. The nucleotide is the basic building block of DNA.
   a. A nucleotide consists of a phosphate, a sugar, and either a purine or pyrimidine nitrogenous base.
   b. Nucleotides are linked into nucleic acid polymers:
      (1) Sugar and phosphates are linked by 3′,5′-phosphodiester bonds.
      (2) The nitrogenous bases project out like stacked shelves.
   c. Nucleic acids have a polarized structure; the ends are marked 5′ and 3′.
3. Chargaff established rules by base composition analysis.
   a. Number of adenines (A) equals number of thymines (T).
   b. Number of cytosines (C) equals number of guanines (G).
   c. $(A) + (T) \neq (G) + (C)$

C. The Watson-Crick proposal:

1. The DNA molecule is a double helix
   a. The DNA molecule is composed of two chains of nucleotides.
   b. The two chains form a pair of right-handed helices.
   c. The sugar and phosphate backbone is located on the outside of the molecule.
   d. The bases are situated inside the double helix; they are stacked perpendicular to the long axis.
   e. Two DNA chains are held together by hydrogen bonds between each base of one chain and a base on the other chain.
   f. The double helix is 20 Å wide.
   g. Pyrimidines are always paired with purines.
   h. Only A-T and G-C pairs fit into the double helix structure.
   i. The two chains are antiparallel.
   j. From the outside, the DNA molecule has a major groove and a minor groove.
   k. The double helix takes a turn every 10 residues.
   l. The two chains are complementary to each other.
2. The importance of the Watson-Crick proposal:
   a. The Watson-Crick model of DNA as the genetic material accounts for its functions:
      (1) Storage of genetic information.
      (2) Self-duplication and inheritance.
      (3) Expression of the genetic message.
3. Alternate conformations of DNA:
   a. B form DNA is fully hydrated, as it is within cells.
   b. A form DNA is drier and resembles some cellular RNA molecules.

c. Z DNA occurs in synthetic nucleotides of only G-C pairs; its occurrence in cells is uncertain.

4. DNA supercoiling:
    a. DNA can be either positively or negatively supercoiled.
    b. Topoisomerases change the level of supercoiling.

IV. The Structure of the Genome

A. The genome of a cell or organism is its unique complement of genetic information.

B. Mapping the Human Genome (The Human Perspective)
    1. The goal of the Human Genome Project (HGP) is to sequence the DNA of all human chromosomes.
    2. The HGP is expected to give valuable information for fighting human diseases.
    3. The HGP has been criticized by many scientists.
        a. The cost of the HGP reduces the money available for other worthy research projects.
        b. Information on a person's genetic make up could be used in a discriminatory manner.

C. The complexity of the genome:
    1. The complexity of the genome refers to the variety of genomic DNA sequences and the numbers of copies of those sequences.
    2. DNA denaturation is the separation of the two DNA strands; the melting temperature of the double helix depends on the G-C content.
    3. DNA renaturation is the reassociation of single strands into a stable double helix.
    4. For bacteria and viruses, the rate of renaturation is directly proportional to the size of the genome.
    5. The complexity of eukaryotic genomes: Reannealing curves show three classes of DNA.
        a. Highly repeated DNA sequences do not code for gene products.
            (1) Satellite DNA has been located at the centromere using in situ hybridization.
            (2) Minisatellite DNAs occupy less of the genome than satellite DNAs.
            (3) Microsatellite DNAs have been implicated in genetic disorders.
        b. Moderately repeated DNA sequences:
            (1) Repeated DNA sequences with coding functions include the genes for histones, transfer RNAs, and ribosomal RNAs.
            (2) Repeated DNAs that lack coding functions include SINEs and LINEs.
        c. Nonrepeated DNA sequences code for the majority of proteins.

V. The Stability of the Genome

A. Information content of the genome can change rapidly.

B. Duplication and modification of DNA sequences:
    1. Genetic exchange between homologous chromosomes can create either a duplication or a deletion of DNA sequences.
    2. Duplication played a major role in the evolution of multigene families, such as globin genes.
    3. Pseudogenes are nonfunctional members of a multigene family.
    4. Evolution of DNA sequences can be observed in cultured cells in the laboratory.

C. Mobile genetic elements:
    1. Genetic elements are capable of moving.
        a. Transposition of a gene happens randomly and can severely affect gene expression.

b. Transposition requires the enzyme transposase coded within the transposable element itself, which facilitates insertion of transposable elements into the target site.
c. Bacterial transposition occurs by replication of the transposable element, followed by insertion.
d. In eukaryotic cells, transposition can occur either by replication via an RNA intermediate or by excision followed by reinsertion.

2. The role of mobile genetic elements in evolution:
   a. Some moderately repeated sequences in human DNA (Alu and L1) are transposable elements.
   b. According to one school of thought, transposable elements have no function and thus are genetic parasites.
   c. Another school of thought proposes that transposition is an important factor in evolution and speciation.

VI. Molecular Maps of the Genome

A. Mapping the genome:
   1. Genetic maps can be made by determining the crossover frequency between genes.
   2. Physical maps of the chromosome are constructed using sequenced fragments of DNA.

B. Restriction endonucleases:
   1. Restriction endonucleases recognize and cut particular nucleotide sequences.
   2. Recognition sites are palindromes; they have twofold rotational symmetry.
   3. Restriction maps can be created by mapping the recognition sites for different restriction enzymes.

C. The formation and use of restriction fragment length polymorphisms (RFLPs):
   1. RFLPs in humans are based on differences in the number of clustered repeats of short sequences.
   2. RFLPs can be used to identify individuals.
   3. RFLPs were used to locate the gene for cystic fibrosis as well as other genetic disorders.

D. Correcting genetic disorders by gene therapy (The Human Perspective):
   1. Gene therapy involves treating a patient by altering his or her genotype.
   2. Foreign genes may be introduced into cells either by infection with a retrovirus or by transfection.
   3. The modern focus of gene therapy has shifted toward such common diseases as cancer and cardiovascular disease.
   4. Gene therapy is only practiced using somatic cells. Germ cell therapy would raise serious ethical questions about tampering with human evolution.

VII. The Chemical Nature of the Gene (Experimental Pathways)

A. The nature of the gene was discovered through a series of initially unrelated studies.
B. Friedrich Miescher first identified "nuclein" in white blood cell extracts and in salmon sperm.
C. The tetranucleotide theory, proposed by Phoebus Levene, stated that DNA was composed of a monotonous repetition of four nucleotide building blocks and could not be the genetic material.
D. Transformation of bacterial cells was first discovered by Frederick Griffith in experiments with pneumococcus bacteria.
E. The role of DNA as a transforming agent was shown in bacterial cell studies by Avery, MacLeod, and McCarty, and in bacteriophage studies by Hershey and Chase.

Key Figure

Figure 10.9 *a.* The double helix.

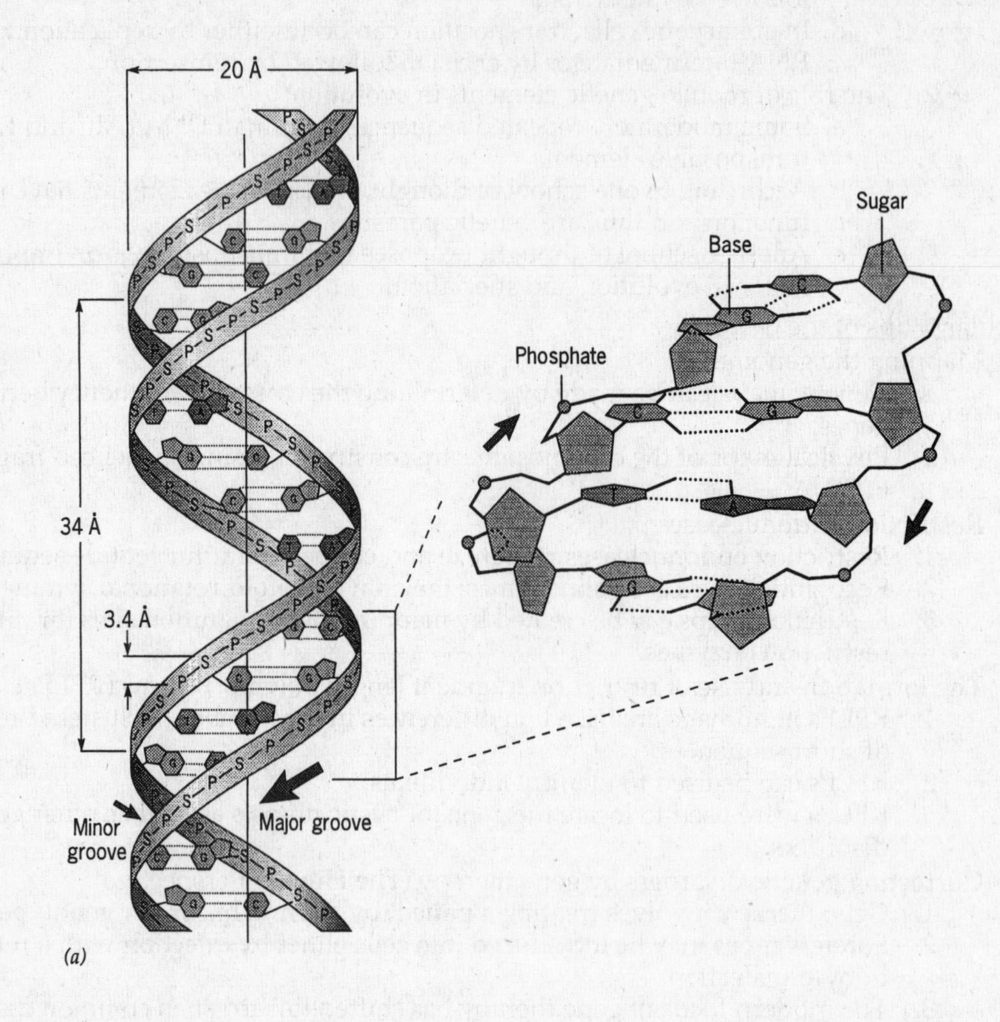

Questions for Thought:

1. The two strands of a double helix of DNA are not identical, yet they contain the same genetic information. Explain.

2. How might this diagram appear if the DNA had been cut by a restriction endonuclease? What is meant by the term "sticky ends"?

3. Can you tell just by looking at this figure why the melting temperature of DNA rich in G-C pairs is higher than that of DNA rich in A-T pairs?

4. Topoisomerases are necessary to unwind supercoiled DNA before it can be replicated or transcribed. What other changes in the structure of the double helix must occur prior to these events? Do you suppose these changes are dependent upon enzymes? The answers can be found in Chapter 13 of your textbook.

Review Problems

Short Answer

1. Give the significance (not the definition) of the following terms or phrases. Say what they do or why they are important. For example:
nucleotide sequence in DNA: *contains the genetic information.*

    a. crossing over

    b. polytene chromosomes

    c. topoisomerases

    d. T. H. Morgan's experiments with *Drosophila*

    e. DNA supercoiling

    f. Barbara McClintock's experiments with maize

    g. restriction fragment length polymorphisms

    h. Avery, Macleod, and McCarty's experiments with pneumococcus

2. Compare and contrast the following:
   a. gene vs. allele

   b. Mendel's law of segregation vs. his law of independent assortment

   c. genome complexity of prokaryotes vs. genome complexity of eukaryotes

   d. genetic map vs. physical map

   e. gene therapy using somatic cells vs. gene therapy using germ cells

Multiple Choice

1. Why were the traits that Mendel chose to study in the pea plant considered lucky choices?
   a. They were the only traits governed by genes.
   b. None of the traits he chose acted as if they were part of the same linkage group.
   c. All the traits he chose could be found on the same plant.
   d. They all obeyed the law of segregation.

2. Alleles are:
   a. different forms of the gene for the same trait.
   b. genes for different traits.
   c. always identical.
   d. inherited in pairs, both members of which come from the same parent.

3. A brown-eyed man has one allele for brown eyes and one for blue eyes. Half his sperm carry the allele for brown eyes, and half carry the allele for blue eyes. This is an illustration of:
   a. Mendel's law of independent segregation.
   b. Mendel's law of independent assortment.
   c. Mendel's law of segregation.
   d. None of the above.

4. Which of the following observations provided evidence that Mendel's "units" were located on the chromosomes?
   a. Chromosomes occur as homologous pairs.
   b. The number of linkage groups and chromosomes is always equal.
   c. One member of each homologous pair goes to each gamete during meiosis.
   d. All of the above.

5. Given that alleles are located on chromosomes, and whole chromosomes are passed on to gametes, why don't offspring get all the alleles on a parental chromosome?
   a. Genetic recombination is a result of gene duplication during evolution.
   b. Genetic recombination is a result of mutation.
   c. Genetic recombination is a result of crossover between members of homologous pairs.
   d. All traits on a given chromosome are always found together in offspring.

6. Watson and Crick used all of the following information in deducing the structure of the genetic material except:
   a. Avery, MacLeod, and McCarty's identification of DNA as the "transforming principle."
   b. x-ray diffraction studies by Astbury, Franklin, Pauling and Wilkins.
   c. Chargaff's rules of DNA base composition.
   d. microscopic examination of DNA structure.

7. Given the structure of DNA as described by Watson and Crick, all of the following must be true of DNA except:
   a. The 3′ end of one of the strands must be aligned with the 5′ end of the other strand.
   b. The total number of purine bases must equal the total number of pyrimidine bases.
   c. The sequence of bases on one strand must be identical to the sequence of bases on the other strand.
   d. The sequence of bases in the genome of one organism must be different from the sequence of bases in the genome of a different organism.

8. The structure of different DNAs can vary in all of the following ways except
   a. The helix may be right-handed or left-handed.
   b. The sequence of nucleotide bases can be different.
   c. The molecule can be fully hydrated as in the B form, or drier as in the A form.
   d. The bases can be attached to the sugars or the phosphates of the backbone.

Use the following diagram to answer questions (9), (10) and (11).

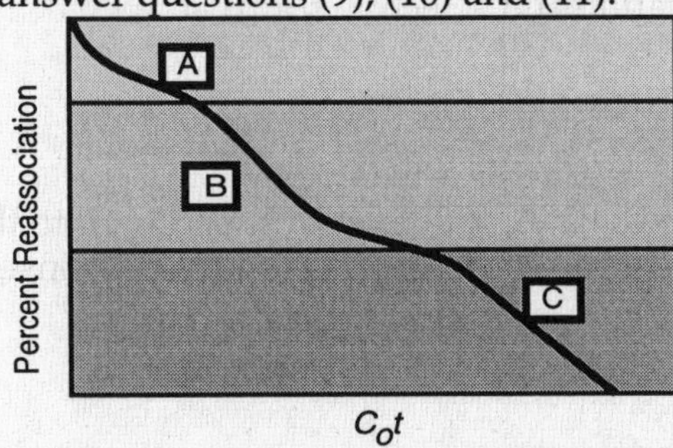

9. Which portion of the curve represents renaturation of the nonhistone coding sequences of the DNA?
   a. A
   b. B
   c. C
   d. cannot determine from this curve

10. Which portion of the curve represents renaturation of the most repetitive DNA?
    a. A
    b. B
    c. C
    d. cannot determine from this curve

11. What is probably the source of the DNA used in this curve?
    a. virus
    b. bacterium
    c. eukaryote
    d. none of the above

12. In situ hybridization is used to:
    a. locate sequences of DNA on a chromosome.
    b. identify mini- and microsatellite DNAs.
    c. identify homologous pairs of chromosomes.
    d. identify the centromere portion of chromosomes.

13. Which of the following statements is true of repeated DNA sequences?
    a. All repeated sequences have coding functions, but not all coding sequences are repeated.
    b. All nonrepeated sequences have coding functions, and all coding sequences are nonrepeated.
    c. All repeated sequences lack coding functions, but not all coding sequences are nonrepeating.
    d. Some coding sequences are repeated; others are nonrepeating.

14. Human cells contain several different proteins that bind calcium. They share similar amino acid sequences but are found in different cell types. What can you infer about the genes for these proteins?
    a. They are all part of the family of globin genes.
    b. They probably arose in evolution by gene duplication.
    c. They are part of a family of pseudogenes.
    d. They must all reside on different chromosomes.

15. Cultured hamster cells treated with methotrexate become resistant to the drug by amplifying the gene for DHFR. This is an example of:
    a. selective gene amplification.
    b. random gene duplication and selective survival.
    c. evolution.
    d. all of the above.

16. In eukaryotes, transposable elements arise by:
    a. reverse transcription of RNA, followed by insertion into the genome.
    b. excision of a portion of the chromosome, followed by insertion at a new place in the genome.
    c. both (a) and (b).
    d. neither (a) nor (b).

17. Which of the following is not a role that has been suggested for transposable elements in the process of evolution?
    a. Mutational effects of transposable elements accumulate in a suppressed form, then are suddenly expressed, resulting in new species.
    b. Insertion of transposable elements into the genome results in gene duplication and the evolution of multigene families.
    c. Transposable elements have no role in evolution; they are genetic parasites.
    d. Natural selection always favors species with many transposable elements.

18. Which of the following kinds of "maps" of genomes rely on the frequency of crossover events?
    a. genetic maps
    b. physical maps
    c. nucleotide sequence maps
    d. restriction fragment maps

19. Restriction endonucleases have become essential tools in the study of molecular biology. Which of the following true statements sums up the significance of these enzymes from the point of view of the bacteria that make them?
    a. Restriction endonucleases cut DNA into fragments that can be easily sequenced.
    b. Restriction endonucleases cut DNA into precisely defined sets of fragments.
    c. Restriction endonucleases cut any type of DNA, as long as it has not been methylated at the cutting sites.
    d. Restriction endonucleases can leave either blunt or sticky ends.

20. Under which of the following conditions would the restriction fragment lengths be identical (i.e., lack polymorphisms)?
    a. when DNA from homologous chromosomes of the same individual is cut with the same restriction endonuclease
    b. when DNA from the same chromosome derived from identical twins is cut with the same endonuclease
    d. when DNA from the same chromosome derived from nonidentical siblings is cut with the same endonuclease
    e. all of the above.

Problems and Essays

1. The cells of human embryos undergo division every 24 hours, and each cell has about ten million ribosomes. A cell with one copy of the ribosomal RNA gene and the maximum number of polymerase enzymes transcribing it can produce a ribosome every 1.16 seconds. What is the minimum number of copies of the rRNA genes required to support human embryonic development?

2. The human genome has hundreds of thousands of Alu sequences scattered throughout it, and many shorter Alu-like sequences that occur between genes, and even within introns. The Alu sequence has remarkable similarities to an RNA called 7SL RNA, a molecule that aids in protein secretion. The 7SL RNA is highly conserved in mice, humans, Drosophila, and a similar gene has even been identified in E. coli, yet the Alu SINE is characteristic of mammals. Give a possible explanation for the appearance of the Alu sequence.

3. Your professor has asked you to take care of two test tubes, each with a different plasmid in it. One contains pFIRST, and the other contains pSECOND. She asks you to get the tube with pFIRST but you forget which tube contains which plasmid. You have a restriction map of both plasmids, as illustrated here, so you take a sample from tube 1 and digest it with the restriction enzymes EcoR1 and BamH1. Next you digest a sample from tube 2 with BamH1 and HindIII. When you separate the DNA fragments using agarose gel electrophoresis, the fragments show the banding patterns below.

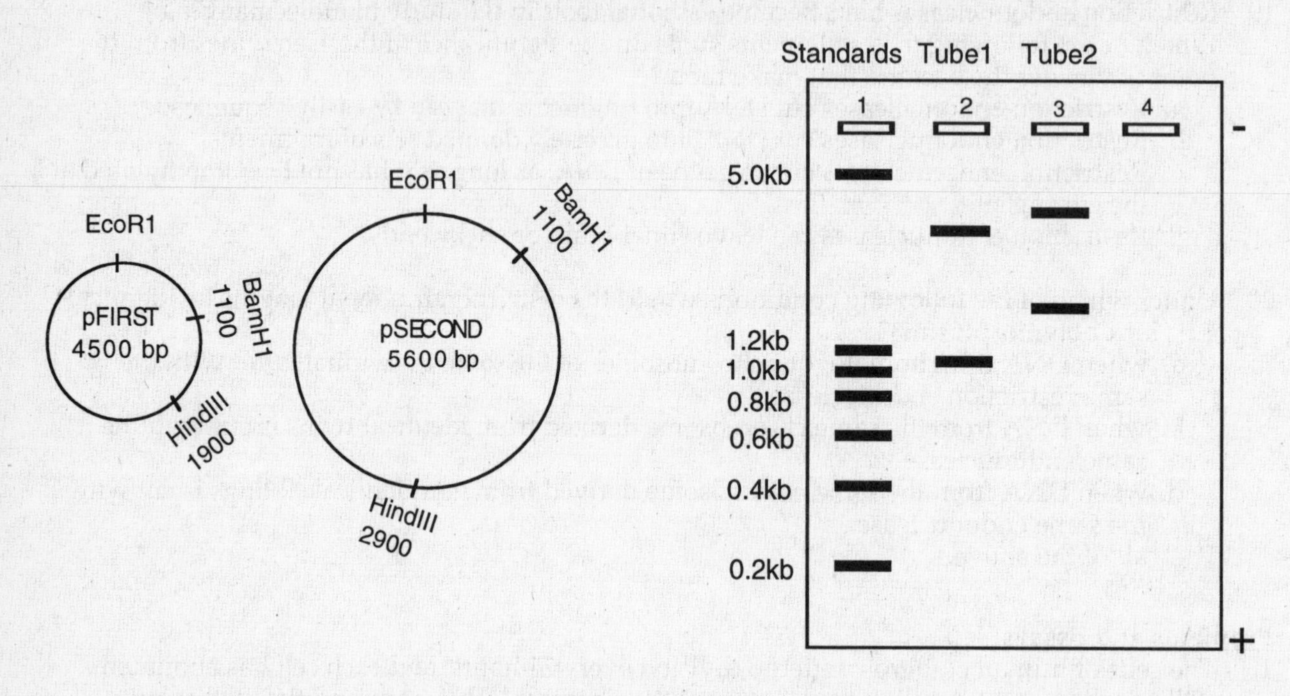

a. Which test tube, 1 or 2, will you give your professor, and why?

b. In the fourth lane of the gel illustrated above, draw the pattern you might see if you had digested the contents of tube 2 with EcoR1 and HindIII.

4. The human genome has three billion base pairs on 23 chromosomes, and an estimated 100,000 genes.
   a. What is the average number of genes on each chromosome?

   b. What is the average number of base pairs for each gene?

   c. If 95% of the base pairs are noncoding DNA, what is the average number of base pairs for each gene?

5. Draw the reannealing curve that might result for DNA extracted from yeast cells. Which part of the curve represents the DNA from the centromere portion of the yeast chromosomes? Which part of the curve represents the actual yeast genes?

6. Two genes, one for seed color and one for seed shape, were studied in dandelion plants and found to obey Mendel's law of independent assortment. When the genes were located by in situ hybridization, however, they were found to reside on the same chromosome. What explanation can you provide for this observation? In the diagram, show where the fluorescence of biotinylated hybrids might be concentrated for these two genes.

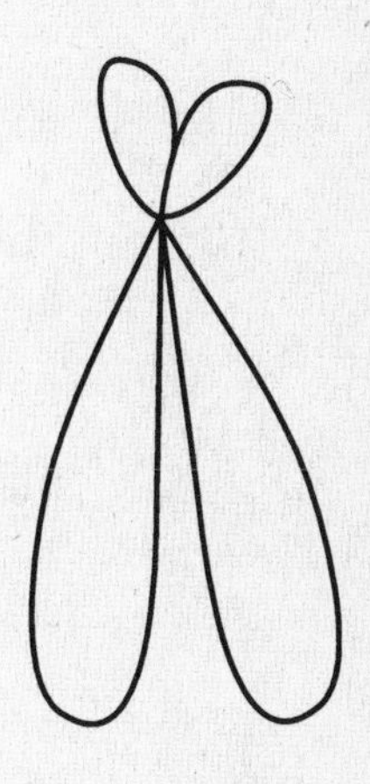

7. You are searching for the gene for a hereditary disease using restriction fragment length polymorphisms. You have isolated DNA from four people with the disease, and four people without the disease, and treated the DNA with a restriction endonuclease. Agarose gel electrophoresis resulted in the banding patterns to the left . Which band probably represents the fragment of DNA closest to the gene for the disease? Why?

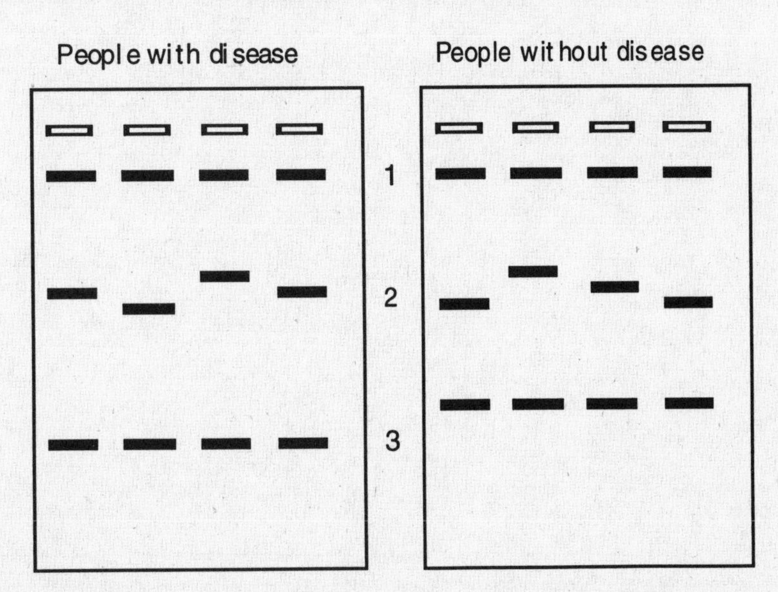

Notes:

CHAPTER ELEVEN

Utilization of Genetic Information: From Transcription to Translation

Learning Objectives

When you have finished this chapter, you should be able to:

1. Understand the concept of information flow in the cell.
2. Define transcription and summarize the main differences between eukaryotic and prokaryotic transcription.
3. Know the roles of each of the three RNA polymerases in eukaryotes.
4. Understand the concept of RNA processing and modification.
   a. Describe the mechanism of rRNA processing in eukaryotes.
   b. Define split gene, intron, exon, and RNA splicing.
   c. Know the importance of mRNA modifications on 5′ and 3′ ends.
5. Know the main features of the genetic code.
6. Explain the role of tRNA in decoding the genetic information.
7. Give an overview of translation and specify the factors involved in each step.

Key Terms and Phrases

gene	transcription	translation
RNA polymerase	template	promoter
processivity	core enzyme	sigma factor (σ)
Pribnow box	rho factor	RNA polymerase I
RNA polymerase II	RNA polymerase III	transcription factors
primary transcript (pre-RNA)	small nuclear RNA (snRNA)	transcription unit
ribosomal RNA (rRNA)	nucleolus	nucleolar organizer
nontranscribed spacer	Svedberg unit (S value)	transfer RNA (tRNA)
messenger RNA (mRNA)	heterogeneous nuclear RNA (hnRNA)	core promoter element
preinitiation complex	TATA-binding protein (TBP)	TATA box
methylguanosine cap	poly(A) tail	split genes
intervening sequences	exon	intron
RNA splicing	splice sites	spliceosome
snRNPs (snurps)	group I intron	group II intron
alternate splicing	exon shuffling	ribozyme
genetic code	codon	reading frame
frameshift mutation	degenerate code	anticodon
wobble hypothesis	aminoacyl-tRNA synthetase	mutagen
translation	initiation	initiation codon
Shine-Dalgarno sequence	initiation factor (IF)	A (aminoacyl) site
P (peptidyl) site	elongation	elongation factors (EF)
peptidyl transferase	translocation	termination
release factors	nonsense mutation	polyribosome (polysome)

Reviewing the Chapter

I. The Relationship between Genes and Proteins

A. Genes store information for producing all cellular proteins.

B. Early observations suggested a direct relationship between genes and proteins.

1. The relationship between a specific gene, a specific enzyme and a specific metabolic condition was first described by Archibald Garrod in 1908 in studies of "inborn errors of metabolism."
2. The "one gene–one enzyme" hypothesis was formulated by Beadle and Tatum. Later it was modified to "one gene–one polypeptide chain."
3. Mutation in a single gene causes a single substitution in an amino acid sequence of a single protein.

C. An overview of the flow of information through the cell:

1. Messenger RNA is an intermediate between a gene and its polypeptide.
2. Transcription is the process whereby RNA is formed from a DNA template.
3. Proteins are synthesized in the cytoplasm by the process of translation from an mRNA template.
4. There are three classes of RNA in a cell: messenger RNA, ribosomal RNA, and transfer RNA.
5. Activity of both rRNA and tRNA depends on their folding into well-defined secondary and tertiary structures.

II. Transcription: The Basic Process

A. DNA-dependent RNA polymerase assembles a linear chain of nucleotides which is complementary to the template DNA strand.

1. The promoter is the site at the 5′ end of the gene, where RNA polymerase binds prior to initiating transcription.
2. The newly synthesized RNA chain grows in a 5′ to 3′ direction, antiparallel to the DNA.
   a. RNA polymerase must be processive– remain attached to DNA over long stretches of template.
   b. RNA polymerase must be able to move from nucleotide to nucleotide.
3. Nucleotides enter the polymerization reaction as trinucleotide precursors. The reaction is driven forward by the exergonic hydrolysis of a pyrophosphate.
4. After the polymerase has passed, the DNA-RNA hybrid dissociates and the DNA double helix reforms.

B. Transcription in prokaryotes:

1. There is only one type of RNA polymerase in prokaryotic cells
2. Transcription-competent prokaryotic RNA polymerase consists of a core enzyme and a sigma factor. The sigma factor dissociates after transcription has begun.
3. Bacterial promoters are located upstream from the site of initiation. They have two conserved regions: the -35 region and the Pribnow box.
4. Differences in the DNA sequences at both -35 and the Pribnow box may regulate gene expression.
5. Termination in prokaryotes can either require a rho factor or be rho-independent. In the latter case, a hairpin loop will be formed on the 3′ end of the RNA.

III. Transcription and RNA Processing in Eukaryotic Cells

A. There are three types of RNA polymerases in eukaryotes:

1. Different classes of RNAs are synthesized by different RNA polymerases.
   a. Most ribosomal RNAs are transcribed by RNA polymerase I.
   b. Messenger RNAs are transcribed by RNA polymerase II.
   c. Transfer RNAs are transcribed by RNA polymerase III.

2. General and specific transcription factors regulate the activity of RNA polymerases.
3. The newly transcribed RNAs are processed.
    a. A primary transcript is the initial RNA molecule synthesized.
    b. A transcription unit is the segment of DNA corresponding to a primary transcript.
    c. Small nuclear RNAs (snRNAs) and their associated proteins are required for RNA processing.

B. Ribosomal RNAs:
1. Ribosomes are produced in nucleoli.
    a. Nucleolar organizers are regions of a chromosome that contain rRNA genes.
    b. The fibrillar core of the nucleolus consists of rDNA and nascent rRNA transcripts.
    c. Both ribosomal subunits are assembled in the nucleolus.
2. Synthesizing the rRNA precursor:
    a. Ribosomal RNA genes are arranged in tandem.
    b. Ribosomal RNA transcription has a "Christmas tree" pattern.
    c. Proteins that convert the rRNA precursor into mature rRNA become associated with pre-rRNA during transcription.
    d. Nontranscribed spacers separate various transcription units in a gene cluster.
3. Processing the rRNA precursor:
    a. By splicing, 28S, 18S and 5.8S rRNAs are derived from a single 45S primary transcript.
    b. Pre-rRNA is heavily methylated.
    c. Pre-rRNA is tightly associated with ribonucleoprotein (RNP) particles.
4. Synthesis and processing of the 5S rRNA:
    a. The 5S rRNA genes are located outside the nucleolus.
    b. A 5S rRNA is transcribed by RNA polymerase III.
    c. RNA polymerase III uses an internal promoter.

C. Transfer RNAs:
1. Transfer RNA genes are located in small clusters scattered around in genome.
2. Transfer RNAs have promoter sequences within the coding region of the gene; the promoter is split into two portions.
3. During processing, tRNA precursor is trimmed and the CCA triplet is added to its 3′ end.

D. Messenger RNAs:
1. Heterogeneous nuclear RNAs are precursors of mature mRNAs.
    a. Heterogeneous nuclear RNAs are found only in nucleus.
    b. Heterogeneous nuclear RNAs have large molecular weights.
    c. Heterogeneous nuclear RNAs are very unstable.
2. The machinery used in mRNA transcription:
    a. RNA polymerase is assisted by general transcription factors (TFII).
    b. The core promoter element for polymerase II lies 24 to32 bases upstream from the initiation site; this region contains the TATA box.
    c. The preinitiation complex of general transcription factors and polymerases assemble at the TATA box.
    d. The preinitiation complex assembly starts with the binding of the TATA-binding protein (TBP) to the promoter.
    e. RNA polymerase is phosphorylated prior to transcription initiation; TFIIH acts as a protein kinase.
3. Messenger RNAs have both noncoding portions and modifications at their 5′ and 3′ ends.

4. Split genes– An unexpected finding:
    a. Eukaryotic genes contain intervening sequences called introns. These sequences are missing from mature mRNAs.
    b. Exons are the parts of a split gene which are included in the mature mRNA.
    c. Hybridization experiments supported the concept of a split gene.
5. The processing of messenger RNA:
    a. During processing, 5' methylguanosine caps and 3′ poly(A) tails are added to mRNAs.
    b. Intervening sequences are removed and exons are connected by splicing.
        (1) Sequences between exon-intron junctions are highly conserved.
        (2) Spliceosomes facilitate mRNA splicing.
        (3) Spliceosomes consist of snRNPs (snRNAs bound to proteins).
    c. Group I and group II introns are capable of self-splicing.
    d. Introns are removed in a preferred order.

E. Evolutionary implications of split genes and RNA splicing:
    1. The theory of "an RNA world" suggests that RNA was the earliest molecule to both store information and catalyze reactions.
    2. RNA splicing via spliceosomes is thought to have evolved from self-splicing RNAs.
    3. Exon shuffling could have played a major role in the evolution of many genes.

F. Creating new ribozymes in the laboratory:
    1. Ribozymes are RNAs that have catalytic activity.
        a. Modified RNAs can catalyze certain reactions.
        b. Synthesis of random RNA molecules is one of the strategies used to look for ribozymes.
    2. Ribozymes have potential uses in medicine, such as fighting certain viral diseases or even cancer (The Human Perspective).

IV. Encoding Genetic Information

A. Information stored in a gene is present in the form of a genetic code.

B. The properties of the genetic code:
    1. The genetic code is triplet, nonoverlapping, and degenerate.
        a. A codon is a nucleotide triplet that codes for a certain amino acid.
        b. Ribosomes start reading mRNA from the initiation codon, AUG.
        c. Frameshift mutations cause the ribosome to move along the mRNA in the incorrect reading frame.
    2. Codon assignment was determined by transcription from artificial mRNAs.
    3. Codon assignment is nonrandom; codons for the same amino acid are similar.
    4. The first two codon bases for a particular amino acid are invariant, whereas the third base may vary.

C. The role of transfer RNA in decoding the codons:
    1. Transfer RNAs are a class of adapter molecules that recognize individual amino acids and the corresponding mRNA codons.
        a. The amino acid is attached to the 3′ end of tRNA.
        b. The anticodon on tRNA complements the codon of the mRNA.
        c. The wobble hypothesis suggests that a single kind of tRNA can recognize codons with variable third bases.
    2. Amino acid activation:
        a. Specific aminoacyl-tRNA synthetases link amino acids with their respective tRNAs.
        b. Energy of one ATP is used to activate the amino acid, which is then transferred to the tRNA molecule.
        c. Codons of the mRNA are ultimately interpreted according to the recognition abilities of the aminoacyl-tRNA synthetases.

V. Translating Genetic Information

A. Translation is the most complex synthetic activity of the cell.

B. Initiation:

1. The ribosome initiates translation at the initiation codon in the mRNA.
2. The small ribosomal subunit identifies the correct AUG codon.
   a. In bacterial mRNAs, the Shine-Dalgarno sequence guides the subunit to the correct initiation codon.
   b. In eukaryotes, the smallest robosomal subunit recognizes the 5′ end of the message and finds the first AUG triplet by scanning.
   c. There are two types of methionyl-tRNAs in a cell; one that recognizes the initiation codon, one that recognizes internal methionine codons. (Prokaryotic proteins start with N-formylmethionine)
3. Initiation factors are required for initiation.
4. Ribosomes have two sites for tRNA molecules: the A (aminoacyl) site and the P (peptidyl) site.

B. Elongation:

1. The elongation cycle is the process of the orderly addition of each subsequent amino acid to the growing polypeptide chain.
2. Several elongation factors (EFs) are required for elongation.
3. Peptidyl transferase, one of the rRNAs of the large subunit, forms a peptide bond between amino acids.
4. In translocation, the ribosome moves one codon in the 3′ direction.

C. Accuracy of translation:

1. Translation has a relatively low error rate. Hypotheses to account for this:
   a. Only complementary codons/anticodons can fit into the ribosomal cleft.
   b. The delay before amino acid transfer ensures accuracy.
2. The antibiotic streptomycin binds to the small subunit of bacterial ribosomes and thus limits accuracy, preventing bacterial translation.

D. Termination:

1. Termination occurs at one of three stop codons: UAA, UAG, or UGA.
2. Release factors interact directly with the stop codons.
3. Nonsense mutations produce stop codons and cause premature chain termination.

E. Polyribosomes are complexes of multiple ribosomes on mRNA, which allow for increased rate of protein synthesis

VI. The Role of RNA as a Catalyst (Experimental Pathways)

A. RNA is capable of catalyzing a complex, multistep reaction of self-splicing.

B. The RNA portion of ribonuclease P can accurately cleave the tRNA precursor.

C. The RNA component plays an important role in the activity of ribosomal peptidyl transferase.

Key Figure

Figure 11.3. The flow of information in a eukaryotic cell.

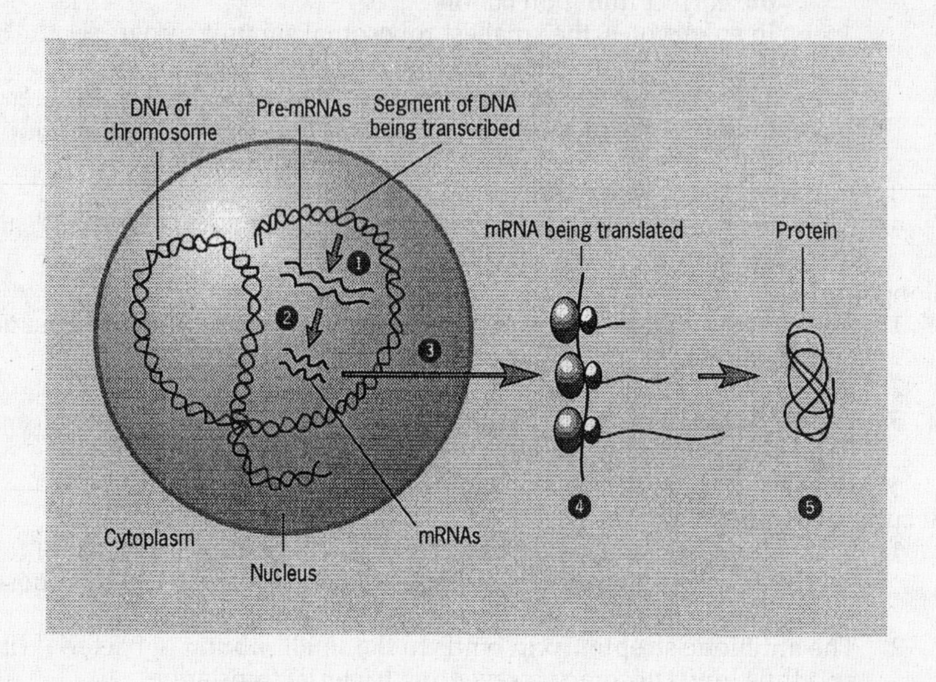

Questions for Thought:

1. How would a diagram depicting the flow of information in a prokaryotic cell differ from this one?

2. In most cases, RNA acts as an intermediary molecule between the genetic information and the final product of that information, proteins. What are the advantages of having an intermediary as opposed to translating the DNA directly into protein?

3. What is the metabolic cost (in either ATP or GTP) of each of these steps? Which step is most "expensive" to the cell?

4. Where in this process are errors likely to occur, and what mechanisms exist to minimize errors?

Review Problems

<u>Short Answer</u>

1. Give the significance (not the definition) of the following terms or phrases. Say what they do or why they are important. For example:
   translation: *interprets the sequence of nucleotides in mRNA into a sequence of amino acids in proteins.*

   a. mRNA as an intermediary between DNA and protein

   b. the highly exergonic nature of RNA polymerization

   c. slight differences in the sequences of nucleotides in promoter regions

   d. methylguanosine cap

   e. alternate splicing

   f. Ames test

   g. Shine-Dalgarno sequence

   h. our changing concept of the gene

2. Compare and contrast the following:
   a. "one gene– one enzyme" vs. "one gene– one transcription unit"

   b. higher-order structure of proteins vs. higher-order structure of RNA

   c. Pribnow box vs. TATA box

   d. upstream promoters vs. internal promoters

   e. frameshift mutations vs. substitution mutations

Multiple Choice

1. Why was Neurospora a good organism for Beadle and Tatum's studies?
   a. Neurospora grows well in rich media and cannot synthesize many enzymes.
   b. Neurospora grows well in minimal media,and is sensitive to enzymatic deficiencies.
   c. Neurospora requires many vitamin additives to grow.
   d. Neurospora was actually a poor choice.

2. The tertiary structure of RNA is mostly dependent upon:
   a. the tendency to bury hydrophobic nucleotides within the core of the molecule.
   b. the formation of additional covalent bonds between nonadjacent nucleotides.
   c. the formation of complementary base pairs between nonadjacent nucleotides.
   d. the formation of salt bridges between the sugar-phosphate backbones.

3. When used to describe the RNA polymerases, the term "processive" refers to:
   a. the ability to recognize and bind to a promoter region.
   b. the ability to remain attached to DNA and form long RNAs.
   c. the ability to add nucleotide triphosphates to a growing RNA chain.
   d. all of the above.

4. Which of the following lists of transcription features applies only to prokaryotes?
   a. Pribnow box, rho factor, core enzyme, -35 region
   b. TATA box, primary transcript, RNA polymerase III, transcription factors
   c. Pribnow box, transcriptional unit, rho factor, RNA polymerase III
   d. TATA box, rho factor, core enzyme, -35 region

5. Of the following statements about the different types of eukaryotic RNA, which one is true?
   a. Over 80% of cellular RNA is ribosomal, yet most newly synthesized RNA is mRNA.
   b. The genes for rRNA and tRNA occur in clusters, whereas most genes for mRNA do not.
   c. All three types of RNAs are processed before reaching their mature form.
   d. All of the above statements are true.

6. The significance of methylation of pre-rRNA is probably
   a. to ensure that the final rRNA will assume the correct tertiary structure.
   b. to prevent the pre-rRNA from being processed.
   c. to prevent degradation of the parts destined to become mature rRNA .
   d. none of the above.

7. If cultured cells are incubated with ^{14}C-methionine, the radiolabel will appear:
   a. in all the rRNAs simultaneously.
   b. first in the 45S RNA, followed by the 32S RNA, followed by the 28S RNA.
   c. first in the 45S RNA, followed by the 28S RNA, followed by the 5S RNA.
   d. first in the 28S RNA, followed by the 32S RNA, followed by the 45S RNA.

8. You have examined nucleoli structures from several sources listed below and find that the cell type with the largest nucleolus is the:
   a. human nerve cell.
   b. frog muscle cell.
   c. frog oocyte.
   d. rat liver cell.

9. The 5S rRNA differs from the other cellular rRNAs in which of the following ways?
   a. The genes are located outside the nucleolus.
   b. The genes are transcribed by RNA polymerase III.
   c. The promoter region of the 5S rRNA gene is internal to the gene.
   d. all of the above

10. Which of the following statements is true of tRNAs?
    a. There are the same number of tRNA genes as there are different tRNAs.
    b. There are more tRNA genes than different tRNAs, and more tRNAs than amino acids.
    c. There are more different tRNAs than there are different triplet codons.
    d. There are the same number of tRNAs as there are amino acids.

11. In a cell that produces only one type of protein, which of the following statements about the relationship between hnRNA and mRNA is true?
    a. Heterogeneous nuclear RNA is always larger than mRNA.
    b. Heterogeneous nuclear RNA is always smaller than mRNA.
    c. Heterogeneous nuclear RNA and mRNA are typically the same size.
    d. There is no consistent relationship between the size of the two molecules.

12. One way to tell if RNA polymerase II is in the process of transcribing an mRNA is to
    a. check for the presence of a repeated 7-amino acid sequence at the C-terminal domain.
    b. check for phosphorylation at the C-terminal domain.
    c. both a and b
    d. neither a nor b

13. Using both the enzyme reverse transcriptase and radiolabeled deoxynucleotides, you make a radioactive DNA copy from an mRNA. When you hybridize that copy DNA to nuclear DNA, you find that there are seven different chromosomal positions where hybridization occurs, and there are three unhybridized loops at each of the seven positions. From this you conclude:
    a. There are seven copies of the gene for that mRNA and three nontranscribed segments in each gene.
    b. There are seven copies of the gene for that mRNA and three exons in each gene.
    c. There are three copies of the gene for that mRNA and seven exons in each gene.
    d. There are seven copies of the gene for that mRNA and three introns in each gene.

14. Why is the methylguanosine cap added to mRNA before the poly(A) tail?
    a. because the cap is added to the 5′ end before the 3′ end is fully transcribed
    b. because the cap is added to the 3′ end before the 5′ end is fully transcribed
    c. because the cap is required for the addition of the poly(A) tail
    d. none of the above

15. The textbook states that the nucleotide sequence of splice sites is "highly conserved" and of "ancient origin." How do we know this sequence is ancient?
    a. It is widespread and occurs in even the most primitive eukaryotes.
    b. The number of mutations that have accumulated in the sequence indicates that it must have been in cells for millions of years.
    c. It is only found in organisms that are highly evolved.
    d. It codes for ancient proteins.

16. About 30% of point mutations (substitutions) that occur in coding regions of DNA are harmless. Why?
    a. because only 30% of the DNA is actually coding for amino acids
    b. because only 70% of the DNA is actually coding for amino acids
    c. because most amino acids are specified by several codons that differ at the third position
    d. all of the above

17. Which type of mutation is likely to be the most harmful?
    a. A base substitution in which a valine is replaced with a leucine.
    b. A +3 frameshift mutation.
    c. A base substitution in which a glutamic acid is replaced with an aspartic acid.
    d. A +1 frameshift mutation.

18. Of the following statements about the aminoacyl tRNA synthetases, which one must be true?
    a. There are at least as many aminoacyl tRNA synthetases as there are amino acids.
    b. There are at least as many aminoacyl tRNA synthetases as there are tRNAs.
    c. There are at least as many aminoacyl tRNA synthetases as there are mRNAs.
    d. There are at least as many aminoacyl tRNA synthetases as there are rRNAs.

19. The following complex of macromolecules is found in a cell: DNA attached to RNA polymerase, attached to a growing chain of RNA, attached to a ribosome, attached to two tRNAs, attached to a growing chain of amino acids. The cell must be:
    a. a plant cell.
    b. an animal cell.
    c. a virus.
    d. a prokaryotic cell.

20. Which of the following observations supports the view that ribosomal RNA has catalytic activity?
    a. The base sequence of rRNAs is highly conserved, whereas the amino acid sequence of ribosomal protein is not.
    b. Bacteria that develop antibiotic resistance have base substitutions on the rRNA, but do not have amino acid substitutions on the ribosomal proteins.
    c. The peptidyl transferase reaction is sensitive to ribonucleases.
    d. all of the above

Problems and Essays

1. Eukaryotic RNA polymerases recognize a promoter that most often has the sequence TATAA. On which strand of DNA, the actual template strand or its complementary partner, does the TATA box sequence occur? Explain.

2. Indicate where in a eukaryotic cell you would be most likely to find each of the following types of RNA.

________ hnRNA	________ tRNA
________ mRNA	________ pre-RNA
________ snRNA	________ nascent rRNA
________ rRNA	________ 45S pre-rRNA

3. A certain gene is 10 kilobases in length, and has two introns and three exons. Within the gene are two restriction sites for EcoR1, as shown. When the mature mRNA is reverse transcribed, and the cDNA is treated with EcoR1, how many restriction fragments result, and what is the length of each?

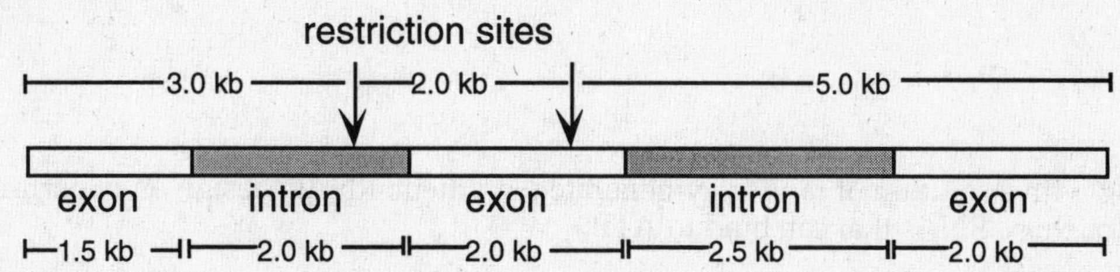

4. Niremberg and Matthaei used the enzyme polynucleotide phosphorylase to assemble RNAs with random sequences in the absence of a template. Suppose they added the enzyme to a solution of 90% CTP and 10% UTP, and then used a cell-free system of protein synthesis to make peptides according to the sequences in their synthetic RNAs.

   a. Which amino acids could possibly appear in the synthetic protein?

   b. Which amino acid would be most abundant in the synthetic protein?

   c. Which amino acid would be least abundant?

5. Starting with a DNA template and the appropriate machinery for both transcription and translation, how many nucleotide triphosphates are needed to synthesize an octapeptide? Why is the recommended caloric intake (per pound) of a growing child higher than that for an adult?

6. A bacterium produces a normal protein with the following amino acid sequence:

   MET-VAL-HIS-LYS-ARG-THR-LEU-VAL

   After irradiation, a mutant strain is produced that makes a mutant protein from the same coding site on DNA:

   MET-VAL-HIS-LYS-GLU-PRO

   Describe the mutation that occurred.

7. Starting with a solution of randomly-generated synthetic RNAs, design an experiment to select for those RNAs that can bind to ATP.

CHAPTER TWELVE

The Cell Nucleus and the Control of Gene Expression

Learning Objectives

When you have finished this chapter, you should be able to:

1. Describe the eukaryotic nucleus, including the morphological features of mitotic chromosomes and nonmitotic chromatin.
2. Know the structure and function of nuclear pores. List the substances that move between the nucleus and the cytoplasm, and the means by which they move.
3. Understand the role of histone proteins in chromatin packaging.
4. Understand the concept of an operon. Give examples of inducible and repressible operons.
5. Explain the concept of differentiated cells in a eukaryotic organism.
6. Define the levels on which regulation of gene expression can take place. Know the role of transcription factors, methylation, DNA packaging, alternative splicing, and mRNA stability in the regulation of gene expression, and give examples of each type of regulation.

Key Terms and Phrases

chromosomes
chromatin
nuclear matrix
nucleolus
nucleoplasm
nuclear envelope
perinuclear space
nuclear lamina
lamins
nuclear matrix
nuclear pore complex
nuclear localization signal
histones
nonhistone chromosomal proteins
nucleosome
nucleosome core particle
linker DNA
packing ratio of DNA
mitotic chromosomes
lampbrush chromosomes
heterochromatin
euchromatin
constitutive heterochromatin
facultative heterochromatin
position effect
Barr body
X inactivation
genetic mosaics
karyotype
Q bands
centromere
telomere
telomerase
chromosome aberrations
inversions
translocation
deletion
cri du chat syndrome
duplication
gene induction
gene repression
positive control
negative control
operon
structural genes
promoter
operator
repressor
inducer
gene regulatory protein
regulatory gene
catabolite gene activator protein (CAP)
repressible operon
inducible operon
trp operon
lac operon
selective gene amplification
transcriptional-level control
transcription factors
general transcription factors
translational-level control
metal response elements
basal level elements
glucocorticoid response elements
glucocorticoid receptor
palindrome
transcription factor motifs
zinc finger
helix-loop-helix (HLH)
genomic imprinting
DNA methylation
processing-level control
high mobility group (HMG) proteins
alternate splicing

DNase I hypersensitive sites
masked mRNA
mRNA editing
homeobox
N-end rule
untranslated regions (UTRs)
poly (A) binding protein
translational bypassing
homeodomain
ubiquitin
enhancers
translational frameshifting
protein splicing
Hox genes
proteosome

Reviewing the Chapter

I. Introduction

A. All cells in a multicellular organism contain the same complement of genes.
B. Cells express their genetic information selectively.
C. Gene expression is controlled by the cell.

II. The Nucleus of a Eukaryotic Cell

A. The contents of the nucleus are enclosed by a nuclear envelope. A typical nonmitotic nucleus includes several major components:
   1. Chromosomes are present as extended fibers of chromatin.
   2. The nuclear matrix is a protein-containing fibrillar network.
   3. Nucleoli are the sites of ribosomal RNA synthesis.
   4. Nucleoplasm is the fluid in which the solutes of the nucleus are dissolved.
B. The nuclear envelope:
   1. The nuclear envelope is a complex structure at the boundary between the nucleus and the cytoplasm of a eukaryotic cell.
      a. A nuclear envelope consists of two membranes separated by a perinuclear space.
      b. The two membranes are fused at sites forming a circular nuclear pore.
      c. The inner surface of the nuclear envelope is lined by the nuclear lamina.
         (1) The nuclear lamina supports the nuclear envelope.
         (2) The nuclear lamina is composed of lamins.
         (3) The integrity of the nuclear lamina is regulated by phosphorylation and dephosphorylation.
   2. The structure and function of the nuclear pore complex:
      a. Proteins and RNAs are transported in and out of the nucleus.
      b. Bidirectional transport occurs through the nuclear pore complex (NPC).
      c. Proteins synthesized in the cytoplasm are targeted to the nucleus by the nuclear localization signal (NLS).
         (1) An NLS usually consists of one or two short stretches of positively-charged amino acids.
         (2) Proteins with an NLS stretch bind to an NLS receptor.
         (3) The NLS receptor docks proteins to the nuclear pore complex.
         (4) Conformation of the NPC changes as the protein passes through.
C. Chromosomes:
   1. Chromosomes are visible only during mitosis.
   2. Packaging the genome:
      a. Chromosomes consist of chromatin fibers, composed of DNA and associated protein.
         (1) Each chromosome contains a single, continuous DNA molecule, packaged to occupy a small space.
         (2) The protein component of chromosomes includes histones and nonhistone chromosomal proteins.
      b. Nucleosomes– the lowest level of chromosome organization:

(1) Histones are a group of very basic proteins that are involved in DNA packaging.
   (a) Histones are highly conserved between different species.
   (b) DNA and histones are organized into repeating subunits called nucleosomes.
   (c) Each nucleosome includes a core particle and a linker.
   (d) DNA is wrapped around the core complex.
   (e) The histone core complex consists of two molecules each of histones H2A, H2B, H3, and H4.
   (f) H1 histone is located outside of the nucleosome core particle.
(2) Positively charged residues on histones interact with negatively charged phosphate groups of the DNA backbone.
   (a) Positively charged residues occur in a spiral along the path of DNA wound around the core.
   (b) DNA-histone interaction is not sequence specific.
(3) Rapidly dividing cells require a large number of histone genes.
(4) Binding of a nonhistone protein to DNA can influence the positioning of nucleosome particles in the adjacent region of the chromosome.

c. Higher levels of chromatin structure:
   (1) A 30 nm filament is the second level of chromatin structure.
   (2) A 30 nm filament packaging is maintained by histone H1.
   (3) Chromatin filaments are organized into large supercoiled loops.
   (4) The presence of loops in chromatin can be seen:
      (a) in mitotic chromosomes from which histones have been extracted.
      (b) in meiotic lampbrush chromosomes of amphibian oocytes.

3. Heterochromatin and euchromatin:
   a. Euchromatin returns to a dispersed state after mitosis.
   b. Heterochromatin is condensed during interphase.
   c. Heterochromatin can be either constitutive or facultative.
      (1) Constitutive heterochromatin remains condensed all the time.
      (2) Constitutive heterochromatin is mostly situated around centromeres and telomeres.
      (3) Constitutive heterochromatin consists of highly repeated sequences and is devoid of protein-coding genes.
      (4) Active genes placed next to constitutive heterochromatin become silenced as a result of their position.
   d. One of the X chromosomes in females is present as facultative heterochromatin and is called a Barr body.
   e. X inactivation:
      (1) Heterochromatization of one X chromosome in each cell of a female mammal occurs at an early stage in development.
      (2) X inactivation is a random process (either of the two X chromosomes can be inactivated), making adult females genetic mosaics.
      (3) X inactivation is initiated by an RNA molecule.

4. The structure of the mitotic chromosome:
   a. Chromatin of a mitotic cell exists in its most highly condensed state.
      (1) The shape of the mitotic chromosome depends on the length of the DNA molecule and the position of the centromere.
      (2) Staining mitotic chromosomes can provide useful information.

(3) A karyotype is a portrayal of pairs of homologous chromosomes ordered according to decreasing size.
    (a) The pattern of Q bands is highly characteristic for each chromosome of a species.
    (b) Specific genes can be located on mitotic chromosomes by in situ hybridization.

b. Centromeres:
    (1) The centromere is located at the site of marked indentation on a chromosome.
    (2) Centromeres contain constitutive heterochromatin.
    (3) Centromeric DNA is the site of microtubule attachment during mitosis.

c. Telomeres:
    (1) The ends of DNA molecules are marked with a set of repeated sequences called telomeres.
    (2) New repeated units are added by a telomerase, which is a reverse transcriptase that assembles DNA from an RNA template.

5. Chromosome aberrations (The Human Perspective):
    a. A chromosomal aberration is a loss or exchange of a segment between different chromosomes, caused by exposure to DNA-damaging agents.
    b. Chromosomal aberrations have different consequences depending on whether they are in somatic or germ cells.
    c. Inversions, translocations, deletions, and duplications are examples of different chromosomal aberrations.
        (1) Inversion involves the breakage of a chromosome and resealing of the segment in a reverse order.
        (2) Translocation is attachment of all or a piece of one chromosome to another chromosome.
        (3) Deletion is the loss of a portion of a chromosome.
        (4) Duplication occurs when a portion of a chromosome is repeated.
    d. Many human diseases are caused by chromosomal aberrations.

D. The nucleus as an organized organelle:
    1. Chromatin fibers are concentrated at specific domains within the nucleus.
    2. Chromosome ordering is directed by the nuclear envelope proteins.
    3. In the nucleus, mRNAs are synthesized at discrete sites.
    4. The nuclear matrix:
        a. The nuclear matrix is a network of protein-containing fibrils.
        b. The nuclear matrix serves as a skeleton, maintaining the shape of the nucleus and anchoring the machinery involved in nuclear activities.
        c. Newly transcribed pre-mRNAs are held in place by elements of the nuclear matrix.

III. Control of Gene Expression in Prokaryotes

A. Bacterial cells selectively express genes to use the available resources efficiently.
    1. The presence of lactose in the medium induces the synthesis of ß-galactosidase.
    2. The presence of tryptophan in the meduim represses the genes that encode enzymes for tryptophan synthesis.

B. The bacterial operon:
    1. An operon is a functional complex of genes containing the information for assembling the enzymes of a metabolic pathway.
        a. A typical operon includes structural genes, a promoter region, an operator region, and a regulatory gene.
        b. Structural genes code for the enzymes themselves. All structural genes are translated from one mRNA.

c. RNA polymerase binds at the promoter.
d. The operator is the site between the promoter and the first structural gene, where a regulatory protein can bind.
e. A repressor is an example of a gene regulatory protein.
f. Regulatory genes code for repressor proteins.
g. When the repressor is bound to an operator, transcription is switched off.

2. The lac operon:
   a. The lac operon is an inducible operon; it is turned on in the presence of lactose.
      (1) The lac operon contains three tandem structural genes.
      (2) Lactose binds to the repressor, changing its conformation and making it unable to attach to the operator.
      (3) Repressor protein can bind to the operator and prevent transcription from the lac operon only in the absence of lactose.
   b. Positive control by cyclic AMP:
      (1) The lac repressor exerts negative control.
      (2) The glucose effect is an example of positive control.
      (3) Cylcic AMP (cAMP) acts by binding to the catabolite gene activator protein (CAP).
      (4) Binding of CAP-cAMP to the lac control region changes the conformation of DNA, which allows RNA polymerase to transcribe the lac operon.
3. The trp operon:
   a. The trp operon is a repressible operon; it is turned off in the presence of tryptophan.
   b. The trp operon repressor is active only when it is bound to a corepressor such as tryptophan.

IV. Control of Gene Expression in Eukaryotes

A. Introduction:
1. Cells of a complex eukaryotic organism exist in many differentiated states.
   a. Differentiated cells retain a full set of genes, as demonstrated in experiments with plant root cells
   b. Nuclei from cells of adult animals are capable of supporting the development of a new individual, as demonstrated in experiments with Xenopus oocytes.
2. Genes are turned on or off as a result of interaction with regulatory proteins.
   a. Each cell type contains a unique set of proteins, characteristic of its state of differentiation.
   b. Regulation of gene expression occurs on three levels:
      (1) Transcriptional-level control mechanisms determine whether a gene will be transcribed, and how often.
      (2) Processing-level control mechanisms regulate the processing of primary RNA transcript into a messenger RNA.
      (3) Translational-level control mechanisms regulate the translation of mRNAs.
3. Selective amplification of DNA templates:
   a. Selective gene amplification can lead to an increase of the copy number of a particular set of genes .
   b. Genes that code for rRNA in amphibian oocytes are selectively amplified.
4. Rearrangement of DNA sequences involved in antibody formation:
   a. An antibody consists of two light (L) and two heavy (H) chains. Each chain includes a variable (V) and a constant (C) portion.
   b. Genes coding for antibodies are subject to rearrangement.

(1) Each antibody chain is coded by at least two separate genes –a C gene and a V gene.
(2) A chromosome usually contains a single C gene and a large number of different V genes.
(3) C and V genes are joined via a J segment

c. A gene-shuffling mechanism creates an additional diversity in protein products.

B. Transcriptional-level control:

1. Differential transcription is the most important mechanism by which eukaryotic cells determine which proteins are synthesized.
   a. Transcription factors are the proteins that either stimulate or inhibit transcription by binding specific DNA sequences.
      (1) A single gene can be controlled by different regulatory proteins.
      (2) A single DNA-binding protein may control the expression of many different genes.
   b. Regulatory sequences are located upstream of the gene.
      (1) The TATA box is where general transcription factors assemble.
      (2) Formation of a preinitiation complex at the TATA site is sufficient for a basal level of gene transcription.
   c. For some genes (such as metallothionein), additional transcription factors bound to basal level elements (BLEs) are required for transcription.
   d. The basal level of transcription is modified (either stimulated or inhibited) by other DNA-binding proteins.
      (1) Metal response elements (MREs) bind transcription factors that are activated by metals, thus increasing transcription of the metallothionein gene above the basal level.
      (2) Binding of activated glucocorticoid receptor protein (in complex with hormone) to the glucocorticoid response elements (GRE) increases transcription.
   e. Enhancers are DNA elements that stimulate transcription above the basal level.
      (1) Enhancers can be located very far upstream from the regulated gene.
      (2) Enhancer-bound proteins stimulate transcription by interacting with components of the basal transcription machinery.
2. The structure of transcription factors:
   a. Transcription factors contain domains that mediate different functions.
   b. The Glucocorticoid Receptor (GR):
      (1) The GR is a nuclear receptor that has three domains: DNA-binding domain, ligand-binding domain, and transcription-activator domain.
      (2) The GR-binding sequence (GRE) is a palidrome. Variations in GRE increase or decrease affinity for the GR.
      (3) Pairs of GR molecules form dimers in which each subunit of the dimer binds to one-half of the DNA palindrome.
      (4) Each DNA-binding domain of the GR contains two α–helical loops, each with a zinc ion at its base, called zinc fingers.
   c. Other transcription factors:
      (1) The DNA-binding domains of most transcription factors possess related structures (motifs) that interact with DNA sequences.
         (a) Most of the motifs contain a segment that binds to the major groove of the DNA.

(b) The zinc finger, the helix-loop-helix, and the HMG box are the most common motifs.

(2) In zinc finger proteins, the zinc ion of each finger is held in place by two cysteines and two histidines.

(3) The helix-loop-helix (HLH) motif has two alpha-helical segments separated by a loop.

(4) The high mobility group (HMG) box motif consists of three alpha helices.

3. Repression of transcription in eukaryotes:
   a. Eukaryotic cells possess both positive and negative regulatory mechanisms.
   b. Genes can be silenced in three ways:
      (1) Specific DNA sequences can interact with repressor proteins.
      (2) DNA can be modified to become less suitable as a template.
      (3) Transcription of a particular gene depends on the balance between positive-acting and negative-acting regulatory factors.
   c. The role of DNA Methylation.
      (1) DNA methylation in eukaryotic cells silences transcription .
      (2) DNA methylation mostly occurs in G-C-rich "islands," in or near transcriptional regulatory regions.
      (3) Methylation patterns of gene regulatory regions change during cellular differentiation.
   d. Genomic imprinting:
      (1) Activity of certain genes, called imprinted genes, depends on whether they originated with the sperm or the egg.
      (2) Inactive and active versions of imprinted genes differ in their methylation patterns.
4. Chromatin structure and transcription:
   a. Nonhistone proteins are able to interact with DNA that is tightly associated with histones.
      (1) Incorporation of DNA into nucleosomes in vitro inhibits transcription.
      (2) DNA can be transcribed even when it is associated with histones.
   b. Nucleosomes are positioned at nonrandom sites along the DNA.
      (1) The wrapping of a DNA segment around the histone core leaves only some parts of it accessible to transcription factors.
      (2) Nucleosomes may play an active role in promoting the interaction between DNA and a regulatory protein.
   c. The binding of a regulatory protein to DNA induces changes in chromatin structure, making other sites more or less accessible.
   d. Transcriptionally active chromatin possesses DNase I hypersensitive sites.
      (1) DNase I hypersensitive sites are usually located in regulatory regions of genes.
      (2) Modification in chromatin structure, triggered by the binding of a transcription factor, is accompanied by the appearance of DNase I hypersensitive sites in and around the binding site.
   e. Transcribed genes retain their nucleosomes, which are likely to be modified to make DNA more accessible to RNA polymerase.
      (1) Histone octamers may be disrupted as theRNA polymerase moves along.
      (2) Histone core proteins may be acetylated, weakening the histone's interaction with DNA.

5. Regulating transcription elongation:
    a. RNA polymerase moves along the template with a changing velocity.
    b. Certain DNA sequences slow down the RNA polymerase.
    c. Some transcription factors promote or inhibit RNA polymerase movement past those sequences.

C. Processing-level control:
1. Protein diversity can be generated by alternative splicing.
2. There may be more than one way to process a primary transcript.
    a. An intron can be spliced out of the transcript or retained.
    b. An mRNA may contain alternate 5′ or 3′ terminal sequences.
3. The mechanism for selecting alternative splice sites is unclear.
4. Cells may control whether or not the primary transcript will be processed.

D. Translational-level control:
1. Translation of mRNAs that have been transported from nucleus to cytoplasm is regulated.
    a. Translational-level control occurs via interactions of specific mRNAs and proteins in the cytoplasm.
    b. Regulatory proteins act on untranslated regions (UTRs) at 5′ and 3′ ends of mRNA.
2. Cytoplasmic localization of mRNAs:
    a. In the fruit fly embryo, the development of the anterior-posterior axis is regulated by the localization of specific mRNAs along this axis in the egg.
    b. Cytoplasmic localization of mRNAs is determined by their 3′ UTRs.
3. The control of mRNA translation:
    a. In an unfertilized egg, mRNAs are stored in inactive form (masked mRNAs).
    b. Masked mRNAs are associated with inhibitory proteins.
    c. The mRNA translation rate can be controlled in different ways:
        (1) phosphorylation of initiation factors
        (2) binding of repressor molecules
        (3) inhibition by an RNA molecule bound to the 3′ UTR of an mRNA.
4. The control of mRNA stability:
    a. The lifetimes of eukaryotic mRNAs vary widely.
    b. Poly(A) tail length may influence the longevity of mRNA.
        (1) The poly(A) tail is bound by a poly(A) binding protein (PABP).
        (2) PABP protects the tail from general nuclease activity.
        (3) PABP increases the sensitivity of the tail to a specific poly(A) ribonuclease.
        (4) While an mRNA is in cytoplasm, its tail its gradually shortened by the poly(A) nuclease.
        (5) When the tail is about 30 residues, it can no longer bind PABP; the mRNA is degraded.
    c. Certain sequences in the 3′ UTR may affect the rate of poly(A) tail shortening.
    d. Oher mechanisms of translational-level control have been discovered recently.
        (1) In translational frameshifting, a change of the reading frame by the ribosome generates two different polypeptides.
        (2) In a termination codon read through, the ribosome continues past a termination codon.
        (3) Editing of an mRNA after it has been transcribed occurs primarily in mitochondria.

(4) In protein splicing, a segment can be excised from a specific polypeptide, and the two ends, joined.

E. Posttranslational control– determining protein stability:
   1. A protein's lifespan is partly determined by the amino acid sequence on the N-terminus (N-end rule).
   2. Condemned proteins are linked to ubiquitin, then enzymatically degraded.

V. Genes That Control Embryonic Development (Experimental Pathways)

A. Homeotic genes have been identified in fruit flies by the effect of mutations on the development of segmentation.
   1. Mutations in homeogenes cause transformation of one body segment into another.
   2. Different homeotic genes have a region of similarity called homeobox.
   3. Homeobox sequences are present in a variety of eukaryotic organisms.

B. In mammals, the linear order of homeotic genes along the chromosome parallels the expression of those genes along the anterior-posterior axis of the embryo

C. Homeotic gene products function as transcription factors

Key Figure

Figure 12-28*a*. Gene regulation by operons.

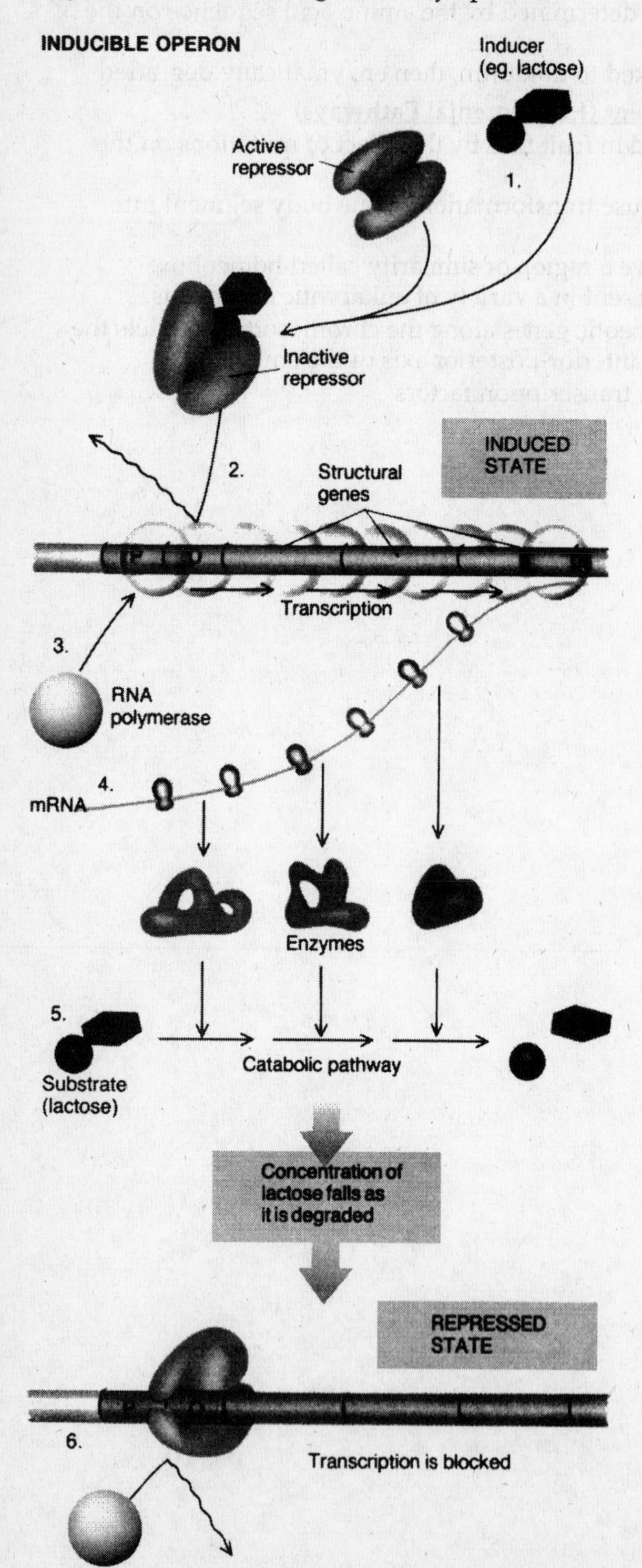

Questions for Thought:

1. In addition to the genes specifically shown in this diagram, what others must exist as part of the operon?

2. What features of this diagram identify it as a prokaryotic system?

3. This diagram shows induction of ß-galactosidase occurring at the level of transcription. Where else might gene expression be regulated?

4. If you knew only that ß-galactosidase was induced in the presence of lactose, but you didn't know whether the mechanism involved initiation of transcription or activation of existing enzyme molecules, what experiment could you perform to determine the exact mechanism of induction?

Review Problems

Short Answer

1. Give the significance (not the definition) of the following terms or phrases. Say what they do or why they are important. For example:
   lac operon: *enables prokaryotic cells to regulate the synthesis of enzymes for lactose metabolism only when it is appropriate to do so, i.e., when there is lactose in the environment.*

   a. nuclear pore complex

   b. nuclear localization signal

   c. evolutionary conservation of histone proteins

   d. H1 histone

   e. nuclear matrix

   f. selective gene amplification

   g. DNA rearrangement of antibody genes

   h. heterodimerization of transcription factors

   i. DNase I hypersensitive sites

   j. acetylated histones

2. Compare and contrast the following:
   a. heterochromatin vs. euchromatin

   b. constitutive vs. facultative heterochromatin

   c. repressible vs. inducible operons

   d. DNA methylation in prokaryotes vs. DNA methylation in eukaryotes

   e. liver fibronectin vs. fibroblast fibronectin

Multiple Choice

1. Which of the following in not true of the nuclear envelope?
   a. The nuclear envelope is exactly like other cellular membranes.
   b. The nuclear envelope separates the genetic material from the cytoplasm.
   c. The nuclear envelope is a pair of concentric membranes.
   d. The nuclear envelope is studded with pores.

2. Which cytoplasmic fibrils are most like the nuclear lamins?
   a. microtubules
   b. microfilaments
   c. intermediate filaments
   d. actomyosin

3. What kind of molecules must pass between the nucleus and the cytoplasm?
   a. DNA
   b. protein
   c. lipids
   d. carbohydrates

4. Which statement best characterizes the nuclear localization signal?
   a. The NLS is typically a small molecular weight metabolic intermediate.
   b. The NLS is a stretch of hydrophobic amino acids on a protein's N terminus.
   c. The NLS is one or two stretches of basic amino acids on a protein's C-terminus.
   d. The NLS is a steroid that binds to DNA.

5. Which sequence represents increasing levels of chromosomal organization, from most dispersed to most condensed?
   a. nucleosomes–30 nm filaments–supercoiled loops–mitotic chromosomes
   b. nucleosomes–supercoiled loops–30 nm filaments–mitotic chromosomes
   c. nucleosomes–30 nm filaments–mitotic chromosomes–supercoiled loops
   d. mitotic chromosomes–30 nm filaments–supercoiled loops–nucleosomes

6. When chromatin is treated with nonspecific nucleases, what is the length of the reulting pieces of DNA?
   a. random numbers of base pairs
   b. about 60 base pairs
   c. about 8 base pairs
   d. about 200 base pairs

7. Which of the following statements about histones is not true?
   a. Histones are very similar between species.
   b. Histones have many basic amino acids.
   c. Histones are rich in lysine and arginine.
   d. Each histone has one single gene that codes for it.

8. In mammalian cells, the DNA of the centromere is characteristic of:
   a. facultative heterochromatin.
   b. constitutive heterochromatin.
   c. euchromatin.
   d. dispersed chromatin.

9. Why are adult women genetic mosaics?
   a. Inactivated X chromosomes are all derived from the male parent.
   b. Inactivated X chromosomes are all derived from the female parent.
   c. Inactivated X chromosomes can be from either the male or the female parent.
   d. none of the above

10. What do telomeres do?
    a. They protect the chromsomes from degradation by nucleases.
    b. They prevent the ends of chromosomes from fusing with one another.
    c. They are required for complete chromosomal replication.
    d. all of the above

11. When all or a piece of a chromosome becomes attached to another chromosome, the aberration is called a(n):
    a. inversion.
    b. translocation.
    c. deletion.
    d. duplication.

12. If there were a mutation in the regulatory gene of an inducible promoter rendering the protein incapable of binding to the repressor, then:
    a. the structural genes would always be expressed.
    b. the structural genes would never be expressed.
    c. the structural genes would only be expressed in the presence of the inducer.
    d. the structural genes would only be expressed in the absence of the inducer.

13. Which of the following is a difference between inducible versus repressible operons?
    a. In an inducible operon, the ligand binds to the regulator protein.
    b. In a repressible operon, the ligand-regulator complex binds to the operator.
    c. In a repressible operon, all the structural gene products have related functions.
    d. In an inducible operon, RNA polymerase binds to the promoter.

14. The experiment, in which root cells from a mature plant could be induced to grow into a fully developed plant, suggested that:
    a. differentiated cells still retained all the genetic information.
    b. root cells are different than leaf cells.
    c. cells loose the chromosomes that are not needed during the process of differentiation.
    d. cell culture is an important technique not only for animal cells, but for plants as well.

15. Antibody diversity is partly due to:
    a. movement of a C gene close to one of over 300 V genes.
    b. movement of a J segment close to a V gene.
    c. variability in the V-J joining site.
    d. all of the above.

14. To stimulate transcription, enhancer sequences:
    a. must be within a few base pairs of the gene they enhance.
    b. must be within a few hundred base pairs of the gene they enhance.
    c. can be tens of thousands of base pairs away from the genes they enhance.
    d. will not function if they are moved experimentally.

15. Which of the following is not a characteristic of transcription factors?
    a. Transcription factors have two fold rotational symmetry.
    b. Active transcription factors are dimers.
    c. Transcription factors are rich in basic amino acids.
    d. Transcription factors have one of a few DNA-binding motifs.

16. Eukaryotes regulate gene expression at all of the following levels except:
    a. transcription.
    b. translation.
    c. mRNA processing
    d. chromosomal loss with differentiation

17. Homeobox genes are involved in:
    a. regulating inducible enzymes.
    b. regulating the appropriate expression of body segmentation.
    c. regulating mRNA longevity.
    d. none of the above.

18. Alternate splicing means that:
    a. the same gene can code for several different proteins.
    b. several different genes can code for the same protein.
    c. gene expression can be regulated at the level of transcription.
    d. pieces of DNA can move around within the genome.

19. Which of the following statements is true?
    a. A cell can potentially make fewer different proteins that the number of different genes it contains.
    b. A cell can potentially make only the same number of different proteins as tthe number of different genes it contains.
    c. A cell can potentially make more different proteins than the number of different genes it contains.
    d. none of the above

20. The longevity of mRNA is related to:
    a. the length of the poly (A) tail.
    b. the presence of PABP.
    c. the nucleotide sequence of the 3' UTR.
    d. all of the above.

Problems and Essays

1. The lac operon in E. coli has three regulatory regions (the gene for the repressor, i; the operator, O; and the promoter, P) and three structural genes (the gene for ß-galactosidase, z; the gene for galactoside permease, y; and the gene for a transacetylase, a).

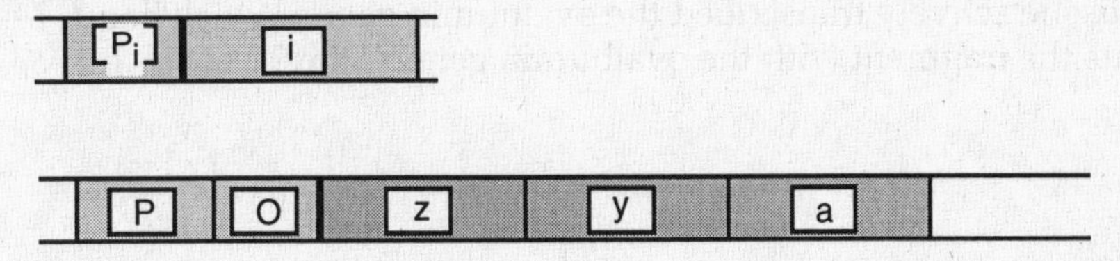

In the absence of mutations, the structural genes, z, y, and a, are all expressed in the presence of lactose and repressed when lactose is not present. The genotype of the wild-type bacterium is

$$i^+ P^+ 0^+ z^+ y^+ a^+$$

a. For the following mutations, would the structural genes be expressed in the presence of lactose and in the absence of lactose? The answer to the first one is given as an example.

	With Lactose			Without Lactose		
Genotype	ß-gal	gal permease	transacet	ß-gal	gal permease	transacet
$i^+ P^+ 0^+ z^+ y^+ a^-$	yes	yes	no	no	no	no
$i^+ P^+ 0^+ z^- y^+ a^+$	no					
$i^- P^+ 0^+ z^+ y^+ a^+$						
$i^+ P^- 0^+ z^+ y^+ a^+$	no	no	no			
$i^+ P^+ 0^- z^+ y^+ a^+$						

b. What would happen to the expression of the structural genes in the presence and absence of lactose if there were a mutation in the promoter region of the i gene, P_i?

2. Chick DNA was isolated from both brain tissue and red blood cells and was treated with varying concentrations of DNase I. The DNA was then cut with a restriction enzyme that releases both the globin gene and the ovalbumin gene in fragments of characteristic size. These fragments were separated on agarose gels, and hybridized with radiolabeled probes for both genes. The results follow:

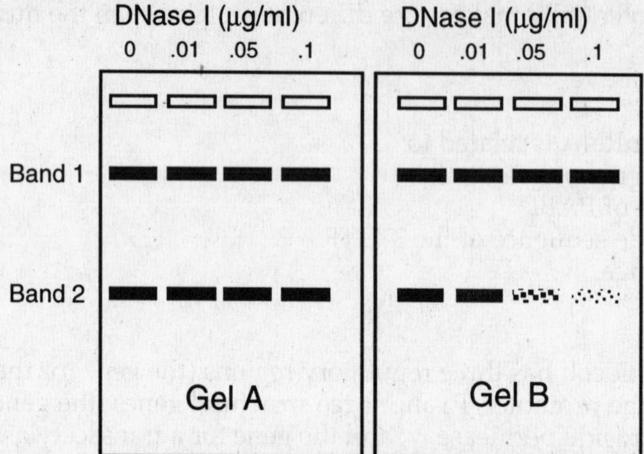

a. Neither tissue actively transcribed the ovalbumin gene. Which band, 1 or 2, most likely represents the fragment with the ovalbumin gene? Why?

b. Which gel, A or B, most likely represents the fragments from red blood cells? Why?

3. What specific problem is presented by the need for copious amounts of rRNA (eg., in the oocytes of amphibians) as opposed to the need for protein? What strategies have cells evolved for providing rRNA?

4. Although many different transcription factors have been isolated, most can be grouped into one of a few families, according to recurring structural motifs. For the following proteins, indicate which structural motif is present.

    ____________________This transcription factor occurs as a heterodimer with two alpha-helical segments separated by a loop. The α-helices are perpendicular to one another.

    ____________________MyoD

    ____________________This transcription factor has several zinc ions affiliated with it.

    ____________________This transcription factor bends DNA, resulting in a spatial arrangement that enhances its access to other regulatory factors.

    ____________________TFIIIA

    ____________________UBF

5. Indicate whether the following describe transcriptionally active or inactive chromatin:

    ____________________Core histones are heavily acetylated.

    ____________________Chromatin is tightly affiliated with histones.

    ____________________Chromatin is hypersensitive to DNase I.

    ____________________Chromatin is methylated.

    ____________________Chromatin is affiliated with histones, but the histone octamers are disrupted.

    ____________________Chromatin is in the form of constitutive heterochromatin.

Notes:

CHAPTER THIRTEEN

DNA Replication and Repair

Learning Objectives

When you have finished this chapter, you should be able to:

1. Define DNA replication. Describe the experiments in both prokaryotes and eukaryotes that established that DNA replication is semiconservative.
2. Give an overview of the replication process in prokaryotes, including the requirements for replication and the components of the replication complex.
3. Know the properties of different DNA polymerases, and the roles of DNA ligase, primase, DNA gyrase, helicase and single-stranded DNA-binding proteins.
4. Describe specific features of replication in eukaryotes. Compare and contrast these with specific features of replication in prokaryotes.
5. Recognize both the necessity for DNA repair and the various ways cells have evolved to repair DNA.
6. Describe some conditions caused by defects in the DNA repair mechanisms of humans. Explain how these conditions have been used to investigate the mechanisms of DNA repair.

Key Terms and Phrases

replication
dispersive replication
nonpermissive temperature
bidirectional replication
primer
lagging strand
primase
replisome
initiator protein
replicase
recognition complex
base excision repair
replication foci
semiconservative replication
temperature-sensitive (ts) mutants
replication fork
DNA polymerase
semidiscontinuous replication
Okazaki fragments
DNA gyrase
single-stranded DNA-binding protein (SSB)
exonuclease
autonomous replicating sequences (ARSs)
conservative replication
Meselson and Stahl
origin
Arthur Kornberg
leading strand
DNA ligase
helicase
oriC site
ter site
replicon
nucleotide excision repair
mismatch repair

Reviewing the Chapter

I. Introduction

A. Self-duplication is a fundamental property of life.
B. DNA duplicates by a process called replication.
C. DNA replication requires a number of auxiliary materials.
D. The DNA replication machinery is also used for DNA repair.

II. DNA Replication

A. DNA replication takes place by separation of the strands of the double helix, and synthesis of two daughter strands complementary to the two parental templates.

B. Semiconservative nature of replication:

1. During DNA replication, half of the parent structure is retained in each of the daughter duplexes.
2. The Meselson and Stahl experiments supported the semiconservative model of replication in bacterial cells.
3. Semiconservative replication was later demonstrated in eukaryotes.

C. Replication in bacterial cells:

1. Temperature-sensitive mutants were used to identify the genes of replication.
2. Replication can be studied using in vitro systems reconstituted from purified cellular compounds.
3. Replication forks and bidirectional replication:
    a. Replication of the bacterial chromosome can be visualized as a theta structure.
    b. Replication forks are the points where a pair of replicating segments come together and join the nonreplicated segments.
    c. Replication starts at the origin site.
    d. Replication proceeds bidirectionally.
3. The properties of DNA polymerases:
    a. DNA polymerase is DNA-dependent, i.e., it uses DNA as a template.
    b. DNA polymerase requires a primer which provides the 3′ hydroxyl terminus on which to add new nucleotides.
    c. Polymerization only occurs in a 5′-to-3′ direction.
    d. There are three classes of DNA polymerases in bacterial cells. None are able to initiate DNA chains.
4. Semidiscontinuous replication:
    a. Both daughter strands are synthesized simultaneously.
    b. The leading strand (in direction of the replication fork movement) is synthesized continuously.
    c. The lagging strand (in the opposite direction of the replication fork movement) is synthesized discontinously.
    d. The lagging strand is constructed of small Okazaki fragments, which are joined by DNA ligase.
    e. Primase is an RNA polymerase that assembles short RNA primers; these primers are later removed and the gaps are sealed.
5. The replication fork:
    a. Movement of the replication fork introduces positive supercoils in DNA.
    b. DNA gyrase changes positive supercoils into negative supercoils.
    c. Helicase and single-stranded DNA-binding proteins unwind the parental duplex and separate the two strands.
    d. Primase and helicase form a primosome, which processively moves along the lagging-strand template.
    e. A single replisome synthesizes both leading and lagging strands.
6. Initiation and termination of replication:
    a. In bacterial cells, replication starts at the oriC site.
    b. Replication begins with binding of an initiator protein.
    c. Replication terminates at the ter site.

C. The structure and functions of DNA polymerases:

1. DNA polymerase III is the primary replication enzyme.
    a. DNA polymerase III is called a replicase.
    b. DNA polymerase III holoenzyme consists of a single catalytic subunit and a number of associated subunits.

c. By forming a sliding clamp, the ß-subunit maintains an association between the polymerase and the DNA template.
d. DNA polymerase I is involved in DNA repair and also removes RNA primers and replaces them with DNA.
e. The function of DNA polymerase II is unclear.

2. Exonuclease activities of DNA polymerases:
    a. Exonuclease degrades nucleic acids by removing the 5′ or 3′ teminal nucleotides one at a time.
    b. DNA polymerase I has 5′-to-3′ and 3′-to-5′ exonuclease activity; each exonuclease activity, as well as the polymerase activity is assigned to a separate domain of the enzyme.
    c. The 5′-to-3′ exonuclease of DNA polymerase I degrades RNA primers.
3. Ensuring high fidelity during DNA replication:
    a. Nucleotides are incorporated into the growing strand if they can form acceptable base pairs with nucleotides of the template strand.
    b. Acceptability of the incoming base is determined by the geometry of the base pair.
    c. During proofreading, mismatched bases are excised.
    d. Careful selection of the nucleotide, immediate proofreading, and mismatch repair account for low error rates in replication (about 10^{-9}).
    e. Replication is rapid (about 1000 nucleotides per second).

D. Replication in eukaryotic cells:
1. Replication in eukaryotes is less well understood than in prokaryotes.
    a. Mutant yeast cells that are unable to produce specific gene products are used to study replication.
    b. In vitro systems can be reconstituted with eukaryotic cellular extracts and/or purified components. Templates from simple sources are usually used.
2. Replication in eukaryotes is similar to that in prokaryotes.
    a. There are five different DNA polymerases in eukaryotes:
        (1) Polymerase α is associated with primase.
        (2) The leading strand and most of the lagging strand are assembled by polymerase δ.
        (3) Polymerase β is involved in DNA repair.
        (4) Polymerase γ replicates mitochondrial DNA.
    b. Eukaryotic DNA polymerases elongate DNA strands in the 5′-to-3′ direction and require a primer. Most possess 3′-to-5′ exonuclease activity.
    c. Eukaryotic DNA is replicated in replicons at many sites simultaneously.
3. Initiation of replication:
    a. Initiation of DNA synthesis in a replicon is regulated.
    b. Replicons located close together in a chromosome tend to undergo replication simultaneously.
    c. More tightly compacted regions are replicated later in the cell cycle.
    d. Origins of replication identified in yeast are called autonomous replicating sequences (ARSs).
    e. A multiprotein origin recognition complex (ORC) is assembled at the ARS. Mutation in the ARS prevents ORC from binding.
    f. Replication origins have not yet been identified in mammals.
4. Replication and nuclear structure:
    1. The replication machinery is immobilized in the nuclear matrix.
    2. Replication forks are located within replication foci.
5. Chromatin structure and replication:

a. Nucleosomes remain associated with the DNA when the replication fork passes through.
b. Histone cores remain intact during replication and each core passes randomly to one daughter duplex.
c. New histone cores are assembled on the other arm of the fork.

III. DNA Repair

A. DNA repair is essential for cell survival.
1. Cells are subjected to a variety of destructive forces.
2. DNA is the cell molecule that is most susceptible to environmental damage.
3. Ionizing radiation, common chemicals, ultraviolet radiation and thermal energy create various spontaneous alterations (lesions) in DNA.
4. Cells have a number of mechanisms to repair genetic damage.
5. Excision is the main mechanism of DNA repair

B. Nucleotide excision repair:
1. A nucleotide excision repair system removes bulky lesions, such as pyrimidine dimers and chemically altered nucleotides.
2. An endonuclease makes nicks on two sides of the altered strand, after which damaged oligonucleotides are removed by helicase and the gap is sealed.
3. Nucleotide excision repair consists of two pathways:
   a. The preferential pathway selectively repairs the genes of greatest importance to the cell.
   b. A slower pathway repairs the rest of the genome.

C. Base excision repair:
1. Base excision repair removes altered nucleotides that produce distortions of the double helix.
2. DNA glycosylase recognizes the alteration and cleaves the base from the sugar.
3. DNA glycosylases are specific for particular types of altered bases.
4. Thymine occurs in DNA instead of uracil because otherwise the repair system would not have been able to distinguish between uracil and altered cytosine.

D. Mismatch repair:
1. Mismatch repair corrects the mistakes that escape the DNA polymerase proofreading.
2. Repair enzymes recognize distortions caused by mismatched bases.
3. The repair system must distinguish between parental and new strands.
4. In bacteria, the two strands are distinguished by the presence of methyl groups on the parental strand.
5. The mechanism of mismatch repair in eukaryotes is not well understood.

E. Excision repair deficiencies are the cause of a number of human disorders, such as xeroderma pigmentosum, Cockayne syndrome and skin cancer (The Human Perspective).

IV. The Role of Human NER Deficiencies in DNA Repair Research (Experimental Pathways)

A. DNA repair deficiency leads to an increased sensitivity to sunlight, in both bacteria and humans.
B. Cells of individuals afflicted with xeroderma pigmentosum are unable to properly excise pyrimidine dimers.
C. Several human genes involved in DNA repair have been isolated using intergenic complementation and transfection-correction techniques.

Key Figure

Figure 13.16. Replication of both the leading and lagging strands in E. coli is accomplished by DNA polymerases working together in a single complex.

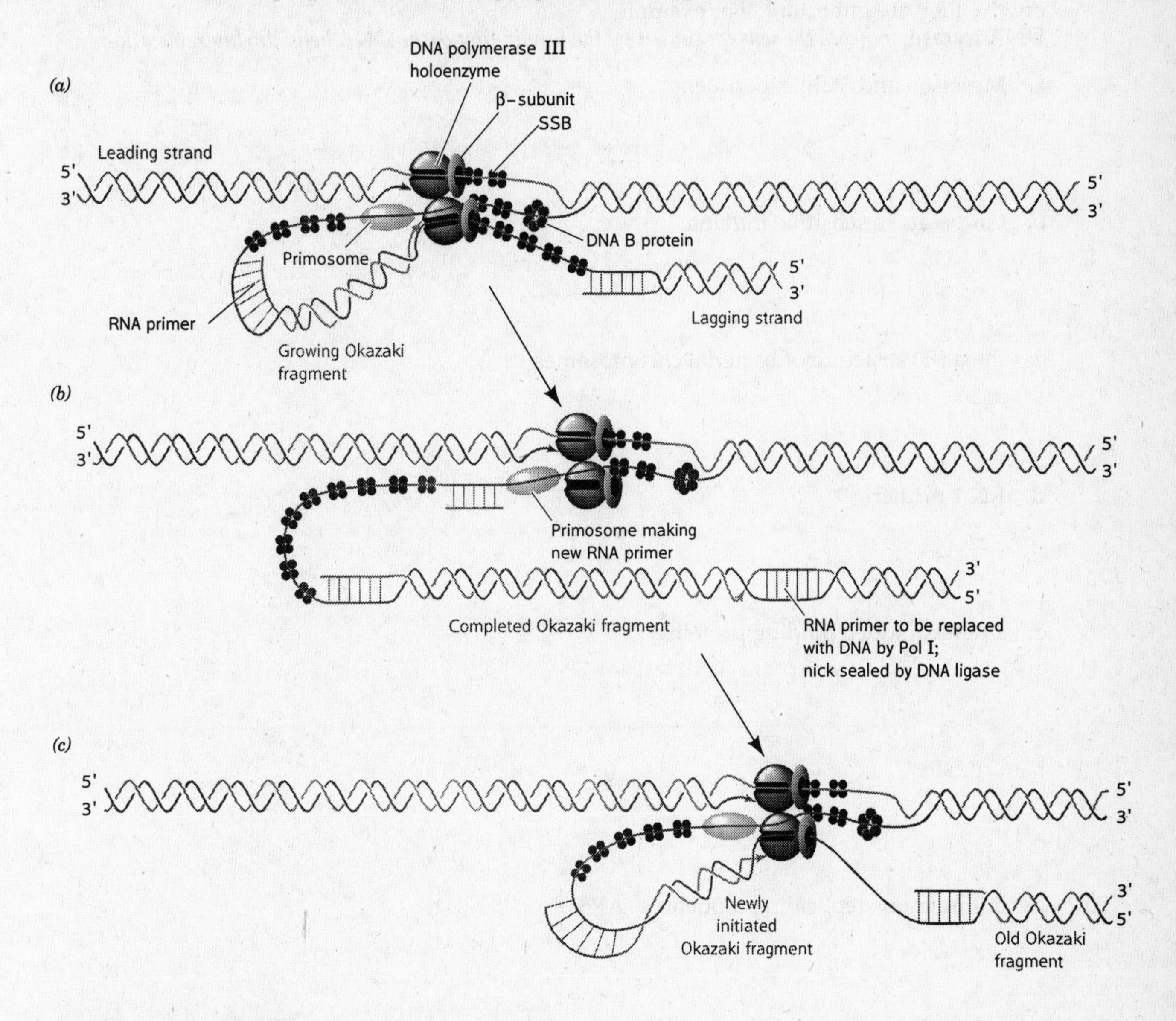

Questions for Thought:

1. Why is it necessary for one of the template strands being replicated to fold back onto the DNA polymerase as the polymerase moves down the DNA?

2. What role do the SSB proteins play in this process? Does one strand require more SSBs than the other? Why or why not?

3. For simplicity, this figure does not show the torsional stress put on DNA by a replication complex, or the enzymes that relieve that stress. Which side of the replication fork shown here would experience the greatest stress?

4. What experimental evidence led to the discovery of Okazaki fragments?

Review Problems

Short Answer

1. Give the significance (not the definition) of the following terms or phrases. Say what they do or why they are important. For example:
   DNA gyrase: *relieves the tension caused by the unwinding of the DNA helix during replication.*

   a. Meselson and Stahl experiment

   b. temperature-sensitive mutants

   c. theta (Θ) structure of bacterial chromosomes

   d. RNA primers

   e. single-stranded binding proteins

   f. oriC site

   g. autonomous replicating sequences (ARSs)

   h. DNA repair enzymes

   i. intergenic complementation technique

2. Compare and contrast the following:
   a. RNA primers of leading strands vs. RNA primers of lagging strands

   b. oriC vs. ter

   c. base substitution errors in transcription vs. base substitutions in replication

   d. replication origins in prokaryotes vs. replication origins in eukaryotes

   e. autonomous replicating sequence vs. origin recognition complex

Multiple Choice

1. Semiconservative replication of DNA means that:
   a. each daughter duplex will have one of the original parental strands and one new strand.
   b. one daughter duplex will be entirely new and the other will have both original parental strands.
   c. both daughter duplexes will be entirely new and the parental duplex will be degraded.
   d. each strand of each daughter duplex will have parts of the parental strands and parts of new strands.

2. In vitro systems used to study replication must include all of the following except:
   a. a DNA molecule to be replicated.
   b. deoxynucleotide triphosphates.
   c. isolated cellular nuclei.
   d. cellular extracts with the factors and cofactors of replication.

3. The theta (Θ) structure that results from autoradiography of replicating bacterial chromosomes illustrates all of the following except
   a. the unreplicated portion of the chromosome.
   b. the pair of daughter molecules in the process of being formed.
   c. replication forks.
   d. the replication origin.

4. A replicating prokaryotic chromosome has __?__ replication forks, and a replicating eukaryotic chromosome has __?__.
   a. one; many
   b. two; two
   c. two; many
   d. many; many

5. Which of the following forms of DNA can serve as a template for DNA polymerase?
   a. partially double-standed DNA
   b. circular double-stranded DNA
   c. intact double-standed DNA
   d. circular single-stranded DNA

6. If replicating bacteria are given a short pulse of radiolabeled thymidine and rapidly killed, the radiolabel shows up as:
   a. large DNA molecules only.
   b. mostly large DNA molecules and very few short DNA molecules.
   c. mostly short DNA molecules and very few large DNA molecules.
   d. free nucleotides only.

7. An input of energy os not requried for which of the following steps of replication?
   a. separating the two DNA strands.
   b. unwinding the coiled DNA.
   c. linking together the individual deoxynucleotides.
   d. All of the above require energy.

8. If a reaction mixture of replicating DNA is suddenly diluted:
   a. the rate of DNA synthesis is slowed.
   b. the rate of DNA synthesis is unchanged.
   c. the rate of DNA synthesis is speeded up.
   d. DNA synthesis stops.

9. Experiments such as that described in number (8) above demonstrated that:
   a. a new DNA polymerase molecule enters the complex to synthesize each Okazaki fragment.
   b. the same DNA polymerase molecule synthesizes each new Okazaki fragment.
   c. DNA polymerase molecules are not involved in DNA synthesis.
   d. none of the above

10. The replisome, which synthesizes DNA on both leading and lagging strands, always moves toward the replication fork. This must mean that:
    a. synthesis on the lagging strand requires that the strand fold back onto the replisome.
    b. synthesis on the lagging strand occurs in the 3'-to-5' direction, and synthesis on the leading strand occurs in the 5'-to-3' direction.
    c. synthesis on the leading strand occurs in the 3' to 5' direction, and synthesis on the lagging strand occurs in the 5'-to 3'-direction.
    d. DNA strands are lined up in the same direction, with 3' ends together and 5' ends together.

11. The exonuclease activity of the DNA polymerases functions to:
    a. remove the RNA primer sequences.
    b. proofread the new DNA strand and remove inappropriate nucleotides.
    c. maximize the fidelity of DNA replication.
    d. all of the above

12. Incorrect pairing of bases during bacterial DNA replication occurs about once every 10^5 to 10^6 nucleotides, but the spontaneous mutation rate is only one error in 10^9 base pairs. What accounts for the difference?
    a. proofreading and correcting by the DNA polymerase
    b. mismatch repair following replication
    c. both (a) and (b).
    d. neither (a) nor (b).

13. The last DNA to be replicated in the eukaryotic chromosome is:
    a. telomeres at the end of the chromosomes.
    b. heterochromatin.
    c. euchromatin in the arms of the chromosomes.
    d. facultative heterochromatin.

14. OriC is to E. coli as___?___is to yeast.
    a. autonomous replicating sequence
    b. origin recognition complex
    c. heterochromatin
    d. DNA polymerase δ

15. DNA polymerase III is to prokaryotes as __?__ is to eukaryotes.
    a. DNA polymerase α
    b. DNA polymerase β
    c. DNA polymerase γ
    d. DNA polymerase δ

16. A newly replicated DNA duplex is likely to have:
    a. nucleosomes both from the parental DNA and newly assembled.
    b. nucleosomes, each with histones both from the parental nucleosomes and newly synthesized.
    c. nucleosomes either all newly synthesized or all from the parental strand.
    d. a random conbination of both nucleosomes and individual histones, both newly synthesized and from the parental strands.

17. How are damaged nucleotides recognized by DNA repair enzymes?
    a. They are tagged with methyl groups.
    b. They are tagged with acetyl groups.
    c. They are recognzied by their bulky geometry.
    d. They are recognized by lesion receptors.

18. If the genes from the following list were damaged, which gene would most likely be repaired first?
    a. the globin genes in a cell destined to become a chick red blood cell.
    b. the heterochromatin in a liver cell.
    c. the ovalbumin genes in a cell destined to become a chick red blood cell.
    d. the antibody genes in a liver cell.

19. Fibroblast cells from several different xeroderma pigmentosa patients were cultured, then fused in intergenic complemetation studies. Which cell fusions would not result in a hybrid capable of repairing damaged DNA?
    a. XP-A and XP-G hybrids
    b. XP-B and XP-B hybrids
    c. XP-C and XP-D hybrids
    d. XP-E and XP-F hybrids

20. Why should patients with xeroderma pigmentosum avoid sunlight?
    a. The UV wavelengths do irreparable damage to DNA.
    b. Sunlight inhibits any residual DNA repair activity in the cell.
    c. These patients lack pigmentation to protect them from burning.
    d. Sunlight inhibits DNA polymerases.

Problems and Essays

1. Onion root tip is a good source of actively growing cells undergoing DNA replication. Suppose you pulse-label some root tip cells with ^{3}H-thymidine, then isolate the chromosomes during metaphase of cell division. Using autoradiography, you examine the pattern of thymidine incorporation into the chromosomes.

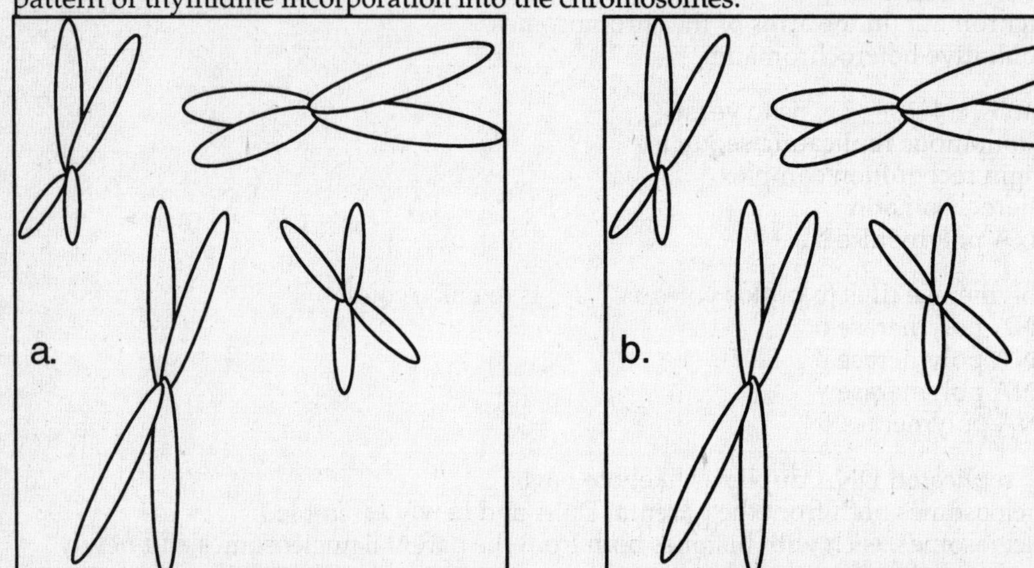

   a. In diagram (a), use dots to indicate where the radiolabeled thymidine appears during the first cell division.

   b. Show the pattern of radioactive labeling during the second cell division on diagram (b).

   c. Explain the labeling pattern that you put in each diagram.

2. On the following diagram of an E. coli replicon, draw leading strands, lagging strands, Okazaki fragments, and primers. Label (or color code) the parts that are DNA and the parts that are RNA. Also label the OriC site.

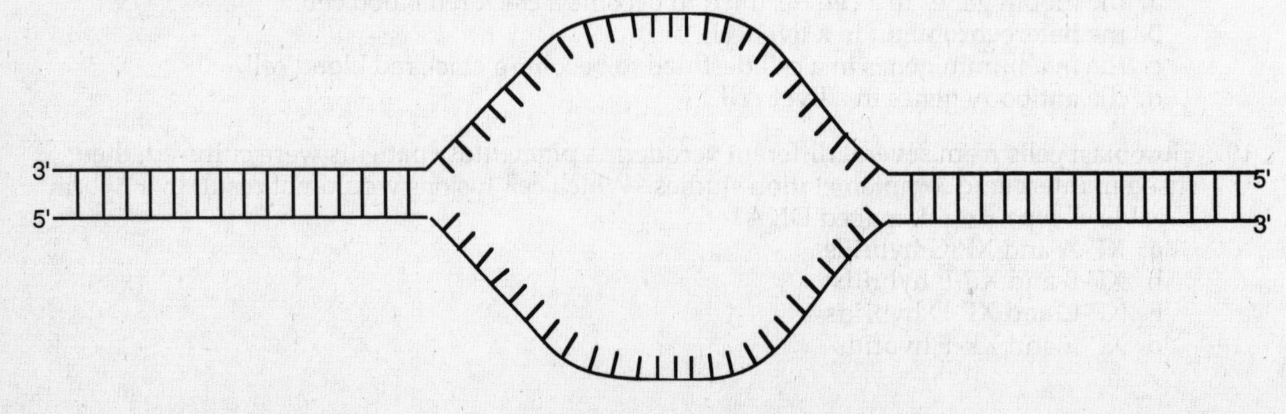

3. You have isolated a temperature-sensitive mutant of E. coli that can replicate its DNA at 20 °C, but not at 37 °C. You are interested in identifying some of the important enzymes involved in DNA replication in normal cells. Using both your temperature-sensitive mutant and your cultures of normal E. coli, design an experiment to identify one such enzyme.

4. The fictional bacterium Abra cadabra has a chromosome with one origin of replication, OriC, and one digeston site for the restriction endonuclease, EcoR1. You have digested many copies of the Abra cadabra chromosome with EcoR1, then provided the linear chromosomes with all the raw materials for DNA replication, including cellular extracts. After various time intervals, you examine some of the chromosomes using electron microscopy. The structures you find before restriction digestion and at time= 0 minutes are shown below. Draw the structures you might find at 5 minutes, 10 minutes, 20 minutes, and 40 minutes.

a. Before restriction enzyme digestion:

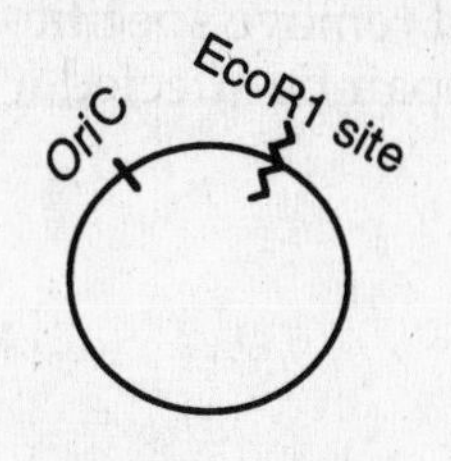

b. time = 0 minutes:

c. time = 5 minutes:

d. time = 10 minutes:

e. time = 20 minutes:

f. time = 40 minutes:

5. One frequent type of DNA lesion involves deamination of the deoxyribonucleotide bases. Next to each DNA base illustrated below, draw the structure (if any) that results from spontaneous deamination.

Natural Bases	Deaminated Bases	Natural Bases	Deaminated Bases
cytosine	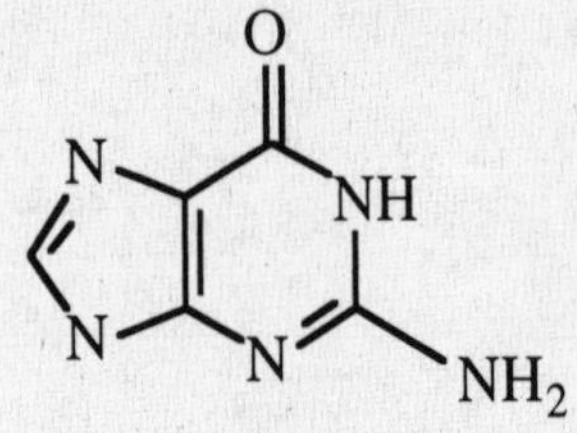	guanine	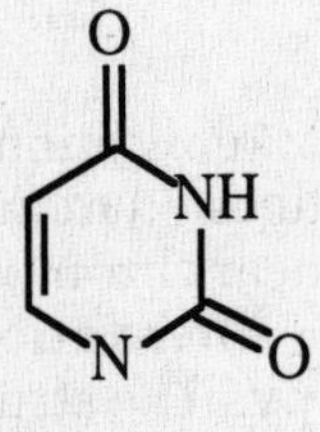
thymine	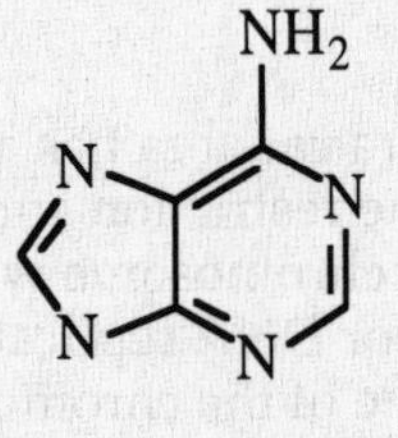	adenine	

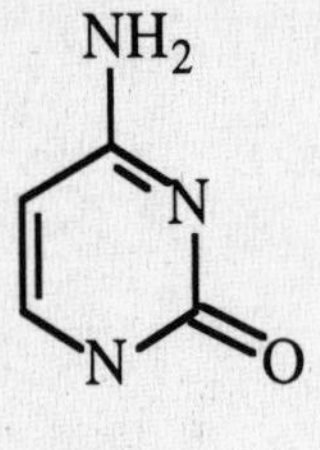

Do any of your structures look familiar?

6. At least 20 different DNA glycosylases recognize and remove specific deaminated DNA bases. How would this important system of DNA repair be affected if uracil, rather than thymine, occurred in DNA as it does in RNA?

CHAPTER FOURTEEN

Cellular Reproduction

Learning Objectives

When you have finished this chapter, you should be able to:

1. Define cellular reproduction and describe its main characteristics.
2. Define the cell cycle, including the different phases. Know in which phases the synthesis of major macromolecules occurs.
3. Understand how the cell cycle is controlled.
   a. Describe the role of cyclins and protein kinases in the cell cycle regulation.
   b. Describe the model systems used for studying the cell cycle, and recognize what makes them particularly suitable for such studies.
4. Give an overview of mitosis and meiosis, and summarize the differences between the two.
   a. Describe the formation of the mitotic spindle and explain how the chromosomes are moved during anaphase.
   b. Understand the difference between chromosomes and sister chromatids.
   c. Define centromere and kinetochore.
5. Describe prophase I of meiosis; define synaptonemal complex, chiasma, bivalent, and tetrad.
6. Understand genetic recombination and its importance in evolution. Know how genetic recombination occurs.

Key Terms and Phrases

cell division	mother cell	daughter cell
cellular reproduction	cell cycle	M phase
interphase	G_1	S phase
G_2	maturation-promoting factor (MPF)	cyclin
cyclin-dependent kinase (Cdk)	checkpoint	p53
mitosis	prophase	chromatid
centromere	kinetochore	centrosome cycle
mitotic spindle	nuclear envelope breakdown	aster
prometaphase	metaphase	anaphase
anaphase A	anaphase B	telophase
cytokinesis	contractile ring theory	phragmoplast
meiosis	gametic meiosis	spermatogonia
primary spermatocytes	spermatids	oogonia
primary oocytes	zygotic meiosis	sporic meiosis
sporophyte	sporogenesis	gametophyte
meiosis I	meiosis II	interkinesis
prophase I	leptotene	zygotene
synapsis	synaptonemal complex (SC)	bivalent (tetrad)
pachytene	diplotene	diakinesis
chiasma	mitotic nondisjunction	aneuploidy
recombination nodules	genetic recombination	Holliday structure

Reviewing the Chapter

I. Introduction

A. Cells reproduce by the process of cell division.
B. A dividing cell is called a mother cell; its descendants are called daughter cells.
C. The mother cell transmits copies of its genetic information to its daughter cells.
D. Cellular gametes are formed by cell division.

II. The Cell Cycle

A. The cell cycle is the succession of events from one cell division to another.
  1. The cell cycle consists of the M phase and interphase.
    a. The M phase includes the processes of mitosis and cytokinesis.
    b. The M phase usually lasts 30 to 60 minutes.
    c. Interphase, constituting the majority of the cell cycle, lasts longer than the M phase.
    d. Both synthesis of macromolecules and DNA replication occur during interphase.
  2. Interphase includes G_1, S, and G_2 periods.
    a. G_1 is the period between the end of mitosis and the beginning of DNA replication.
    b. DNA replication occurs during the S phase.
    c. The G_2 occurs between the end of DNA synthesis and the beginning of mitosis.

B. Cell cycles in vivo:
  1. Three cell types are distinguished bssed on their capacity to grow and divide.
    a. Extremely specialized cells (nerve cells) have lost the ability to divide.
    b. Cells that normally do not divide can be induced to begin DNA synthesis (liver cells).
    c. Cells that have a high level of mitotic activity (stem cells).
  2. There is great variability in the length of the cell cycle in different cells.
    a. G_1 is most variable in length.
    b. Cells that stay indefinitely in G_1 are considered to be in the G_o state.

C. Cell cycle synthetic activities:
  1. Overall rates of RNA and protein synthesis are constant during interphase.
  2. During mitosis, protein synthesis drops and RNA synthesis stops.
  3. Histones are synthesized almost exclusively during S phase; histone messenger RNAs are destroyed after replication is complete.
  4. DNA replication and synthesis of histone proteins are regulated together.

D. Control of the cell cycle:
  1. Cell cycle control is focused on initiation of DNA replication and initiation of mitosis.
  2. The cytoplasm contains factors that regulate the state of the nucleus, as evidenced from cell fusion experiments.
    a. The cytoplasm of a replicating cell contains factors that stimulate initiation of DNA synthesis.
    b. The cytoplasm of a cell undergoing mitosis contains factors that trigger chromosomal condensation.
    c. G_1-S and G_2-M transitions are both under positive control.
  3. The role of protein kinases:
    a. Entry of the cell into M phase is triggered by activation of a protein kinase called maturation-promoting factor (MPF).
      (1) MPF consists of two subunits: a kinase and a regulatory subunit, cyclin.

(2) Increased concentration of cyclin activates the kinase.
(3) The cyclin level fluctuates predictably during the cell cycle.

b. MPF-like, cyclin-dependent kinases (Cdks) occur in yeast cells.
   (1) The product of the cdc2 gene is a cyclin-dependent kinase responsible for passage through both control points.
   (2) For cells to pass through a point of commitment Cdks must be transiently activated by specific cyclins.
   (3) Activated Cdk phosphorylates the sets of proteins specific for the particular transition.
c. G_1 cyclins are responsible for G_1-S transition (START point).
d. Mitotic cyclins are responsible for the G_2-M transition.
e. Cyclins and cyclin-dependent kinases are highly conserved between species.

4. Checkpoints, kinase inhibitors and cellular responses:
   a. Cells are able to control their progress through the cell cycle according to the events inside and outside of the cell.
   b. Progress of a cell through the cell cycle can be arrested at a checkpoint if certain conditions have not been met.
      (1) In yeast cells, a certain mass must be reached to enter S phase.
      (2) DNA damage or inhibition of DNA synthesis causes the cell to arrests its entry into mitosis.
   c. The rate of cell division is controlled by growth-altering agents.
      (1) Cells stimulated by external agents to pass G_1-S transition proceed towards mitosis.
      (2) Growth-altering agents can stimulate or inhibit cell division.
   d. Cells may synthesize proteins that inhibit progression through the cell cycle.
      (1) Inhibitor proteins may act on the Cdk-cyclin complex or the DNA replication machinery.
      (2) Certain transcription factors can either stimulate or repress the transcription of genes involved in cell cycle activities.
      (3) Inhibitor proteins often act as tumor suppressors.
      (4) Mutations in the p53 gene are present in approximately 50 percent of all human cancers. The p53 gene product is a transcription factor.

III. M Phase: Mitosis and Cytokinesis

A. Mitosis is a process of nuclear division in which two cells with identical genetic content are produced.
   1. Mitosis is usually accompanied by cytokinesis.
   2. Cytokinesis partitions the cytoplasm into two roughly equal parts.
   3. Mitosis maintains the chromosome number.
   4. Mitosis can occur in either diploid or haploid cells.
   5. Mitosis consists of prophase, prometaphase, metaphase, anaphase, and telophase.

B. Prophase:
   1. In prophase, duplicated chromosomes are prepared for segregation and mitotic machinery is assembled.
   2. Formation of the mitotic chromosome:
      a. Mitotic chromosomes are compacted into short, thick fibers.
         (1) Each mitotic chromosome consists of two chromatids.
         (2) Chromatids are connected to each other at their centromeres.
         (3) Proteins that mediate packaging of mitotic chromosomes form a scaffold to which DNA loops are attached.

(4) Chromosome compaction is triggered by activation of cyclin-dependent kinase, which phosphorylates H1 histone molecules.

b. Centromeres and kinetochores:
    (1) Centromeres occur at primary constrictions on mitotic chromosomes.
    (2) Kinetochores are on the outer surfaces of centromeres.
    (3) Kinetochores are the sites where chromosomes attach to the microtubules of the mitotic spindle.

3. Formation of the mitotic spindle:
    a. The mitotic spindle is made of microtubules.
    b. The centrosome cycle (doubling of centrioles in the cytoplasm) progresses simultaneously with the cell cycle.
    c. In the mitotic spindles of animal cells, microtubules are arranged in an aster formation around each centrosome.
        (1) Microtubules of the aster emerge from the pericentriolar material, which acts as a nucleating center.
        (2) Certain types of animal cells and most of higher plants lack centrioles.

4. The dissolution of the nuclear envelope and the fragmentation of cytoplasmic organelles:
    a. Nuclear envelope breakdown is caused by Cdk kinase.
    b. Cdk kinase phosphorylates nuclear lamins, which leads to the disassembly of the nuclear lamina.
    c. The endoplasmic reticulum and Golgi complex are fragmented during mitosis, while mitochondria, lysosomes, and chloroplasts stay intact.

C. Prometaphase:
1. During prometaphase the definitive mitotic spindle is formed and chromosomes are moved by microtubules into the center of the cell.
    a. The microtubules grow into the region around a chromosome
    b. The ends of the microtubules penetrate into the body of the kinetochore.
    c. A single kinetochore becomes attached to microtubules from both spindle poles.
2. Chromosomes arrive at the center of the cell by differential growth of microtubules at opposite poles.

D. Metaphase:
1. In metaphase, chromosomes are aligned at the spindle equator on the metaphase plane.
2. Microtubules in a mitotic spindle are highly organized.
    a. Astral microtubules radiate from the centrosome to the region outside the body of the spindle and determine the plane of cytokinesis.
    b. Chromosomal (or kinetochore) microtubules extend from the centrosome to the kinetochores of the chromosomes; they move chromosomes to the poles.
    c. Polar (interpolar) microtubules extend from the centrosome past the chromosomes and maintain the integrity of the spindle.
3. Misalignment of one or several chromosomes during metaphase delays the progression of the cell through mitosis.

E. Anaphase:
1. Entry to anaphase is triggered by a drop in Cdk activity and degradation of mitotic cyclins.
2. Sister chromatids split from each other during anaphase; this separation requires topoisomerase II activity.

3. Chromosomes are split in synchrony; chromatids begin migrating to the poles.
    a. As chromosomes move toward a pole, microtubules attached to kinetochores are shortened.
    b. Movement of chromosomes toward the poles is called anaphase A.
    c. Two spindle poles move in opposite directions due to elongation of polar microtubules; this process is called anaphase B.
4. Movement of chromosomes toward the poles is very slow.

F. Telophase:
1. The cell returns to interphase condition during telophase.
    a. Nuclear envelopes of two nuclei are reassembled.
    b. Chromosomes become dispersed.
    c. The endoplasmic reticulum and Golgi complex re-form.
2. The cytoplasm is partitioned into two cells.

G. Forces required for mitotic movements:
1. Mitotic movement is powered by microtubule motors (dynein and kinesin-related proteins).
    a. Microtubule motors are located at the spindle poles and kinetochores.
    b. Motor proteins can be studied in two ways:
        (1) by analyzing a mutant cell without a motor.
        (2) by using antibodies against motor proteins.
    c. Motor proteins have a number of general features.
        (1) Motor proteins along polar microtubules keep the poles apart.
        (2) Motor proteins in the kinetochore bring chromosomes to the metaphase plate and keep them there.
        (3) Motor proteins in the region of spindle equator elongate the spindle during anaphase B.
2. Forces required for chromosome movement at anaphase:
    a. According to the now abandoned traction fiber hypothesis, shortening of the chromosomal microtubules during anaphase is caused by loss of subunits at the poles.
    b. Chromosomes possess all the motor proteins they need to move; this machinery is located in the region of the kinetochores.
    c. According to Inoué's theory, disassembly of the microtubules at the pole generates sufficient force to pull the chromosomes.

H. Cytokinesis:
1. Cytokinesis in animal cells:
    a. Cytokinesis starts with an indentation of the cell surface, lying in the same plane as chromosomes of the metaphase plate.
    b. The contractile ring theory suggested that a thin cortical band composed of actin and myosin filaments generates the force to cleave the cell.
        (1) Sliding of actin filaments pulls the cortex and attached plasma membrane to the center of the cell.
        (2) The cortical ring is assembled rapidly prior to cytokinesis and dismantled immediately after it.
        (3) The cortical ring forms midway between two poles.
    c. Dissociation of both mitosis and cytokinesis is possible.
2. Cytokinesis in plant cells– formation of the cell plate:
    a. Plant cells build a cell membrane and cell wall in the cell center.
    b. The cell plate begins with the appearance of the microtubular phragmoplast.
    c. Material for the cell wall is brought to the phragmoplast by ER-Golgi vesicles.
    d. Cell plate formation proceeds laterally from the center of the cell.

IV. Meiosis

A. During meiosis, chromosome number is halved and haploid cells are formed
  1. Each daughter cell has one member of each pair of homologous chromosomes.
  2. Meiosis consists of two subsequent divisions.
     a. Homologous chromosomes pair, ensuring that daughter cells receive a full haploid complement of chromosomes.
     b. Genetic recombination occurs during pairing of homologues.
     c. Genetic recombination increases genetic variability and thus enhances the likelihood of survival of cellular progeny.
  3. In different eukaryotes meiosis occurs at different points in the life cycle.
     a. In gametic (terminal) meiosis, meiosis is linked to gamete formation.
        (1) Gametic meiosis is typical for all multicellular animals.
        (2) In vertebrates, formation of spermatozoa from spermatogonia and eggs from oogonia takes place by gametic meiosis.
     b. In zygotic (initial) meiosis, meiosis occurs just after fertilization; zygotic meiosis is typical for fungi.
     c. In sporic (intermediate) meiosis, meiosis is independent of gamete formation and fertilization.
        (1) Sporic meiosis is typical for higher plants.
        (2) Sporogenesis (production of spores) includes meiosis.
        (3) Gametes are produced by haploid gametophytes via mitosis.

B. The stages of meiosis:
  1. DNA is replicated prior to meiosis. The premeiotic S phase lasts longer than a premitotic S phase.
  2. Prophase I consists of several stages:
     a. In leptotene, the first stage of prophase, chromosomal condensation starts.
     b. During zygotene, the second stage of prophase I, homologous chromosomes pair.
        (1) Genetic recombination begins before the chromosomes are visibly paired.
        (2) In synapsis, homologues associate via the synaptonemal complex.
           (a) The synaptonemal complex is a ladderlike structure.
           (b) The synaptonemal complex functions as a scaffold to hold the DNA molecules in place during the crossover.
           (c) A pair of synapsed homologous chromosomes form a bivalent (tetrad) composed of four chromatids.
     c. In pachytene, synapsis ends.
        (1) Recombination nodules appear on chromosomes.
        (2) Pachytene may last for weeks.
     d. During diplotene, the synaptonemal complex is disassembled and homologous chromosomes start moving apart.
        (1) Chiasmata are the remaining points of attachment between homologous chromosomes.
        (2) Chiasmata visually portray the extent of recombination.
        (3) Diplotene chromosomes can be observed as lampbrush chromosomes with characteristic loops of transcriptionally active DNA.
     e. During diakinesis, the last stage of prophase I, chromosomes are prepared for attachment to the meiotic spindle fibers.
        (1) Diakinesis ends with the disappearance of the nucleolus and the disassembly of the nuclear envelope.
        (2) Diakinesis is triggered by an increase in MPF activity.

3. During metaphase I, homologous chromosomes are aligned at the metaphase plate so that both chromatids of one chromosome face the same pole.
    a. Homologous chromosomes are held by one or several chiasmata.
    b. Absence of a chiasma can lead to abnormal segregation of chromosomes.
4. Homologous chromosomes separate during anaphase I.
    a. Maternal and paternal chromosomes of each tetrad segregate into the two daughter cells independent of other chromosomes.
    b. Behavior of chromosomes during anaphase I corresponds to Mendel's law of independent assortment.
5. Interkinesis is a short stage between two meiotic divisions.
    a. Cells at interkinesis have a haploid number of chromosomes.
    b. Cells have a diploid amount of nuclear DNA.
6. Meiosis II is simpler than meiosis I.
    a. During metaphase II, chromosomes are aligned so that the kinetochores of sister chromatids face opposite poles.
    b. Sister chromatids separate during anaphase II.
    c. Meiosis II produces cells haploid in both the amount of nuclear DNA and chromosome number.

C. Meiotic nondisjunction and its consequences (The Human Perspective):
1. Meiotic nondisjunction occurs when homologous chromosomes do not separate during meiosis I or sister chromatids don't separate during meiosis II.
2. Meiotic nondisjunction leads to the formation of gametes (and later zygotes) with abnormal numbers of chromosomes.
3. Aneuploidy is a condition in which an abnormal number of chromosomes occurs in cells.
    a. Trisomy is the presence of one extra chromosome.
    b. Monosomy is the lack of one chromosome.
4. Persons with an abnormal number of somatic chromosomes seldom survive.
5. Persons with an abnormal number of sex chromosomes are sterile.
6. A reduction in genetic recombination increases the risk of meiotic nondisjunction.
    a. Genetic recombination is reduced in the oocytes of older women.
    b. Infants of older mothers have a higher incidence of aneuploidy.

D. Genetic recombination during meiosis:
1. Meiosis increases genetic variability by mixing maternal and paternal alleles between homologous chromosomes.
2. According to early views, recombination took place by the actual physical breakage and reunion of an entire compacted chromatid.
3. According to modern views, recombination occurs via breakage and reunion of single DNA molecules.
    a. Recombination occurs without the addition or loss of a single base pair.
    b. Because of compaction, the breakage and reunion of an entire chromosome would randomly and irretrievably split hundreds of genes.
    c. Prior to recombination, DNA strands are aligned via homology search.
        (1) Breaks are introduced into one strand of each duplex at corresponding sites.
        (2) Broken strands invade the other DNA molecule.
        (3) The two duplexes are joined to each other by a Holliday structure.
    d. The point of crossover moves via branch migration.
4. Recombination intermediates in bacterial and viral genomes can be observed using the electron microscope.
5. Recombination in eukaryotes is more complex than in prokaryotes.

V. The Discovery and Characterization of MPF (Experimental Pathways)

A. MPF was first observed in studies of the effect of cytoplasm on the state of the nucleus in oocytes.

B. In developing embryos, MPF activity fluctuates following the stages of the cell cycle.

C. Levels of cyclin were found to correlate with levels of MPF activity.

D. Purified MPF was shown to possess kinase activity and stimulate nuclei to prepare for entry into mitosis

Key Figure

Figure 14.1. The eukaryotic cell cycle.

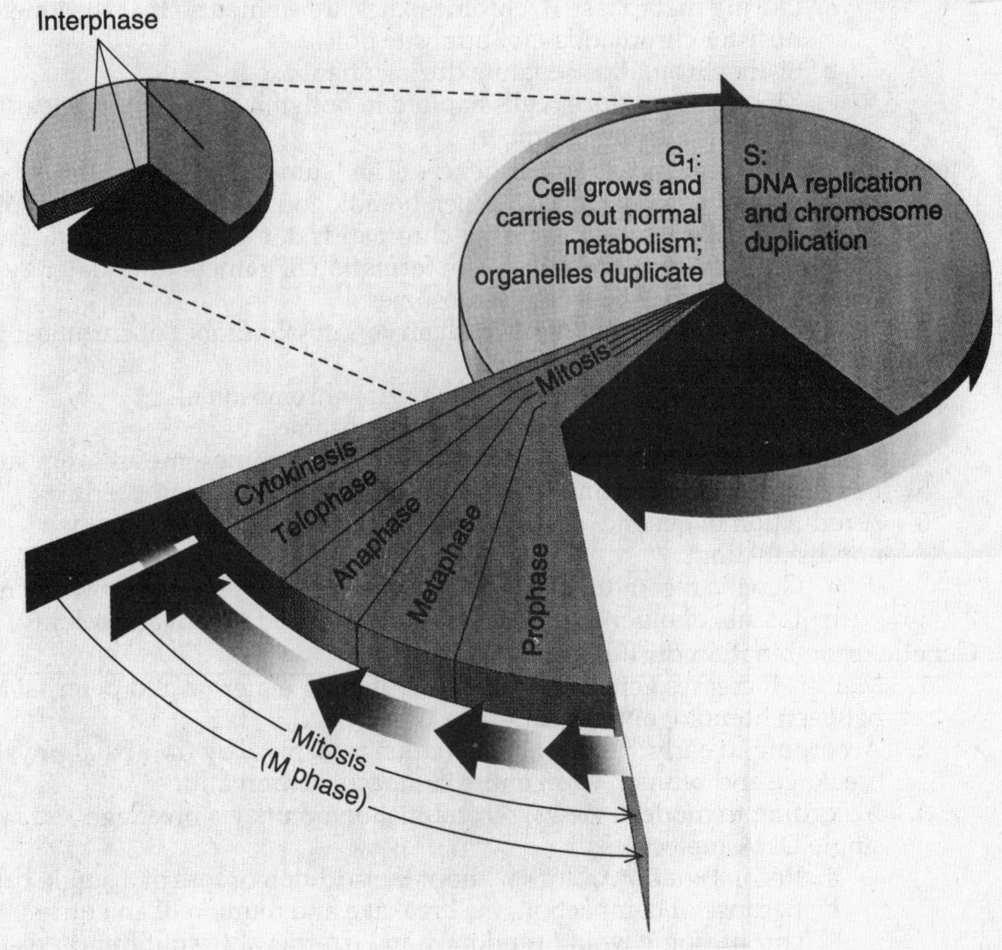

Questions for Thought:

1. In which phase of the cycle is protein synthesis occurring?

2. How would this diagram differ if it portrayed the cell cycle of a primary oocyte or primary spermatocyte? Why?

3. How would this diagram differ if it portrayed the cell cycle of a fully mature nerve cell? Why?

4. If the cycle is compared to a clock with the G_1-S transition at 12:00, at what hours is cylcin at its highest level?

Review Problems

Short Answer

1. Give the significance (not the definition) of the following terms or phrases. Say what they do or why they are important. For example:
   meiosis: *reduces the chromosome number by half, a necessary step for species that undergo sexual reproduction via fertilization.*

   a. MPF (maturation promoting factor)

   b. cell cycle checkpoints

   c. p53 protein

   d. formation of the mitotic chromosome

   e. phosphorylation of histones

   f. mitotic spindle

   g. fragmentation of the endoplasmic reticulum

   h. polar microtubules

   i. slow rate of chromosome movement during anaphase

   j. chiasmata

2. Compare and contrast the following:
   a. G_0 vs. G_1 periods of the cell cycle

   b. centromere vs. kinetochore

   c. prometaphase vs. metaphase

   d. anaphase A vs. anaphase B

   e. phragmoplast vs. contractile ring

Multiple Choice

1. A cell that is entering the M phase of the cell cycle is:
   a. always diploid, either with duplicated or unduplicated chromosomes.
   b. always haploid, always with duplicated chromosomes.
   c. either haploid or diploid, always with duplicated chromosomes.
   d. either haploid or diploid, either with duplicated or unduplicated chromosomes.

2. A culture of cells is given ^{3}H-thymidine for 30 minutes; then a sample of the cells is fixed and examined by autoradiography. It is found that:
   a. all of the cells engaged in mitosis have incorporated the labeled thymidine into their chromosomes.
   b. none of the cells engaged in mitosis have incorporated the labeled thymidine into their chromosomes.
   c. all of the cells, both mitotic and nonmitotic, have incorporated the labeled thymidine into their chromosomes.
   d. none of the cells, neither mitotic nor nonmitotic, have incorporated the labeled thymidine into their chromosomes.

3. During which phases of the cell cycle do the chromosomes exist in a doubled state?
   a. G_2 and early M
   b. G_1 and S
   c. late M and G_1
   d. G_0 and G_1

4. Liver cells are highly specialized, but when the liver is damaged or parts of it are removed surgically, the tissue grows. Liver cells fall into which category of cells?
   a. cells permanently locked in G_o
   b. cells that can be induced to enter the S phase
   c. cells subject to continual renewal
   d. cannot be determined from the information given

5. Throughout interphase, __?__ are synthesized at a relatively constant rate, whereas __?__ are synthesized only during the S phase.
   a. most major proteins, DNA replication enzymes
   b. histones, DNA replication enzymes
   c. structural proteins, functional proteins
   d. most major proteins, histones

6. If a cell in the G_1 phase is fused with a cell in the S phase, then
   a. the G_1 nucleus will enter the S phase.
   b. the S nucleus will enter the G_1 phase.
   c. both nuclei will enter the G_2 phase.
   d. both nuclei will shut down.

7. Cyclin concentrations are highest during which periods of the cell cycle?
   a. late G_1 and early S
   b. late G_2 and early M
   c. late G_1 and late G_2
   d. late M and late S

8. Cyclin is to MPF as __?__ is to cyclin-dependent kinases.
   a. cyclin
   b. MPF
   c. cdc2
   d. activator

9. The product of the cdc2 gene is
   a. cyclin.
   b. p53.
   c. cyclin-dependent kinase.
   d. CAK.

10. A mutation in the gene that codes for p53 can result in
    a. loss of an inhibitor protein, p21.
    b. loss of a transcription factor.
    c. uncontrolled cell growth and cancer.
    d. all of the above.

11. Which of the following structures is not a feature of the mitotic chromosome?
    a. compacted 30 nm chromatin fiber
    b. topoisomerase II
    c. centrosome
    d. kinetochore

12. During which of the following mitotic phases are chromosomes dispersed the most?
    a. prophase
    b. prometaphase
    c. metaphase
    d. anaphase

13. Which type of spindle microtubules are attached to chromosomes?
    a. astral microtubules
    b. kinetochore microtubules
    c. polar microtubules
    d. All of the above spindle microtubules are attached to chromosomes.

14. What is the best evidence that anaphase chromosomes are not pulled to the poles like fish on a line?
    a. Dyneins and kinesin-related proteins are associated with kinetochores.
    b. Depolymerization of spindle microtubules occurs at the kinetochores.
    c. Microtubule disassembly establishes the rate of chromosomal movement.
    d. In fact, it is believed that chromosomes are pulled in like fish on a line.

15. Which of the following experimental observations (some real, some contrived) would not be consistent with our current understanding of cytokinesis?
    a. Anti-myosin antibodies injected in a dividing cell cause cytokinesis to stop.
    b. When the myosin II gene is mutated, cells undergo nuclear division, but not cytokinesis.
    c. Anti-actin antibodies injected into dividing cells have no effect on cytokinesis.
    d. Actin filaments seen in micrographs are aligned parallel to the cleavage furrow.

16. The cleavage furrow is to animal cells as the___?___is to plant cells.
    a. cell plate
    b. cell membrane
    c. cell wall
    d. microtubule

17. The life cycle of mosses is characterized by an "alternation of generations," where the small, haploid gametophyte produces gametes, which fertilize and give rise to the tall sporophyte generation. What type of life cycle is this?
    a. gametic
    b. sporic
    c. zygotic
    d. asexual

18. If diplotene lampbrush chromosomes are incubated with radiolabeled uracil and then examined using autoradiography, most of the radiolabel will appear
    a. on the lampbrush loops.
    b. along the lampbrush backbone.
    c. on both the lampbrush loops and the lampbrush backbone.
    d. unaffiliated with the chromosomes.

19. Which stage of cell division provides visual evidence of Mendel's law of independent assortment?
    a. mitotic anaphase
    b. mitotic metaphase
    c. meiotic anaphase I
    d. meiotic anaphase II

20. The following diagram represents a recombination event between homologous chromosomes (one homoloque in bold and the other in fine). Did branch migration occur in this crossover, and if so, in which direction?

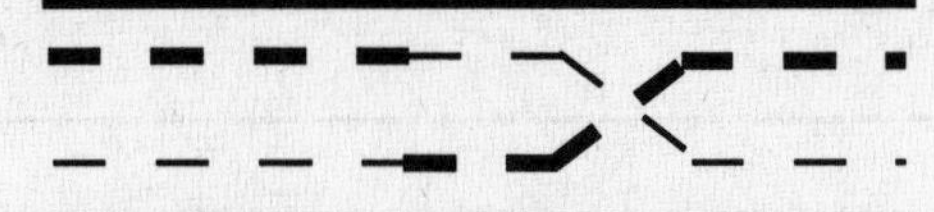

   a. No branch migration occurred.
   b. Yes, branch migration occurred from right to left.
   c. Yes, branch migration occurred from left to right.
   d. It cannot be determined from the information given.

Problems and Essays

1. You are interested in studying the cytosolic factors that control the phases of the cell cycle. You have set up two fusion experiments. In the first experiment, you fuse G_1 cells with S cells. In the second, you fuse G_2 cells with S cells. Both hybrids are given ^{3}H-thymidine, and then the nuclei from the G_2 and G_1 cells are examined by autoradiography. Using the diagrams below, indicate where radioactivity is found.

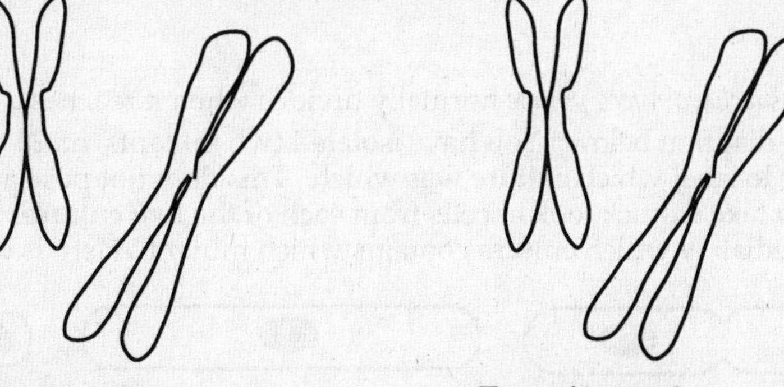

2. Your lab is searching for human genes that code for proteins playing a role in regulating the cell cycle. Because many of the cell cycle proteins are highly conserved, human genes can produce proteins that function perfectly well in yeast cells. You have many resources with which to study this phenomenon, including a culture of human cells, several yeast cell cultures (including some temperature-sensitive mutants that prevent steps of the cell cycle at nonpermissive temperatures), restriction enzymes, and techniques for introducing foreign DNA into yeast cells. Design an experiment to identify and isolate a human gene encoding a cell cycle control protein.

3. The figure below represents the activity curve of MPF isolated from the cytoplasm of frog embryonic cells. Superimpose on this figure a curve representing the concentration of cyclin in the cytoplasm. Now add a curve representing the activity of MPF in the presence of RNase.

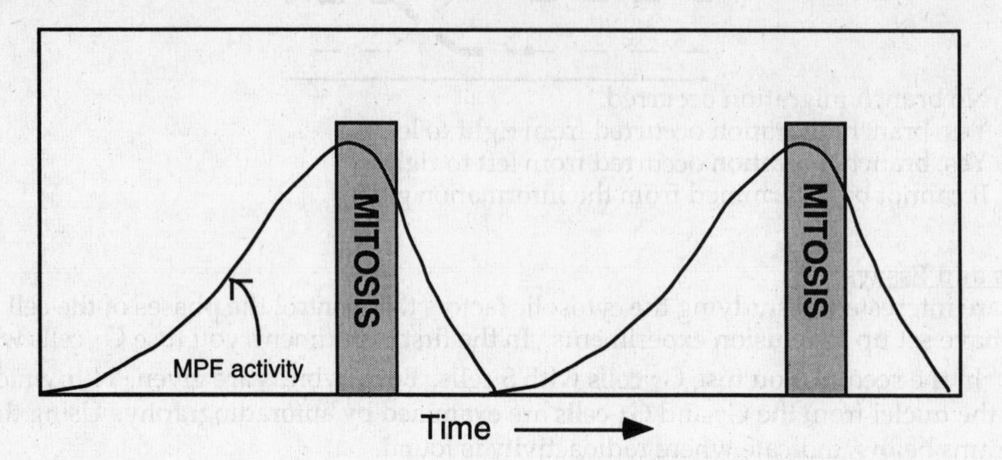

4. The yeast *Schizosaccharomyces pombe* normally divides when it reaches a certain size, as shown in the wild type diagram below. You have isolated two mutants, $cdc25^-$ and $wee1^-$, but your professor forgot to label which culture was which. This does not pose a problem. Using the microscope, you take a quick look at cells from each of the two cultures of mutants, and determine immediately which culture contains which mutant. Here is what you see:

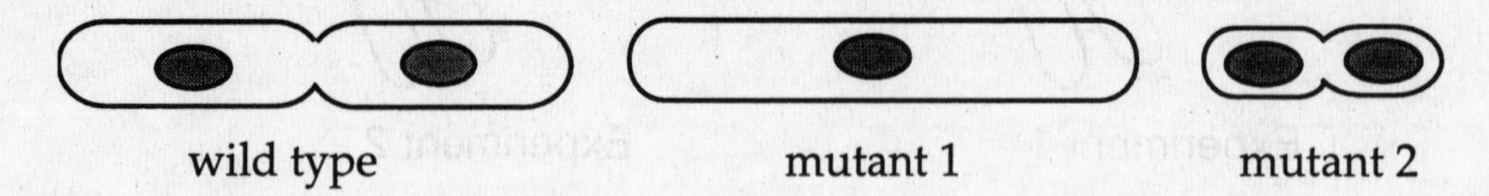

wild type mutant 1 mutant 2

   a. Which mutant, 1 or 2, corresponds to the $cdc25^-$ strain, and which corresponds to the $wee1^-$ strain?

   b. Which phase of the cell cycle is altered by these mutations? Is the phase increased or decreased in mutants 1 and 2?

5. The p53 protein is a transcription factor that initiates production of a cyclin-dependent kinase inhibitor (among other things) when there is damage to DNA. Thus, cells are arrested in G_1 until the damage is repaired. A second checkpoint, at the end of G_2, acts independently of p53. The first graph on the left shows how p53 influences the percentage of cells in each of the three phases, G_1, S, and G_2, both before and 8 hours after DNA-damaging gamma irradiation. In the second graph, draw the bars representing the approximate proportion of cells in each of the three phases following irradiation– phases in which the p53 gene has been mutated and is nonfunctional.

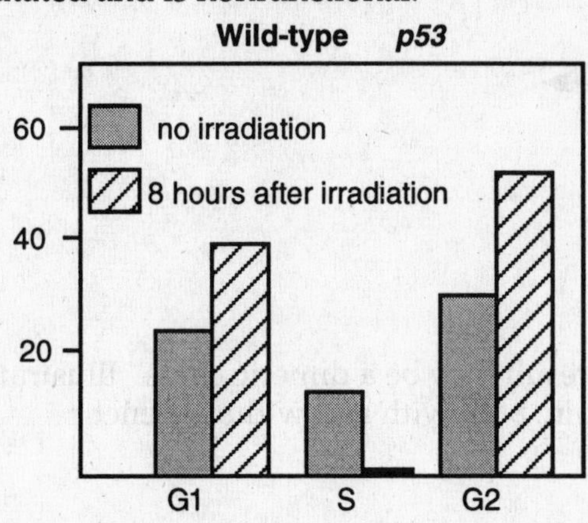

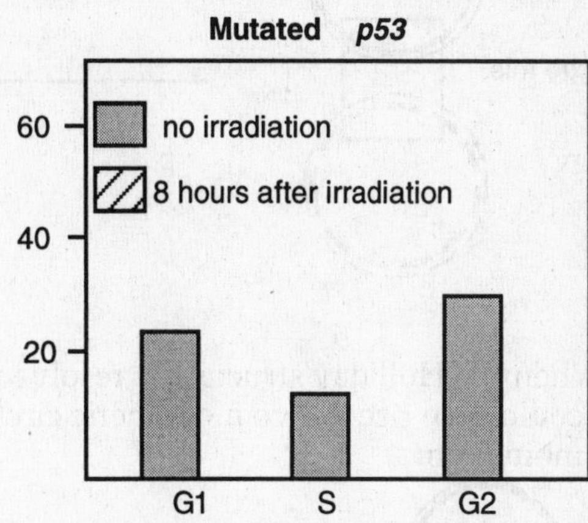

6. For each of the following phases of the cell cycle, indicate whether the chromosome number is haploid (H), diploid (D), or could be either (E). Also indicate whether chromosomes are replicated (R) or unreplicated (U).

Chromosome Number	Chromosome Status	
________	________	a. G_2.
________	________	b. meiotic metaphase
________	________	c. mitotic prometaphase
________	________	d. meiotic pachytene
________	________	e. meiotic interkinesis
________	________	f. mitotic cytokinesis
________	________	g. G_o

7. When genetic recombination occurs in circular chromosomes, as it does in prokaryotic cells or in plasmids, the sorts of intermediate structures that result differ from those in linear molecules.
   a. Imagine two circular chromosomes undergoing recombination. Draw the Holliday structure that results.

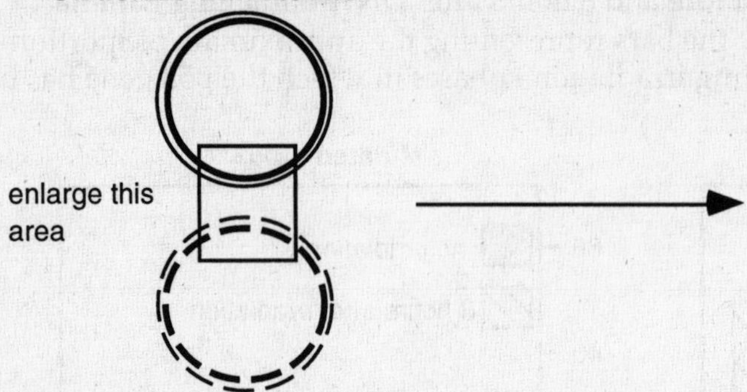

   b. When the Holliday structure is resolved, the result may be a dimeric circle. Illustrate how this could give rise to two monomeric circles again, both with and without genetic recombination:

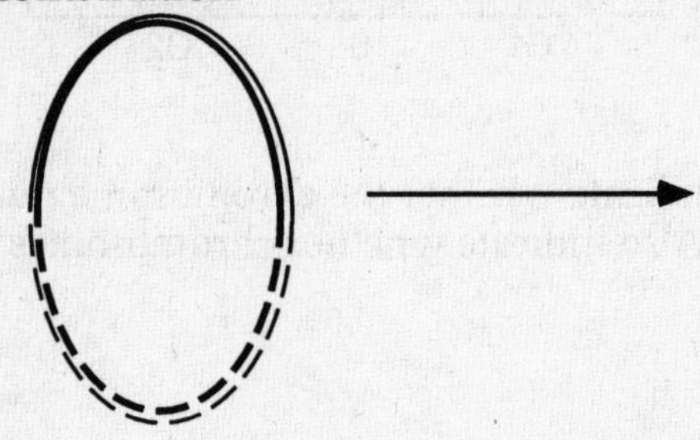

CHAPTER FIFTEEN

Cell Signaling: Communication Between Cells and Their Environment

Learning Objectives

When you have finished this chapter, you should be able to:

1. Identify the steps common to signal transduction pathways.
2. Define first messenger, second messenger, and effector.
   a. Give some examples of first messengers and effectors.
   b. Name some of the second messengers that have been identified to date.
3. Describe specific ways in which second messengers and effectors influence cellular activity.
4. Distinguish between heterotrimeric G proteins and receptor tyrosine kinases.
   a. Compare signal cascades that alter metabolism with those that affect cell growth and division. (Name specific components of both types of pathways.)
   b. Discuss ways in which mutations in genes coding for signal pathway proteins can lead to cancer and other disorders.
5. Give some examples of convergence, divergence, and crosstalk between different signaling pathways.
6. Cite the major experimental findings that have contributed to our understanding of signal transduction.

Key Terms and Phrases

cell signaling	ligand	signal transduction
signal transduction pathways	G protein	cyclic AMP (cAMP)
second messenger	adenylyl cyclase	effector
reaction cascade	protein kinase A (PKA)	heterotrimeric G protein
G protein-coupled receptors	phospholipase C	protein kinase C
calcium-binding proteins	calmodulin	protein tyrosine kinase
receptor tyrosine kinase (RTK)	phosphatidylinositol 3-hydroxy kinase (PI(3)K)	
insulin receptor substrates (IRSs)	SH2 domains	ras oncogene
apoptosis	programmed cell death	

Reviewing the Chapter

I. Introduction

A. Cells must respond appropriately to external stimuli to survive.
B. Cells respond to stimuli via cell signaling, which includes:
   1. recognition of the stimulus by a specific plasma membrane receptor.

2. transfer of a signal across the plasma membrane.
3. transmission of the signal to effector molecules within the cell, which causes a change in cellular activities.
4. cessation of the cellular response due to inactivation of the signal molecule.

C. Some signal molecules enter cells; others bind to cell-surface receptors.

II. Some of the Basic Characteristics of Cell Signaling Systems

A. The stimulus at the surface differs from the signal in the cell interior.

B. Signal transduction pathways consist of a series of proteins.
   1. Each protein in a pathway alters the conformation of the next protein.
   2. Protein conformation is usually altered by phosphorylation.
      a. Signal transduction pathways have protein kinases and phosphatases.
      b. Phosphorylation stimulates or inhibits the activity of the protein.
   3. GTP-binding proteins (G proteins) are often used as on-and-off switches.

III. G Protein-Coupled Receptors and Second Messengers

A. Different types of stimuli acting on the same target cell may induce the same response.
   1. Glucagon and epinephrine each bind to different receptors on the same cell.
   2. Both hormones stimulate glycogen breakdown and inhibit its synthesis.

B. The discovery of a second messenger– cyclic AMP:
   1. Second messengers are released in the cytoplasm after binding of a ligand.
      a. Cyclic AMP (cAMP) is one type of second messenger.
      b. Second messengers amplify the response to a single extracellular ligand.
   2. Glucose mobilization– an example of a response induced by cAMP:
      a. Cyclic AMP is formed by adenylyl cyclase, an effector.
      b. Binding of a ligand activates adenylyl cyclase.
      c. Cyclic AMP triggers a reaction cascade which amplifies the message.
      d. The cascade starts with the binding of cAMP to cAMP-dependent protein kinase A (PKA).
      e. PKA inhibits glycogen synthase and activates phosphorylase kinase.
   3. Other aspects of cAMP signal transduction pathways:
      a. Some PKA molecules phosphorylate nuclear proteins.
      b. Phosphorylated transcription factors regulate gene expression.
      c. Phosphatases halt the reaction cascade by removing phosphate groups.
      d. Regulation can occur at any stage of the reaction cascade.
      e. Cyclic AMP is produced only as long as the external stimulus is present.

C. The structure and function of G protein-coupled receptors:
   1. Glucagon and epinephrine activate the same effector (adenylyl cyclase).
      a. Their receptors are members of the same family of proteins.
      b. Both receptors have seven membrane-spanning α-helices.
      c. Both receptors bind G proteins that transmit the signal from the receptor to the effector.
   2. Heterotrimeric G proteins:
      a. G proteins bind either GDP or GTP.
      b. G proteins consist of α, β and γ subunits.
      c. When the α subunit is bound to GTP, the G protein is active; it is inactive when bound to GDP.
   3. Stimulation of G protein-coupled receptors follows a sequence of events:
      a. activation of the G protein by the receptor.
      b. relay of the signal from G protein to effector.
      c. dissociation of the alpha subunit; receptor loses its affinity for the ligand.
      d. Bound GTP is hydrolyzed by the α subunit, which ends the response.
   4. G protein-coupled receptors are very widespread.

5. Disorders associated with G protein-coupled receptors (The Human Perspective):
    a. Mutations in genes coding for effectors (such as adenylyl cyclase) have a wider effect on bodily functions than mutations in genes coding for receptors.
    b. Several disorders are caused by defects in receptors or G proteins.
        (1) Congenital nephrogenic diabetes insipidus is caused by the inability of the vasopressin receptor to activate the G protein.
        (2) Certain benign thyroid tumors (adenomas) are caused by a mutation in a receptor. As a result, the G protein is constantly activated.
6. The specificity of G protein-coupled responses:
    a. Different stimuli can use a similar mechanism for signal transduction.
    b. Specificity is in both the ligand-binding site and different effectors.
    c. G proteins can be either stimulatory or inhibitory.

D. Other second messengers:
1. Phosphatidylinositol-derived second messengers:
    a. Phospholipase C produces second messengers derived from phosphatidylinositol– inositol triphosphate (IP_3) and diacylglycerol (DAG).
    b. DAG activates protein kinase C, which phosphorylates serine and threonine residues on target proteins.
    c. One IP_3 receptor is a calcium channel located at the surface of the smooth endoplasmic reticulum. Binding of IP_3 opens the channel and allows Ca^{2+} ions to diffuse out.
2. Cytoplasmic calcium levels are determined by events within a membrane.
    a. In the cell, calcium is stored within certain cytoplasmic organelles (smooth endoplasmic reticulum or mitochondria).
    b. Calcium channels can be transiently opened by action potential (in muscle cells) or calcium itself.
    c. Calcium binds to calcium-binding proteins (such as calmodulin), which affects other proteins.
    d. Calcium ions are important second messengers in plant cells.

IV. Receptor Tyrosine Kinases (RTKs): A Second Major Type of Signaling Pathway

A. Some receptors, eg., the insulin receptor, use a different signaling pathway.

B. The mechanism of action of insulin– signaling by RTK:
1. The insulin receptor is a tyrosine kinase called a receptor tyrosine kinase (RTK).
2. Insulin binding activates its tyrosine kinase.
    a. Tyrosine kinase phosphorylates another subunit of the receptor (autophosphorylation).
    b. Tyrosine kinase also phosphorylates insulin receptor substrates (IRSs).
    c. RTKs phosphorylate tyrosines within "phosphotyrosine motifs."
3. Phosphorylated IRSs bind effector proteins that have SH2 domains (such as PI(3)K), each of which may activate a separate signaling pathway.

C. The role of RTKs in other cellular activities:
1. RTK signaling pathway regulates growth-related functions, among other things.
2. Most RTKs are monomers that dimerize when the ligand binds.
3. Ras protein is an important component of many RTK reactions.
    a. Ras was originally discovered as a viral oncogene.
    b. Ras is present in normal genome of animals; many human tumors have mutated ras.
    c. The product of the ras gene, Ras, is a small monomeric G protein.

d. Mutations in the ras gene prevent the Ras protein from hydrolyzing GTP, thus always keeping it turned on.

4. The MAP kinase cascade:
   a. The MAP kinase cascade regulates both transcription and translation.
      (1) The binding of a ligand to the RTK activates Ras (via several other proteins), which recruits the Raf protein to plasma membrane.
      (2) Raf initiates the MAP kinase cascade.
      (3) The last protein kinase in the cascade phosphorylates specific transcription factors in the nucleus.
      (4) A second parallel cascade activates translation.
   b. The MAP kinase cascade has been found in all eukaryotes studied.
   c. Oncogenes are mutated genes whose proteins act in mitogenic pathways.

V. Convergence, Divergence, and Crosstalk Among Different Signaling Pathways

A. Information circuits within cells have a number of common features.
   1. Unrelated growth factor signals can converge to activate a common effector.
   2. Identical signals can diverge to activate a variety of effectors.
   3. Signals can be passed back and forth between pathways via crosstalk.
   4. The Ras pathway plays a central role in routing information through the cell.

B. Insulin, epidermal growth factor (EGF), and PDGF initiate signaling pathways which converge to activate Ras and the MAP kinase cascade.

C. Binding of EGF or PDGF to its receptor can activate effectors of several divergent pathways, such as Grb2 (MAP kinase cascade), PI(3)K, and phospholipase C.

D. Crosstalk between signaling pathways:
   1. Cyclic AMP can block signals transmitted through the MAP kinase cascade.
   2. Ca^{2+} and cAMP can influence each other's pathways.

VI. Other Signaling Pathways

A. The role of NO and CO as cellular messengers:
   1. Nitric oxide (NO) is both an intercellular and intracellular messenger with a variety of functions.
      a. NO is produced by nitric oxide synthase.
      b. NO stimulates the cell's phagocytic activity.
      c. NO mediates dilation of blood vessels and can regulate blood pressure.
      d. NO can act as a neurotransmitter.
   2. Carbon monoxide (CO) acts as a messenger to stimulate the production of cGMP.

B. Signals that originate from cell surface contacts:
   1. Cells have specialized contacts with other cells and with the extracellular matrix.
   2. Membrane proteins and cytoskeletal elements transmit signals from the environment to the cell interior.
      a. Integrins are receptors at sites of cell-substrate and cell-cell contact.
      b. The interaction of integrin with a ligand generates signals in the cell.
      c. Focal adhesion kinase (FAK) is an effector in integrin-mediated responses.
         (1) The action of FAK can cause reorganization of the cytoskeleton.
         (2) FAK can trigger aggregation of platelets in blood.

C. The role of phosphatases in cell signaling:
   1. Kinases and phosphatases have opposite effects on their substrate.
      a. Some phosphatases are multifunctional; others are specific.
      b. Most phosphatases remove phosphates from either serine and threonine residues or tyrosine residues.
   2. Most phosphatases are soluble in the cytoplasm.

3. Receptorlike protein tyrosine phosphatases (RPTP) span the plasma membrane and act as cell-surface receptors involved in cell signaling and cell adhesion.

D. Pathways that lead to cell death:

1. Cells that are no longer needed or can harm the individual undergo apoptosis, programmed cell death.
    a. Apoptosis requires the activation of a specific set of genes.
    b. Apoptosis occurs in embryonic development and in some adult tissues.
2. Cells can be triggered to enter apoptosis when growth factors are removed.

E. Signaling pathways in plants:

1. Some signaling mechanisms in plants and animals are similar; others are different.
    a. Both plants and animals use Ca^{2+} and phosphoinositide messengers.
    b. Cyclic nucleotides are not used in plant cell signaling.
2. Bacteria, yeast, and plants have receptor histidine protein kinase not found in animal cells.
3. Salicylic acid is a messenger in a pathway initiated by an attack of a pathogen.

VII. The Discovery and Characterization of GTP-Binding Proteins (G Proteins) (Experimental Pathways)

A. Certain hormones, when bound to a receptor, activate adenylyl cyclase on the inner side of the membrane.

1. Various receptors stimulate a common population of adenylyl cyclases.
2. Receptors and adenylyl cyclase effectors exist separately.

B. Guanyl nucleotides regulate the response of adenylyl cyclase to the hormone.

1. GTP changes the affinity of hormone-binding sites for their hormones.
2. GTP stimulates adenylyl cyclase activity.
3. Binding of GTP, but not its hydrolysis, stimulates cAMP formation.
4. Hydrolysis of GTP inactivates adenylyl cyclase.
5. Activation of the receptor leads to the release of bound GDP and binding of GTP.

C. There is a regulatory component between the hormone receptor and the adenylyl cyclase effector. This component is inactivated by cholera toxin.

D. GTP-binding protein consists of three subunits.

E. The α subunit of the G protein has the GTP-binding site and activates adenylyl cyclase.

Key Figure

Figure 15.10. The membrane-bound machinery for transducing signals via a seven-helix transmembrane receptor and a heterotrimeric G protein.

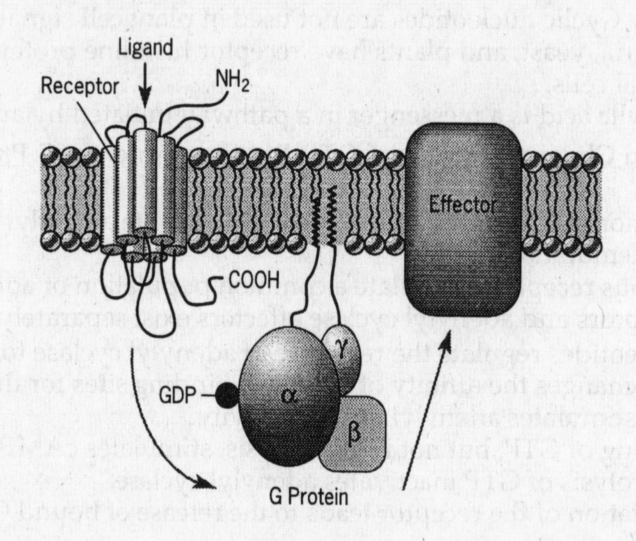

Questions for Thought:

1. Different versions of the type of receptor illustrated here will bind different ligands, and generate different intracellular responses. How can the same type of receptor act in such diverse ways?

2. What characteristics must the amino acids in the membrane-spanning regions of the receptor have? How might they differ from those of the amino acids in the cytoplasmic and extracellular loops?

3. Identify the sites in this receptor complex where signal amplification can occur. Where else in a signal transduction pathway can the signal be amplified?

4. In a gene for any of the protein components of this complex, imagine how a mutation might alter the function of the receptor. Now refer to the "The Human Perspective" in your textbook, and see what types of mutations in G protein-coupled receptors have been identified in some human disorders.

Review Problems

Short Answer

1. Give the significance (not the definition) of the following terms or phrases. Say what they do or why they are important. For example:
   G proteins: *carry the signal generated by ligand binding at the receptor to an effector molecule.*

   a. the many steps of a signal transduction cascade

   b. cAMP

   c. phosphatases

   d. calmodulin

   e. SH2 domain

   f. receptor tyrosine kinases

   g. Ras protein

   h. MAP kinase cascade

   i. crosstalk

   j. apoptosis

2. Compare and contrast the following:
   a. kinase vs. phosphatase

   b. heterotrimeric G protein vs. monomeric G protein

   c. phospholipase C vs. protein kinase C

   d. tyrosine kinases vs. serine/threonine kinases

   e. signaling pathways in plants vs. signaling pathways in animals

Multiple Choice

1. Cell signaling means all of the following except:
   a. intercellular communication.
   b. export of cellular waste products.
   c. environmental monitoring.
   d. response to stimuli.

2. In a signal transduction pathway, what is passed from the upstream member of the pathway to the downstream member?
   a. electrons
   b. protons
   c. phosphates
   d. information in the form of alterations in protein conformation

3. Both epinephrine and glucagon bind to G protein-coupled receptors and activate glycogen breakdown. Therefore, epinephrine and glucagon must:
   a. have very similar structures and bind to the same receptor.
   b. bind to receptors with different ligand-binding sites but similar functions.
   c. bind to different receptors and activate different second messengers.
   d. bind to the same receptors, one being intracellular and the other, extracellular.

4. If you break liver cells by homogenization, separate the broken cell membranes (particulate) from the cytoplasm (soluble), and add epinephrine to the soluble portion, the result will be:
   a. the production of cAMP.
   b. binding of the epinephrine to its receptor.
   c. activation of adenylyl cyclase.
   d. none of the above.

5. Glucagon binds to the glucagon receptor, which interacts with a membrane-bound G protein, whose α subunit releases GDP and binds GTP. The α subunit then dissociates from the G protein and diffuses to a membrane-bound adenylyl cyclase which produces cAMP. In this scenario, name the (1) transducer, (2) effector, (3) first messenger, and (4) second messenger.
   a. (1) glucagon (2) cAMP (3) G protein (4) GTP
   b. (1) G protein (2) adenylyl cyclase (3) glucagon (4) cAMP
   c. (1) G protein (2) GTP (3) glucagon (4) cAMP
   d. (1) glucagon (2) G protein (3) adenylyl cyclase (4) cAMP

6. Epinephrine binds to the same type of receptors in liver, fat, and smooth muscle cells. Yet in liver, glycogen breaks down; in fat, triacylglycerols break down; and smooth muscle cells relax. How can the same hormone produce three such different responses?
   a. The same receptors activate three different second messengers.
   b. The same second messengers activate three different intracellular cascades.
   c. The same enzyme that breaks down glycogen also breaks down fat and relaxes muscle.
   d. none of the above

7. G proteins are said to be self-inactivating. What G-protein function accounts for this?
   a. GTPase activity of the α subunit
   b. activation of the effector
   c. binding to the receptor
   d. dissociation of the subunits

8. People who suffer from thyroid adenomas have constitutively activated G proteins in some cells of the thyroid gland. Tumor cells not only secrete copious quantities of thyroid hormones but also divide and grow excessively. Based on these observations, in what kind(s) of signal transduction pathways do the mutant G proteins participate?
   a. cAMP-regulated pathways
   b. MAP kinase pathways
   c. both cAMP and MAP kinase pathways
   d. neither cAMP nor MAP kinase pathways

9. What factors are responsible for the specificity of G protein-coupled responses?
   a. Protein kinases can phosphorylate different proteins in different cells.
   b. Receptors can exist as isoforms.
   c. G-protein subunits can exist as isoforms.
   d. all of the above

10. Acetylcholine on a smooth muscle cell is to phospholipase C as epinephrine on a liver cell is to ___?___.
   a. protein kinase C
   b. cAMP
   c. adenylyl cyclase
   d. GTP

11. Serine and threonine are to ___?___ as tyrosine is to receptor tyrosine kinases.
   a. phosphatases
   b. G proteins
   c. cAMP
   d. protein kinases A and C

12. Ryanodine and IP_3 receptors are found on __?__ membranes, whereas epinephrine and glucagon receptors are found on __?__ membranes.
    a. plasma, intracellular
    b. intracellular, plasma
    c. mitochondrial, endoplasmic reticulum
    d. outer, inner

13. Depending on the cell, increases in intracellular Ca^{2+} can do all of the following except:
    a. stimulate the increase of intracellular calcium.
    b. stimulate the removal of intracellular calcium.
    c. stimulate the breakdown of cAMP.
    d. stimulate the breakdown of calmodulin.

14. The sequence of events that occurs after insulin binds to its receptor tyrosine kinase is:
    a. binding of IRS —> phosphorylation of proteins with SH2 domains —> effect
    b. binding of proteins with SH2 domains —> phosphorylation of IRS —> effect
    c. autophosphorylation and phosphorylation of IRS —> binding of proteins with SH2 domains —> effect
    d. autophosphorylation and binding of IRS —> phosphorylation of proteins with SH2 domains —> effect

15. Proteins with SH2 domains are to __?__ as adenylyl cyclase and phospholipase C are to G protein-coupled receptors.
    a. IRSs
    b. receptor tyrosine kinases
    c. insulin
    d. autophosphorylation

16. Which mutant form of ras is likely to cause malignancy?
    a. ras that cannot hydrolyze GTP
    b. ras that cannot bind to GTP
    c. ras that cannot bind to Grb2 or Sos
    d. ras that cannot bind to Raf

17. The MAP kinase cascade stimulates transcription via specific transcription factors, and stimulates translation via
    a. SREs
    b. MAPK
    c. PHAS-I and eIF4E
    d. all of the above

18. What second messenger has been implicated in blood pressure regulation?
    a. cAMP
    b. NO
    c. CO
    d. IP_3

19. A mutation in the FAK gene would likely result in:
    a. abnormal cell shape.
    b. abnormal wound healing.
    c. abnormal cell growth.
    d. all of the above.

20. A cell responds to an external stimulus by increasing the intracellular Ca^{2+} concentration, releasing IP_3 from the membrane, and producing salicylic acid. What kind of cell does this describe?
    a. bacterial
    b. viral
    c. plant
    d. animal

Problems and Essays

1. In your laboratory, you have a culture of human cells that express the receptors for both platelet-derived growth factor (PDGF) and epidermal growth factor (EGF). You can add either PDGF or EGF to your culture and the cells enter the S phase and begin DNA replication (treatment a). You have done a series of experimental manipulations (treatments b through f), and tested for DNA replication after each one. Write your results in the third column of the table below.

Treatment	Additions	DNA Replication
a. none	PDGF or EGF	yes
b. microinjection (intracellular) of constitutively active Ras protein	none	
c. microinjection of saline ("mock" injection)	PDGF or EGF	
d. microinjection of saline ("mock" injection)	none	
e. microinjection of anti-Ras antibodies	PDGF or EGF	
f. microinjection of anti-Ras antibodies	none	

2. The nicotinic acetylcholine receptor is membrane-bound protein that also functions as a ligand-gated ion channel. Upon binding acetylcholine, the channel opens and sodium ions pass into the cell, flowing in the direction of their electrochemical gradient. Put a check next to all of the features listed below that apply to the nicotinic acetylcholine receptor.

    a. __________Ligand binding to the receptor is reversible.

    b. __________Low concentrations of GTP stimulate the dissociation of the ligand from the receptor.

    c. __________Ligand binding results in increased intracellular concentrations of cAMP.

    d. __________Ligand binding changes the voltage measured across the cell membrane.

    e. __________The effects of ligand binding cannot be reconstructed in homogenized cells, even if both soluble and particulate fractions are included.

    f. __________Mutations in the ras gene will influence the effect of ligand binding.

3. A single mutation in the ras gene (resulting in the substitution of a valine for glycine at amino acid position 12 of the protein) reduces the GTPase activity of the protein and transforms the cell from normal to malignant. Explain.

4. When one molecule of epinephrine binds a ß-adrenergic receptor, 500,000 protein kinase A molecules are activated per second. One stage of the amplification occurs at the level of adenylyl cyclase, which can produce 1000 cAMPs per second. You are interested in finding where the other 500-fold amplification occurs, so you perform the following experiment:: Using radioactive GppNp, a GTP analog that cannot be hydrolyzed, you determine that when 0.8% of a mole of receptors are bound to epinephrine, four moles of GppNp are bound to G proteins. Based on your data, where does the additional 500-fold amplification occur?

5. Serotonin is a small molecular weight amine that can act as a neurotransmitter, conveying a signal between adjacent nerve cells, or as a hormone, entering the blood stream and conveying a signal to cells in nonadjacent tissues. It has been implicated in personality, mood and sleep, and also in blocking pain at the level of the central nervous system. You are interested in establishing the mechanism by which serotonin affects target cells. Your lab has made the following observations:
   i. Serotonin causes an increase in cAMP in target cells.
   ii. The increase in cAMP can be observed in suspensions of homogenized cells, but not if the particulate fraction has been removed by centrifugation.
   iii. Dissociation of serotonin from membrane fragments in homogenized cells requires the presence of GTP.
   iv. Membranes of target cells have GTPase activity. When target cells that are sensitive to both serotonin and epinephrine are given both hormones, the total effect is not additive.

   Propose a mechanism of action for serotonin.

CHAPTER SIXTEEN

Cancer

Learning Objectives

When you have finished this chapter, you should be able to:

1. Define cancer. Discuss the characteristics of benign tumors, malignant tumors, and secondary tumors (metastases).
2. Describe the phenotypic characteristics of transformed cells and how these characteristics differ from those of normal cells.
3. Give an overview of known causes of cancer.
4. Describe the initiation and the promotion phases of tumorogenesis. Explain tumor progression on the basis of cancer genetics.
5. Distinguish between tumor suppressor genes and oncogenes.
   a. Describe how mutations in these genes can affect cellular activity.
   b. Explain mutations in these two types of genes cause respective differences in the heritability of cancers.
   c. Give specific examples of these two types of genes.
   d. Describe some cancers, including their underlying cellular causes, that have been attributed to these genes.
6. Discuss the relationship of DNA and RNA tumor viruses to cellular genes; recognize the role of these viruses in advancing our understanding of cancer.

Key Terms and Phrases

cancer	malignant tumor	benign tumor
secondary tumor (metastasis)	transformed cells	tumor-associated antigens
anchorage dependence	DNA tumor viruses	RNA tumor viruses
monoclonal	tumorigenesis	initiation
promotion	tumor-suppressor genes	RB gene
p53 protein	oncogenes	proto-oncogenes

Reviewing the Chapter

I. Introduction

A. Cancer results from changes in cellular behavior caused by underlying alterations in the genome of the cells.
   1. Gene expression in cancer cells is altered.
   2. Cancer cells proliferate in an uncontrolled manner, forming malignant tumors.
   3. Malignant tumors tend to metastasize (establish secondary tumors).

B. Cancer is the focus of a massive research effort.
   1. Current treatments for cancer simultaneously kill cancer cells and normal cells.

2. Genes responsible for human cancers have been identified, but gene therapy as a treatment for cancer is not yet possible.
3. The immune system is not very effective in destroying cancer cells.

II. The Biology of Cancer

A. Malignant cells are not responsive to influences that cause normal cells to cease growth and division.
   1. The capacity for growth and division is similar between cancer cells and normal cells.
   2. When there are no growth factors in the medium or when cells contact surrounding cells:
      a. Normal cells stop growing.
      b. Malignant cells continue to grow.

B. The phenotype of a cancer cell:
   1. In culture, normal cells can be transformed by chemicals or viruses.
   2. Different types of cancer cells share a number of similarities.
      a. Cancer cells often have aberrant chromosome numbers (aneuploidy).
      b. The cytoskeleton of cancer cells is disorganized and reduced.
      c. Cancer cells possess certain cell surface proteins called tumor-associated antigens.
      d. Cancer cells are less adhesive.
      e. Cancer cells are motile.
      f. Cancer cells ignore signals transmitted by neighboring cells.
      g. Cancer cells do not need growth factors for proliferation.
      h. Cancer cells do not need a solid surface for growth (lack of anchorage dependence).

III. The Causes of Cancer

A. Carcinogenic chemicals can cause cancer by causing changes in the genome.

B. DNA tumor viruses and RNA tumor viruses carry genes whose products interfere with cell growth regulation.

C. Causes of most types of human cancer are still unknown.

IV. The Genetics of Cancer

A. The development of a malignant tumor is a multistep process.
   1. Cancer results from the uncontrolled proliferation of a single cell.
   2. Tumorigenesis occurs by a cumulative progression of genetic alterations.
      a. During tumorigenesis, cells become increasingly less responsive to normal growth regulation and better able to invade normal tissues.
      b. The formation of a benign tumor often precedes the development of malignancy.
      c. The products of the genes involved in carcinogenesis are usually responsible for cell cycle regulation, cell adhesion, and DNA damage repair.
   3. Tumorigenesis consists of initiation and promotion.
      a. Tumor initiators cause mutations in genes.
      b. Tumor promoters stimulate cell growth and proliferation, causing additional cancer-causing mutations.
   4. Cancers often arise in tissues that are engaged in a high level of cell division.

B. Tumor-suppressor genes and oncogenes:
   1. Tumor-suppressor genes encode proteins that restrain cell growth.
      a. A normal cell fused to a cancer cell can suppress malignant characteristics of the latter.
      b. Specific regions of chromosomes are deleted in cells of certain cancers.
      c. Tumor-suppressor genes act recessively.

2. Oncogenes encode proteins that promote the loss of growth control and the conversion of a cell to a malignant state.
   a. Oncogenes were first discovered in genomes of tumor viruses.
   b. An oncogene is an altered cellular gene (proto-oncogene).
   c. Proto-oncogenes encode proteins that function in a cell's normal activities.
   d. Oncogenes act dominantly.
3. For a cell to become malignant, both alleles of a tumor-suppressor gene must be lost, and a proto-oncogene must be converted into oncogene.

C. Tumor-Suppressor Genes
1. Introducing a normal copy of tumor-suppressor gene into tumor cells may be a possible treatment for cancer.
2. RB gene was the first tumor-suppressor gene to be discovered.
   a. Retinoblastoma is inherited in certain families, and occurs sporadically in the population at large.
   b. The cells of children with inherited retinoblastoma have a deletion in one copy of the RB gene.
   c. The development of retinoblastoma requires both copies of RB to be altered or eliminated.
   d. The absence of a functional pRb is sufficient to promote the development of retinoblastoma.
   e. The reintroduction of a wild-type RB gene is sufficient to suppress a cell's cancerous phenotype.
3. The role of pRb in regulating the cell cycle:
   a. The protein encoded by the RB gene (pRb) regulates the passage of cells from G_1 to S phase.
   b. Transcription factors of E2F family are targeted by pRb.
   c. The arrest of the cell cycle in G1, required for normal cell differentiation, is directed by pRb.
   d. Animals with one mutated copy of the RB gene have an elevated risk of developing cancer.
4. The role of p53 in human carcinogenesis:
   a. The p53 protein suppresses the formation of tumors and maintains genetic stability.
      (1) The p53 gene may be the most important tumor suppressor gene in human genome.
      (2) Fifty percent of human cancers have mutations in p53.
      (3) Tumor cells with p53 mutations are highly invasive.
   b. The p53 protein acts as a transcription factor, activating the expression of a gene that inhibits the passing of the cell through the G_1-S transition.
   c. The p53 protein triggers apoptosis in cells whose DNA is damaged beyond repair.
5. Cooperative functions of RB and p53:
   a. Many tumor cells have mutations in both p53 and RB .
   b. Mutations in both copies of both RB and p53 cause cells to proliferate and develop into tumors.
   c. Cells transformed with virus that cannot inactivate p53 undergo apoptosis.
6. Other tumor suppressor genes:
   a. Mutations of tumor-suppressor genes that are not RB or p53 are detected in only a few types of cancer.
   b. Colon cancer is often caused by an inherited deletion in a tumor-suppressor gene APC.

c. Inherited breast cancer is caused by mutations in BRCA tumor suppressor genes, which probably act as transcription factors.

D. Oncogenes:
1. Most of the known oncogenes are derived from proto-oncogenes that are involved in signal transduction pathways.
2. Oncogenes that encode growth factor receptors:
   a. Simian sarcoma virus contains the oncogene (sis) which is derived from a cellular gene which codes for the growth factor PDGF.
   b. Oncogene erbB directs the formation of an altered EGF receptor that stimulates the cell regardless of the presence of growth factor.
   c. Some malignant cells contain more surface receptors than normal cells and are therefore sensitive to lower concentrations of growth factors.
3. Oncogenes that encode cytoplasmic protein kinases:
   a. Raf (a serine/threonine protein kinase in the MAP kinase cascade) can be converted into an oncogene by mutations that turn it into an enzyme that is always "on."
   b. The oncogene product of src is a protein tyrosine kinase which phosphorylates proteins involved in signal transduction, control of cytoskeletal configuration and cell adhesion.
4. Oncogenes that encode nuclear transcription factors:
   a. A number of oncogenes encode proteins that may act as transcription factors.
   b. Myc protein stimulates cells to reenter cell cycle from G_O stage.
   c. Overexpression of myc may cause cells to proliferate uncontrollably.
5. Oncogenes that Encode Proteins that Affect Apoptosis.
   a. Alterations that tends to diminish cell's ability to self-destruct increase likelihood of a cell giving rise to a tumor.
   b. The overexpression of blc-2 gene leads to suppression of apoptosis, allowing abnormal cells to proliferate into tumors.

E. Genes that promote tumor formation through secondary effects on other genes:
1. Mismatch repair defects predispose cells to abnormally high mutation rates, which increases the risk of malignancy.
2. A deficiency in the mismatch repair system may be involved in a hereditary form of colon cancer.
3. Tumor cells of this form of cancer often have variations in the length of microsatellite sequences, due to errors during replication.

D. Concluding remarks on cancer genetics:
1. Because many different types of cancers share the same genetic defects, ta common approach to treatment may be possible.
2. The identification of defects in specific genes can lead to screening procedures for individuals predisposed tocancer.

V. The Discovery of Oncogenes (Experimental Pathways)

A. Experiments of Rous (1911) showed that a tumor could be transmitted from one animal to another by a virus.

B. RNA-dependent DNA polymerase (reverse transcriptase) is the enzyme that replicates the genetic material of RNA tumor viruses.

C. The capacity to transform a cell resides in a restricted portion of the viral genome.

D. The transforming genes of the viral genome (oncogenes) are not true viral genes but are cellular genes previously picked up by RNA tumor viruses.

E. The src gene was the first oncogene identified; it is present in all vertebrate classes.
1. Both viral and cellular src genes (v-src and c-src, respectively) code for protein kinases.
2. The viral version of the gene has higher activity than the cellular version.

3. Increased activity of an oncogene product can transform a normal cell into a malignant cell.

F. Cellular genes can be converted into oncogenes by carcinogenic chemicals or by a mutation in sequences regulating their expression.

G. Single base substitution mutations can convert proto-oncogenes into oncogenes.

Key Figure

Figure 16.10. Contrasting effects of mutations in tumor-supressor genes and in oncogenes.

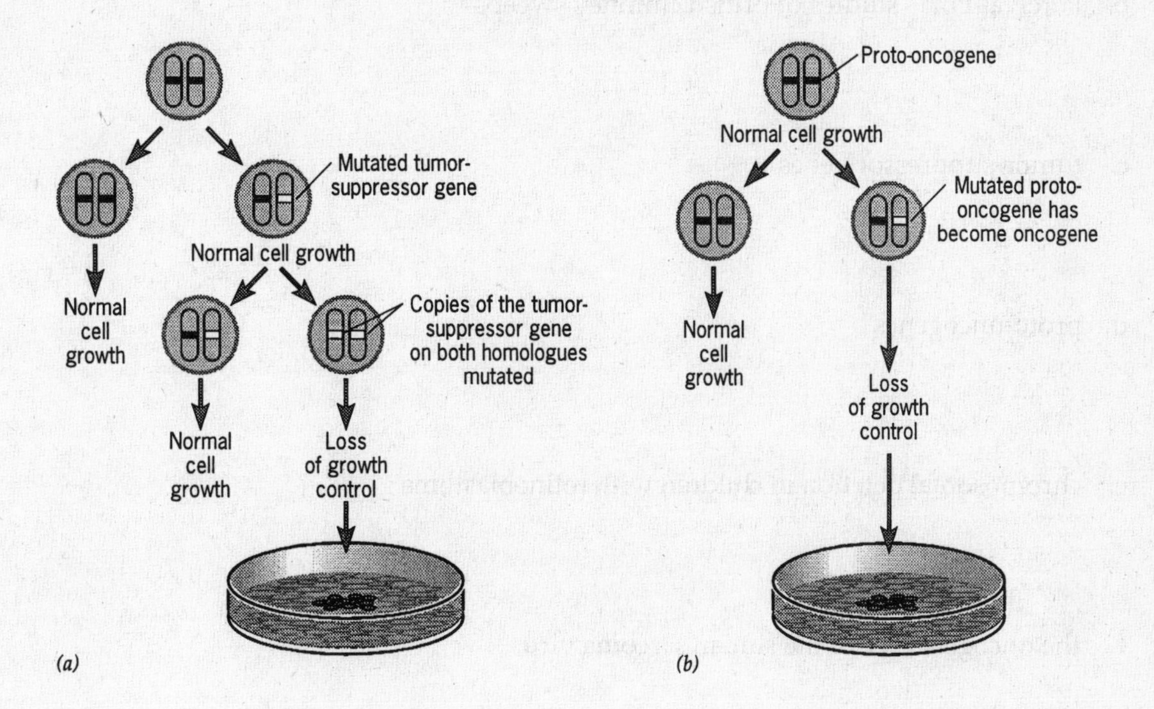

Questions for Thought:

1. Which type of mutation would be expressed as a genetic dominant and which as a recessive? Why?

2. The RNA and DNA tumor viruses carry oncogenes that are highly homologous to normal cellular genes. Why wouldn't a virus with a tumor suppressor gene homologue be a tumor virus?

3. What kinds of cellular functions are carried out by the gene products of normal tumor suppressor genes? The gene products of normal proto-oncogenes?

4. Assuming you could treat cancers with gene therapy, how would the therapy in cancers involving tumor-suppressor genes differ from that in cancers involving oncogenes?

Review Problems

Short Answer

1. Give the significance (not the definition) of the following terms or phrases. Say what they do or why they are important. For example:
   metastasis: *responsible for the invasive and potentially deadly nature of cancer.*

   a. aneuploidy in transformed cells

   b. Percival Pott's studies of British chimney sweeps

   c. tumor-suppressor genes

   d. proto-oncogenes

   e. chromosomal deletion in children with retinoblastoma

   f. the oncogene, sis, of the simian sarcoma virus

2. Compare and contrast the following:
   a. benign vs metastatic tumor

   b. tumor virus genes vs. oncogenes

   c. initiators vs. promoters

d. tumors in children with familial retinoblastoma vs. those with sporadic retinoblastoma

e. oncogenes that encode growth factors vs. those that encode growth-factor receptors

Multiple Choice

1. How might you distinguish between a malignant tumor and a benign tumor?
   a. Using microscopy, compare cell shape.
   b. Using microscopy, compare the chromosome integrity.
   c. Using cell culture, compare cell life span.
   d. all of the above

2. Which of the following proteins has been implicated in aneuploidy of cancer cells?
   a. p53
   b. pRB
   c. Myc
   d. bcl-2

3. All cancer-causing agents identified thus far have what feature in common?
   a. They are all found in cigarette smoke.
   b. They are all mutagenic.
   c. They are all chemicals.
   d. They do not have any features in common.

4. Cancer is said to be monoclonal. What does that mean?
   a. Cancer arises only once in an individual's lifetime.
   b. Cancer stimulates the production of antibodies.
   c. Cancer can only be cured using one treatment.
   d. Cancer results from the uncontrolled proliferation of one wayward cell.

5. In what way is tumor progression like biological evolution?
   a. Survival of cancer victims is like survival of the fittest.
   b. Cancer cells that proliferate fastest become the most abundant in the tumor.
   c. Tumor progression requires both initiation and promotion.
   d. Cells of benign tumors cannot disperse to distant tissues.

6. Of agents that can act as either initiators or promoters, which can be mutagenic?
   a. initiators
   b. promoters
   c. both initiators and promoters
   d. neither initiators nor promoters

7. In some breast cancers, estrogen acts as a(n):
   a. initiator.
   b. promoter.
   c. progressor.
   d. mutagen.

8. Cancers that arise due to mutated tumor supressor genes require that
   a. one allele of the gene undergoes mutation.
   b. both alleles of the gene undergo mutations.
   c. neither allele of the gene undergoes mutation.
   d. none of the above

9. In children with retinoblastoma, the chromosomal deletion that is present in all of their cells indicates that the chromosomal aberration:
   a. arose somatically in retinal cells.
   b. first occurred in the retina, and then spread to other cells.
   c. arose during early embryonic development.
   d. was inherited from a parent.

10. Would you expect to find that the cells of children with nonfamilial (i.e., sporadic) retinoblastoma have the chromosomal deletion referred to in number (9) above?
    a. yes
    b. no
    c. no way of telling
    d. not enough information given

11. The normal protein product of the RB gene acts as an:
    a. inhibitor of transcription.
    b. activator of transcription.
    c. inhibitor of translation.
    d. activator of replication.

12 Which of the following is not a normal function of p53?
    a. It activates transcription of a cell cycle inhibitor.
    b. It plays a role in DNA repair.
    c. It triggers apoptosis in damaged cells.
    d. It plays an essential role in the embryonic development of the mouse.

13. When a culture of rodent cells is infected with a DNA tumor virus whose gene products bind to p53 and RB, all the cells take on characteristics of malignancy. If the viral gene coding the p53-binding protein is mutated, but that coding for the RB-binding protein is not mutated, how will infected cells respond?
    a. They will become malignant.
    b. They will show no effects of the virus.
    c. They will undergo apoptosis.
    d. They will enter the S phase.

14. The textbook states that cells of patients with adenomatous polyposis coli (APC) "contain a deletion of a small portion of chromosome #5, which was subsequently identified as the site of a tumor-suppressor gene called APC." In which cells of these patients would you expect to find this chromosomal deletion?
    a. all cells
    b. cells of the polyps of the colon epithelium
    c. germ cells
    d. cancer cells

15. When the genes responsible for familial breast cancer were sequenced, it was found that the polypeptides they encoded contained a zinc finger motif. What does this suggest to you about the normal functions of the products of these genes?
    a. The genes code for hormones.
    b. The genes code for growth factors.
    c. The genes code for growth factor receptors.
    d. The genes code for transcription factors.

16. Tumor-suppressor genes usually act as __?__ genes, and oncogenes usually act as __?__ genes.
    a. dominant, dominant
    b. recessive, recessive
    c. dominant, recessive
    d. recessive, dominant

17. Which of the following types of genes have not been identified as oncogenes?
    a. genes that encode growth factor receptors
    b. genes that encode membrane-bound ion channels
    c. genes that encode cytoplasmic protein kinases
    d. genes that encode proteins involved in mismatch repair

18. An oncogene codes for a growth factor receptor that is constitutively active. Of the following statements about a culture of cells expressing this gene, which would you expect to be true?
    a. The cells would be stimulated to divide by addition of growth factors to the medium.
    b. Cells in the culture would have no response to the additon of growth factors.
    c. Cells in the culture would be arrested in the G_0 phase of the cell cycle.
    d. none of the above

19. The genes responsible for cancer in RNA and DNA tumor viruses are:
    a. unique sequences found only in the viral genomes.
    b. also found in cells of the host, but only after the cells have been transformed.
    c. also found in cells of the host, even when the cells are normal.
    d. also found in many pathogenic bacteria.

20. Of the following statements explaining how alterations of the ras gene cause cancer, which is true?
    a. A base substitution in the c-ras gene that replaces a critical glycine with another amino acid.
    b. Infection of the cell by a tumor virus carrying a v-ras gene.
    c. Linkage of the c-ras gene to a viral promoter.
    d. All of the above statements are true.

Problems and Essays

1. The following have been used as treatments for cancer patients, but none of these approaches has been entirely effective. For each one, briefly explain the theory of how the treatment fights cancer cells, and why the treatment has not cured cancer.
    a. chemotherapy

    b. gene therapy

c. surgery

d. stimulating the immune system

2. The human population of the world is about five billion (5×10^9). Each person has an average of 100 trillion cells (100×10^{12}) and a life span of 50 years. About one third of the population develops cancer in a lifetime. Calculate the total number of human cells that become malignant per year.

3. You have isolated an RNA tumor virus that is capable of inducing leukemia in mice, but you do not know the mechanism, or the nature of the viral gene product. You also have a mutant strain of the virus that cannot induce murine leukemia and is missing 15% of the genome.
   a. Design an experiment to isolate the portion of the normal viral genome that causes the cancer.

   b. How might you go about identifying both the gene product of the viral oncogene and making an educated guess about its function?

4. You are testing the growth of both normal and transformed cells, and the effects of the growth factor EGF on both. The following is a graph of the growth curve of normal cells, both with and without added EGF.

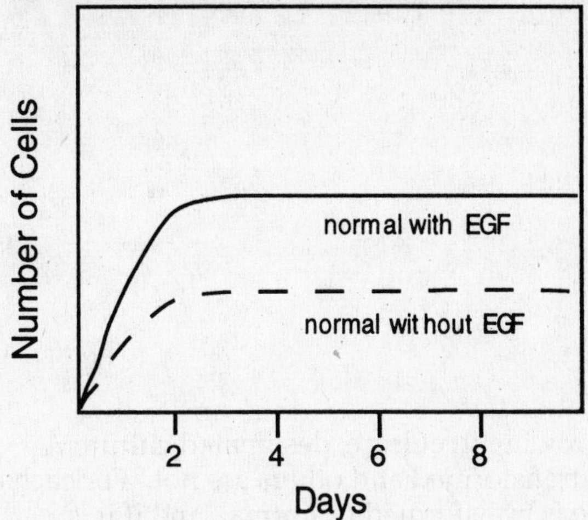

a. On the graph, superimpose two curves respectively representing transformed cells both with EGF and without EGF.

b. Now add a curve showing how the addition of EGF on day six would alter the normal, "without EGF" curve.

c. Cells in the normal +EGF and -EGF cultures are in what phase of the cell cycle after day four

5. Why do tumor-suppressor genes act as recessives, and oncogenes act as dominants?

6. Some cancers develop in two phases, initiation and promotion. Explain at a cellular level why the promoting influence can occur months or even years after initiation and still be effective in promoting cancer?

7. In your lab you have several mouse cell lines growing in culture, designated culture A through culture D. Some of these cell lines are transformed and others are not. For each of the descriptions below, tell whether the cell line is transformed or normal, and if it is transformed, make an educated guess about the gene involved in the transformation.

   a. Cell line A is insensitive to the addition of growth factors to the medium. If the cells are exposed to damaging radiation, the growth curve is unaffected. Many of the cells have abnormal numbers of chromosomes.

   b. Cell line B will not grow suspended in liquid media. The cells require a hard substrate, and then they only grow until one cell layer covers the substrate; they may grow a bit denser if growth factors are added to the medium. The cells lose their ability to move once the substrate is covered with a monolayer of cells.

   c. Cell line C grows quickly and is quite successful when the cells are suspended in soft agar. The cells are impervious to the addition of serum to their medium. Upon exposure to damaging radiation, most of the cells undergo apoptosis.

   d. Cell line D grows quickly and is impervious to DNA-damaging radiation, entering cycle after cycle of mitosis even after large doses of radiation. If transfected with DNA containing a gene for p53, these cells will undergo apoptosis.

CHAPTER SEVENTEEN

Techniques in Cell and Molecular Biology

Learning Objectives

When you have finished this chapter, you should be able to:

1. Describe the features of microscopy.
   a. Understand resolution, magnification, and visibility (contrast), as well as the technologies that enhance each.
   b. Know which types of microscopy are best suited for different experimental applications.
2. Understand how radioisotopes are used as tracers. Describe the methods of tracing radioactivity in cells, including autoradiography and liquid scintillation spectrometry. Know what types of insights each method provides.
3. Define cell culture; distinguish between different types of cell cultures. Understand the advantages of working with cells in culture, as well as some of the disadvantages.
4. Describe some techniques used for the separation and purification of proteins and nucleic acids.
   a. Understand the principles of centrifugation and know when to apply the different types of centrifugation.
   b. Understand the principles of chromatography and gel electrophoresis, and know when these techniques are used.
5. Know how x-ray diffraction provides insights into the three-dimensional structure of macromolecules.
6. Give an overview of the most commonly used recombinant DNA techniques.
   a. Describe DNA cloning; know what vectors are, and how they are used.
   b. Define DNA library, cDNA, site-directed mutagenesis, and PCR. Appreciate the importance of these techniques in the development of modern molecular biology.
   c. Know how transgenic animals are used as models.
7. Understand how antibodies are used for visualization of different materials; define hybridoma, monoclonal antibody, immunofluorescence.

Key Terms and Phrases

light microscopy
objective lens
empty magnification
bright-field microscope
polarization microscopy
shadow casting
freeze-fracture replication
confocal scanning microscopy
alpha particle
autoradiography
refractive index
ocular lens
fluorescence miscroscopy
phase-contrast microscopy
differential interference contrast
transmission electron microscope (TEM)
scanning electron microscope (SEM)
isotopes
beta particle
liquid scintillation spectrometry
angle of incidence
resolution
numerical aperture
serial sections
negative staining
freeze-etching
critical-point drying
radioactive isotopes
gamma radiation
half-life

cell culture
mass culture
clonal culture
protoplast
differential centrifugation
homogenizer
cell-free systems
protein purification
fractionation
specific activity
selective precipitation
chromatography
isoelectric point
high-performance liquid chromatography (HPLC)
cation exchanger
ion-exchange chromatography
anion exchanger
gel filtration chromatography
affinity chromatography
electrophoresis
tracking dye
polyacrylamide gel electrophoresis (PAGE)
SDS-PAGE
isoelectric focusing
x-ray diffraction
x-ray crystallography
spectrophotometry
ultracentrifugation
Svedberg (S) unit
nucleic acid hybridization
recombinant DNA
DNA cloning
vector DNA
replica plating
in situ hybridization
yeast artificial chromosomes (YACs)
DNA libraries
cDNA
site-directed mutagenesis
expression cloning
transfection
transgene
transgenic animals
animal models
knockout mice
embryonic stem cells
antiserum
polymerase chain reaction (PCR)
DNA sequencing
monoclonal antibody
Sanger sequencing method
hybridoma
indirect immunofluorescence
Gilbert and Maxam sequencing method
direct immunofluorescence
dideoxyribonucleoside triphosphate (ddNTP)

Reviewing the Chapter

I. Introduction

A. Research in cell biology requires complex instrumentation and techniques.
B. Understanding the technology aids in understanding the cell.

II. The Light Microscope

A. The light microscope uses the refraction of light rays to magnify an object.
   1. The specimen is illuminated by rays from a light source.
   2. A condenser directs light toward the specimen.
   3. The objective lens collects light from the specimen to form an enlarged, real image.
   4. The ocular lens forms an enlarged, virtual image used by the eye.
B. Resolution:
   1. Resolution is the ability to see two neighboring points as distinct entities.
   2. Resolution is limited by the wavelength of light.
   3. The numerical aperture is a measure of the light-gathering qualities of a lens.
   4. The resolution of a light microscope is 200 nm, sufficient for seeing large organelles.
   5. Lens aberrations affect resolving power.
C. Visibility:
   1. Visibility requires that the specimen and the background have different refractive indices.
   2. Translucent specimens are stained with dyes.
D. Phase-contrast microscopy:
   1. The phase-contrast microscope utilizes interference to convert the differences in the refractive index of some parts of a specimen into differences in light intensity.
   2. Differential interference contrast (DIC) optics give a three-dimensional quality.
E. Fluorescence microscopy:
   1. Fluorochromes release visible light upon the absorption of UV rays.
   2. Fluorochrome stains cause cell components to glow.

F. Polarization microscopy:
   1. Submicroscopic structures with birefringence rotate plane-polarized light.
   2. Birefringent structures appear bright on a dark background.

G. Video microscopy is used to observe living cells; data can be processed by a computer.

H. Confocal scanning microscopy:
   1. A laser beam is used to examine planes at different depths in a specimen.
   2. "Sections" can be viewed without cutting, as in CAT scans.

I. Preparation of specimens for light microscopy:
   1. A whole mount is an intact object, either alive or dead.
   2. Sections are very thin slices of an object.
      a. The tissue is fixated, embedded, and sliced.
      b. Procedures minimize alterations from the living state.

III. Transmission Electron Microscopy

A. Transmission electron microscopes (TEMs) use electrons instead of light to form images.
   1. The limit of resolution of standard TEMs is 3 to 5 Å.
   2. The components of an EM:
      a. An electron beam from a tungsten filament is accelerated by high voltage.
      b. The electron beam is focused with a magnetic field.
      c. Electrons pass through the specimen and strike a fluorescent screen, creating an image.
      d. Differential scattering of electrons by the specimen creates the image.
         (1) Electron scattering (and visibility of an object) is proportional to the mass thickness of that part of the specimen.
         (2) Tissues are stained with heavy metals to obtain contrast.

B. Specimen preparation for electron microscopy:
   1. Specimens must be fixed, embedded, and sectioned thinly.
      a. Glutaraldehyde and osmium tetroxide are common fixatives.
      b. Specimens are dehydrated prior to embedding.
      c. Epon is a common embedding material.
      d. Thin sections cut with glass or diamond knives are collected on grids.
   2. In negative staining, heavy metal diffuses into spaces between specimen molecules.
   3. Shadow casting coats a specimen with metal to produce a three-dimensional effect.
   4. Freeze-fracture replication and freeze-etching:
      a. In freeze-fracture replication, frozen tissue is fractured with a knife.
         (1) A heavy-metal layer is deposited on a fractured surface.
         (2) A cast of the surface is formed with carbon.
         (3) The metal-carbon replica is viewed in the TEM.
      c. In freeze-etching, a layer of ice is evaporated from the surface of the specimen prior to coating it with heavy metal.

IV. Scanning Electron Microscopy

A. SEMs form images from electrons that have bounced off the surface of a specimen.

B. Specimens for SEM are dehydrated by critical-point drying.

C. Specimens are coated with a layer of carbon, then gold.

D. The image in SEM is formed like an image on a television screen– it is indirect.

E. SEM has a large range of magnification and depth of focus.

V. The Use of Radioisotopes

A. Radioisotopes can be easily detected and quantified.
   1. A tracer is a substance that can be followed in cells.
   2. Properties of radioisotopes:

a. Atoms that differ only in numbers of neutrons are isotopes of each other.
b. Isotopes with an unstable combination of protons and neutrons are radioactive.
c. Three forms of radiation can be emitted during atomic disintegration.
(1) Alpha particles consist of two protons and two neutrons.
(2) Beta particles are equivalent to electrons.
(3) Gamma radiation is electromagnetic radiation or photons.
d. The half-life of a radioisotope measures its instability; half of radioactive material disintegrates during its half-life.

3. Liquid scintillation spectrometry:
a. Scintillants absorb the energy of an emitted particle and release it in the form of light.
b. Radiation of a tracer in a sample can be detected by measuring light emitted by a scintillant.

B. Autoradiography:
1. Autoradiography localizes radioisotopes within cells and tissues.
2. A particle emitted from a radioactive atom activates a photographic emulsion.
3. The location of the radioisotopes in the specimen is determined by the positions of the overlying silver grains in a photographic emulsion.

VI. Cell Culture

A. Much of cell biology research is carried out on cultured cells.
1. In a cell culture, cells can be obtained in large quantity.
2. Most cultures contain only a single type of cell.
3. Many different types of cells can be grown in culture.
4. Cell differentiation can be studied in a cell culture.

B. Cells in a culture require media that includes hormones and growth factors.

C. Culture can be primary or secondary, depending on the origin of the cells
1. A primary culture consists of cells from the organism, usually an embryo.
2. Cells in a secondary culture are derived from a previous culture.

D. There are two types of primary culture– mass cultures and clonal cultures.

E. Nonmalignant cells are capable of a limited number of cell divisions.

F. Many types of plant cells can be grown in culture.

VII. The Fractionation of a Cell's Contents by Differential Centrifugation

A. Differential centrifugation facilitates the isolation of particular organelles in bulk .
1. Prior to centrifugation, cells are broken by mechanic disruption in a buffer solution.
2. The homogenate is subjected to a series of sequential centrifugations.
3. Organelles sediment into a pellet according to their size.
a. Nuclei and remaining whole cells sediment first.
b. Mitochondria, chloroplasts, lysosomes, and peroxisomes sediment next, followed by microsomal fragments..
c. Ribosomes can be sedimented only in an ultracentrifuge.

B. Cellular organelles isolated by centrifugation retain a high level of activity.

VIII. Isolation, Purification, and Fractionation of Proteins

A. Protein purification involves the stepwise removal of contaminants.
1. Protein purification can be either analytic or preparative.
2. Purification is measured as an increase in specific activity of a protein.

B. Selective precipitation:
1. At low ionic strength, proteins tend to remain in solution (they are salted-in).
2. At high ionic strength, protein solubility decreases.
3. Ammonium sulfate is the most commonly used salt for protein precipitation.

C. Chromatography:
    1. A mixture of dissolved components can be chromatographically fractionated as it moves through porous material.
        a. Components are fractionated between mobile and immobile phases.
        b. The molecule's rate of progress through the immobile phase is proportional to its affinity for the matrix.
        c. High-performance liquid chromatography (HPLC) has a high resolution.
    2. Ion-exchange chromatography:
        a. Proteins can be separated according to ionic charge.
        b. The isoelectric point is the pH value at which the number of positive and negative charges on the protein are equal.
        c. In ion-exchange chromatography, proteins with charged groups associate with the column matrix.
    3. Gel filtration chromatography:
        a. Gel filtration separates proteins by molecular weight.
        b. A gel filtration column is packed with cross-linked polysaccharides of different porosity.
        c. Proteins that are small enough to enter the pores are eluted from the column last.
    4. Affinity chromatography isolates one protein from a mixture using a specific ligand.

D. Polyacrylamide gel electrophoresis:
    1. Electrophoresis is based on the fact that proteins can migrate in an electric field.
        a. In polyacrylamide gel electrophoresis (PAGE), proteins are driven by a current through a gel composed of cross-linked acrylamide.
        b. Movement of proteins depends on molecular size, shape, and charge density.
        c. The position of the proteins in the gel can be visualized by staining, autoradiography, or Western blot.
        d. SDS-PAGE:
            (1) PAGE is often done in the presence of the negatively charged detergent sodium dodecylsulfate (SDS).
            (2) In SDS-PAGE, proteins are separated on the basis of their molecular weight only.
    2. Isoelectric focusing:
        a. In isoelectric focusing, the gel consists of a mixture of ampholytes that have varying ratios of positive and negative charges.
        b. Electrophoretic movement of ampholytes establishes a pH gradient.
        c. A protein stops when it reaches the pH equal to its isoelectric point.
    3. Two-dimensional gel electrophoresis separates proteins on the basis of both isoelectric point and molecular weight.

IX. Determination of Protein Structure by X-ray Diffraction Analysis

A. X-ray diffraction requires protein crystals.
B. Crystals are hit with x-rays; scattered radiation is collected by a photographic plate.
C. The diffraction pattern provides information about the structure of the protein.
D. x-ray diffraction is useful in the study of both proteins and nucleic acids.

X. Purification and Fractionation of Nucleic Acids

A. DNA purification procedures differ from protein purification procedures.
    1. To obtain DNA, nuclei are isolated and lysed.
    2. DNA is separated from contaminating materials (RNA and protein).
B. Separation of DNAs by gel electrophoresis:
    1. PAGE is used for the separation of small DNA and RNA molecules. Large ones are separated in agarose gels.

2. Nucleic acids are separated on the basis of their respective molecular weights.

XI. Measurements of Protein and Nucleic Acid Concentrations by Spectrophotometry

A. The amount of protein or nucleic acid in a solution can be found by measuring the amount of light absorbed by the solution.
   1. The light-absorbing capacity of a solution is read using a spectrophotometer.
   2. Absorbance of light is directly proportional to the number of absorbing molecules present in the light path.

B. Tyrosine and phenylalanine absorb light in the ultraviolet range.

XII. Ultracentrifugation

A. Sedimenting particles are affected by two opposing forces– diffusion and centrifugal force (which is dependent on particle density)
   1. Large proteins diffuse slower than smaller ones.
   2. Sizes of organelles and macromolecules can be expressed in Svedberg (S) units.
   3. There are two types of ultracentrifuges.
      a. Preparative centrifuges are used to purify components for further use.
      b. Analytic ultracentrifuges allow one to follow the progress of the substances during the centrifugation.

B. Sedimentation behavior of nucleic acids:
   1. Nucleic acid analysis involves the study of sedimnetation patterns in an ultracentrifuge.
      a. In velocity sedimentation, nucleic acid molecules are fractionated in a sucrose gradient on the basis of nucleotide length .
      b. In equilibrium (isopycnic) sedimentation, nucleic acids are fractionated according to their buoyant density in a cesium gradient.

C. Nucleic acid hybridization:
   1. Nucleic acid hybridization is based on the ability of two complementary DNA strands to form a double-stranded hybrid.
   2. Nucleic acid hybridization methods:
      a. Two populations of single-stranded nucleic acids are incubated together.
      b. One population of single-stranded nucleic acids is usually immobilized.
      c. Nucleic acids that are complementary to the immobilized ones form double-stranded complexes.
      d. Nonhybridized single-stranded nucleic acids are washed away.
   3. The Southern blot technique is based upon nucleic acid hybridization.
   4. Nucleic acid hybridization can be used to determine the degree of similarity in nucleotide sequence between two samples of DNA.

XIII. Recombinant DNA Technology

A. Analysis of eukaryotic genomes involves recombinant DNA molecules.
   1. DNA is first cut with restriction enzymes.
   2. Recombinant DNAs can be formed in several ways, such as creating "sticky ends" with restriction enzymes.
   3. The two components of a recombinant DNA are linked using DNA ligase.

B. DNA cloning:
   1. DNA cloning produces large quantities of a specific DNA segment within host cells.
   2. Bacterial plasmids and bacterial virus lambda (λ) are two commonly used vectors.
   3. Cloning eukaryotic DNAs in bacterial plasmids:
      a. Plasmids used for DNA cloning are modified forms of the wild type.
         (1) Cloning plasmids contain a replication origin.
         (2) Cloning plasmids usually carry genes for antibiotic resistance.
      b. Recombinant plasmids are introduced into bacterial cells by transformation.

c. Plasmid-containing bacteria are selected by treatment with antibiotics.
d. Cells containing various plasmids are grown into separate colonies which can be screened for the presence of a particular DNA sequence.
   (1) Replica plating produces dishes with representatives of the same bacterial colonies in the same position on each dish.
   (2) In situ hybridization uses a labeled DNA probe to localize the colony possessing the desired DNA fragment.

4. Cloning eukaryotic DNAs in phage genomes:
   a. Recombinant DNA molecules are packaged into lambda phage heads.
      (1) Lambda phage can take inserts up to the size of 25 kb.
      (2) Site of phage infection are identified as a plaques in a bacterial "lawn."
      (3) DNA fragments are identified by the same techniques as those used for cloning of recombinant plasmids.
   b. Formation of a DNA library:
      (1) A DNA library is a collection of DNA fragments representing the entire genome of an organism.
      (2) DNA fragments for the library can be obtained by cutting genomic DNA with restriction enzymes.
      (3) Cleaving of genomic DNA is random, which generates overlapping fragments.
      (4) Overlapping fragments are useful for chromosome walking.
5 Yeast artificial chromosomes (YACs) can accommodate large DNA inserts.
6. Cloning cDNAs:
   a. The isolation of genomic fragments allows study of the genome. Topics include:
      (1) regulatory sequences flanking the coding portion of a gene.
      (2) noncoding intervening sequences.
      (3) the structure of multigene families.
      (4) the evolution of DNA sequences.
      (5) transposable genetic elements.
   b. Coding regions of a gene can be studied using cDNAs, synthesized by reverse transcriptase using mRNA as a template.
7. In expression cloning, cDNA is inserted next to a strong bacterial promoter, which allows foreign DNA to be transcribed and translated during the infection process.

C. Chemical synthesis and site-directed mutagenesis:
   1. Polynucleotides can be chemically synthesized.
   2. Site-directed mutagenesis inserts genetic elements containing modified nucleotide sequences into host cells.

D. Gene transfer into eukaryotic cells and mammalian embryos:
   1. DNA can be introduced into eukaryotic cells by the process of transfection.
   2. Transgenic animal and plants:
      a. Transgenic animals allow scientists to determine the effects of overexpression of a particular DNA sequence.
      b. Genetic engineering can produce animal models used to study human diseases.
      c. The main goal of genetic engineering in plants is to improve the efficiency of both photosynthesis and nitrogen fixation.
      d. Knockout mice are obtained by transfecting embryonic stem cells, introducing them into an embryo, and implanting the embryo into a female mouse.

E. Enzymatic amplification of DNA by PCR:
    1. Polymerase chain reaction (PCR) cheaply, rapidly, and extensively amplifies a single region of DNA, from a very small amount of template.
    2. PCR utilizes the DNA polymerase from bacteria living in hot springs.
    3. PCR can be used in criminal investigations.

F. DNA sequencing:
    1. Techniques developed by Sanger, and Glibert and Maxam are widely used for sequencing nucleic acids.
    2. Sanger-Coulson (dideoxy) sequencing technique:
        a. Four samples of identical single-stranded DNA molecules are obtained.
        b. DNA of each sample is incubated with a primer, DNA polymerase, four dNTPs, and a low concentration of dideoxyribonucleoside triphosphate (ddNTP), different one in each sample.
        c. DNA fragments of different lengths are synthesized in each sample, with synthesis terminating where ddNTP has been (randomly) incorporated.
        d. DNA fragments are separated by gel electrophoresis and the DNA sequence is read from the gel bands.
        e. The amino acid sequence of the protein is deduced from the nucleotide sequence.

XIV. The Use of Antibodies

A. Antibodies are highly specific proteins produced by lymphoid tissues in response to the presence of foreign materials.

B. Preparation of antibodies:
    1. A population of polyvalent antibodies can be obtained by repeated injections of a purified antigen into an animal; the blood of that animal serves as a source of the antiserum.
    2. Univalent antibodies are produced by descendants of a single antibody-producing cell.
        a. Antibody-producing cells do not grow and divide in culture.
        b. Fusion of a normal antibody-producing lymphocyte and a malignant myeloma cell will create a viable hybridoma cell that can produce large amounts of a monoclonal antibody.

C. Antibodies can be conjugated with a fluorescent substance that allows visualization of antigens.
    1. In direct immunofluorescence, antibodies with bound fluorescent molecules bind to antigens. The binding sites are then visualized under a fluorescence microscope.
    2. In indirect immunofluorescence, cells are first incubated with unlabeled antibodies, and then with labeled secondary antibodies which are directed against the primary antibody.

Key Figure

Figure 17.33. Formation of a recombinant DNA molecule.

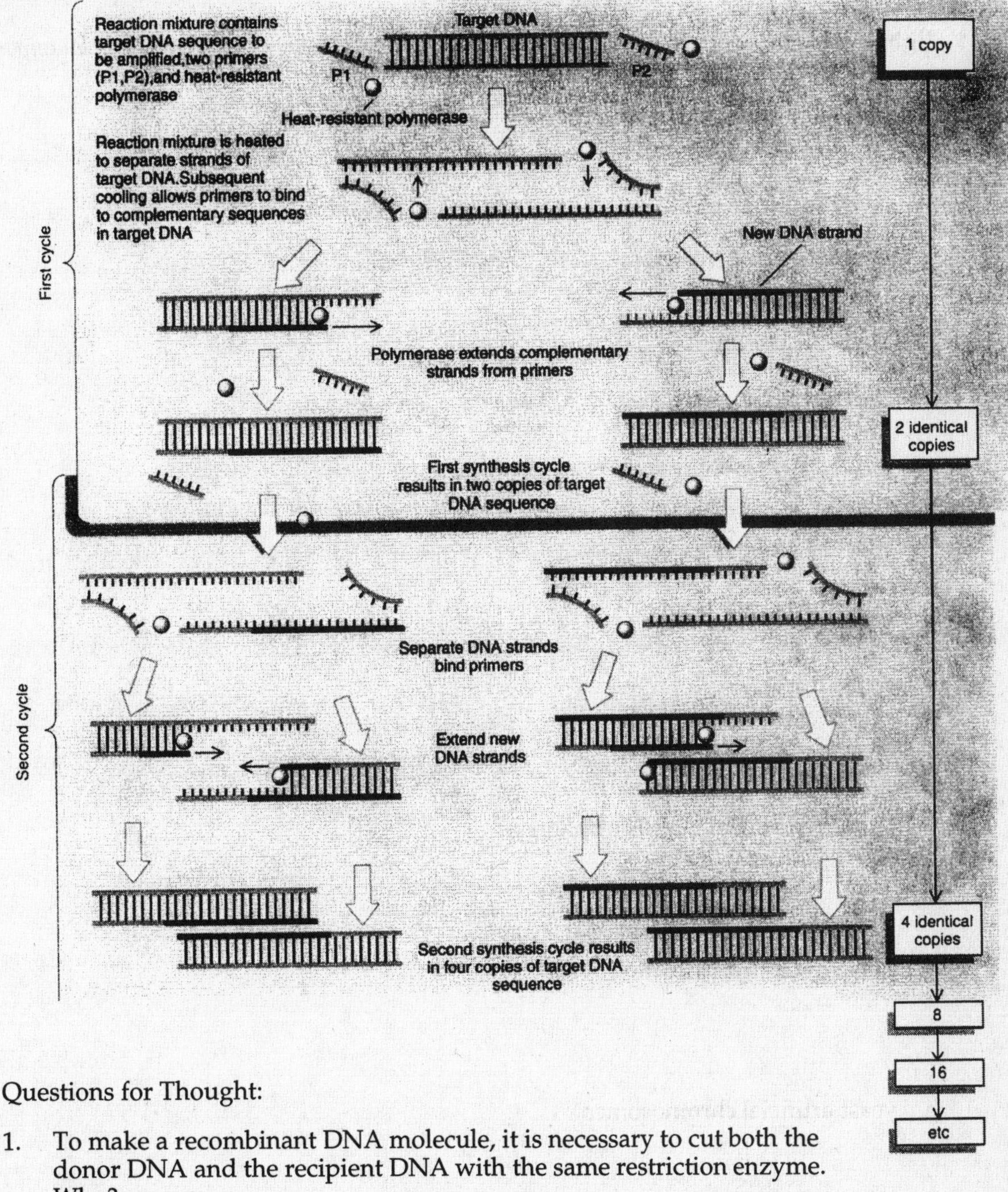

Questions for Thought:

1. To make a recombinant DNA molecule, it is necessary to cut both the donor DNA and the recipient DNA with the same restriction enzyme. Why?

2. What genetic elements must reside on the plasmid DNA in order for this technique to succeed? Why?

3. What steps follow the formation of the recombinant plasmid?

4. Are there ethical questions raised by recombinant DNA technology? What are they?

Review Problems

Short Answer

1. Give the significance (not the definition) of the following terms or phrases. Say what they do or why they are important. For example:
   light microscope: *enabled biologists to discover the existence of cells. This is still an important instrument for cell biologists.*

   a. microscopic resolution

   b. interference microscopy

   c. electron microscopy

   d. autoradiography

   e. cell culture

   f. two-dimensional gel electrophoresis

   g. nucleic acid hybridization

   h. yeast artificial chromosomes

   i. animal models for human disease

   j. polymerase chain reaction

2. Compare and contrast the following:
   a. magnification vs. resolution

   b. transmission electron microscopy vs. scanning electron microscopy

   c. differential centrifugation vs. density-gradient centrifugation

   d. cloning eukaryotic DNA in bacterial plasmids vs. in phage genomes

   e. cloning genomic DNA vs. cloning cDNA

Multiple Choice

1. A light microscope has an objective lens with a magnification of 40x and an ocular lens with a magnification of 10x. What is the total magnification of the image?
   a. 40x
   b. 50x
   c. 400x
   d. 450x

2. Which of the following does not contribute to the resolution attainable by a microscope?
   a. the wavelength of the light
   b. the magnification of the objective lens
   c. the refractive index of the material between the specimen and the lens
   d. the numerical aperture of the lens

3. Under which of the following conditions would the visibility of an object under a microscope be lowest?
   a. The object has the same refractive index as does the medium.
   b. The object and the background each bend light differently.
   c. The object diffracts some, but not all, of the light rays that hit it.
   d. The object absorbs some, but not all, of the light rays that hit it.

4. You are interested in reconstructing the three-dimensional shape of chromosomes in living cells that are in the process of mitosis. Which microscopic technique will you choose?
   a. freeze-fracture microscopy
   b. confocal scanning microscopy
   c. light microscopy
   d. scanning electron microscopy

5. You wish to find the intracellular compartments that contain high calcium concentrations. Which of the following microscopic techniques will you choose?
   a. fluorescence microscopy
   b. polarization microscopy
   c. phase-contrast microscopy
   d. transmission electron microscopy

6. Why are transmission electron micrographs never in color?
   a. Cellular structures have no color. They are all shades of gray.
   b. Color micrographs would be too expensive.
   c. Color film has not yet been invented.
   d. The photographic film detects electrons that are transmitted through the specimen rather than the different wavelengths of light that determine color.

7. You think you might have discovered a new organelle using electron microscopy, but you are not sure if it's real or an artifact of your technique. What might you do to prove it is real?
   a. Look for the structure in many similarly prepared specimens from the same source.
   b. Lood for the structure in many similarly prepared specimens from different sources.
   c. Look for the same strucure in specimens prepared using different staining protocols.
   d. Look for the same structure in specimens prepared using different (or no) fixatives.

8. The half-life of tritium is 12 years. If you had some tritium and kept it for 24 years, what percentage of your original radioactive sample would be left?
   a. 12%
   b. 25%
   c. 75%
   d. none

9. You want to know if a culture of cells is in the process of DNA synthesis. You incubate your cells in the presence of radioactive thymidine to see if it is being incorporated into the DNA. What is the best technique to detect the labeled deoxynucleotide in nuclear DNA?
   a. autoradiography
   b. polyacrylamide gel electrophoresis
   c. agarose gel electrophoresis
   d. two-dimensional gel electrophoresis

10. The major difference between defined and undefined cell media used in cell culture is
    a. undefined medium is a recent development in cell culture technology.
    b. undefined medium is completely artificial.
    c. defined medium includes serum, lymph or other fluids from living sources.
    d. defined medium is free of serum, lymph or other fluids from living sources.

11. Which of the following would be the least promising source of cells for a primary cell culture?
    a. embryonic tissue
    b. actively growing malignant tumor tissue
    c. adult rat brain tissue
    d. plant protoplasts

12. You have homogenated and fractionated rat liver cells using differential centrifugation. Which fraction would have the fewest mitochondria?
    a. whole homogenate before any centrifugation steps
    b. the supernatant that results after the first low-speed centrifugation step
    c. the supernatant that results after a high-speed ultracentrifugation step
    d. None of the above fractions would have any mitochondria.

13. Which type of column chromatography separates proteins on the basis of molecular weight?
    a. ion-exchange chromatography
    b. gel filtration chromatography
    c. affinity chromatography
    d. isoelectric focusing

14. In gel electrophoresis, the tracking dye moves:
    a. more slowly than all the different molecules of the sample.
    b. at the same rate as all the different molecules of the sample.
    c. more quickly than all the different molecules of the sample.
    d. not at all.

15. Nucleic acid hybridization can be used as a measure of evolutionary relationships between species. Of the following statements about the DNA of related species, which is true?
    a. Closely related species form hybrid DNAs with relatively low melting temperatures.
    b. Closely related species form hybrid DNAs with relatively high melting temperatures.
    c. There is no correlation between DNA hybrid melting temperature and relatedness of species.
    d. One cannot create DNA hybrid molecules from DNAs of closely related species.

16. DNA fragments and DNA probes are to Southern blotting as __?__ are to Northern blotting.
    a. DNA fragments and RNA probes
    b. RNA fragments and RNA probes
    c. RNA fragments and DNA probes
    d. protein fragments and DNA probes

17. The role of the vector DNA in DNA cloning is to:
    a. identify the source of DNA as foreign.
    b. identify the host cell that has taken up the one specific gene of interest.
    c. make the foreign DNA susceptible to digestion with endonucleases.
    d. carry the foreign DNA into the host cell.

18. Why does one need to make replica plates when screening for a specific DNA sequence among a large number of recombinant bacterial colonies?
    a. It may take several tries to positively identify the specific sequence of interest.
    b. The screening process requires several different steps, each of which must be done on a new colony of recombinants.
    c. One wants a living culture of recombinant cells available after screening, a process that destroys the cells.
    d. It is good science to replicate all experimental results.

19. What is a "gene gun"?
    a. a gun that fires DNA-coated pellets into plant cells
    b. a new technique using "explosively fast" enzymes to form recombinant DNA
    c. a weapon of mass destruction that exploits recombinant DNA technology for warfare
    d. none of the above

20. Why are heat-stable DNA polymerases from thermophilic bacteria required for the polymerase chain reaction?
    a. The heat-stable forms are the only ones that recognize all four deoxyribonucleotides.
    b. These enzymes amplify DNA in a reasonable amount of time.
    c. These enzymes are the most readily available forms of DNA polymerase in the world.
    d. These enzymes are stable enough to withstand the temperatures required to melt DNA.

Problems and Essays

1. You are studying subcellular structure with a microscope that has an angular aperture of 70° ($\theta = 70°$) and you want to maximize your resolution.
   a. First calculate the resolution of your scope, using white light in air.

   b. Now place a blue filter between the light source and the specimen. The wavelength of blue light is 450 nm. Does this improve your resolution? If so, by how much?

   c. Add immersion oil (refractive index = 1.5), still using your blue filter. Now what is the resolution? Given your equipment, can you improve on this level of resolution?

2.

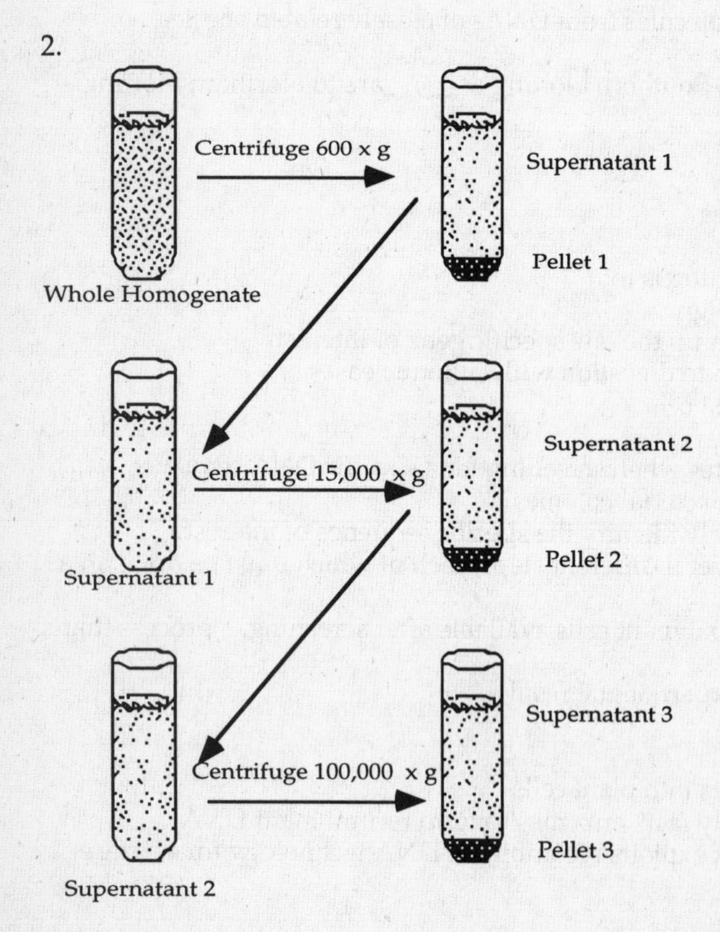

You have homogenized liver tissue and separated the subcellular fractions according to the centrifugation scheme shown to the left. You have saved small aliquots of each fraction.

a. Using the aliquots of each fraction, you do enzyme assays specific for the enzyme succinic dehydrogenase (SDH). Recall that SDH is one of the enzymes of the TCA cycle. Assuming a perfect cell fractionation, which fractions will test positive for SDH activity?

b. Which fraction will have the highest specific activity (activity per unit of protein) of SDH?

c. Which fraction will have the most total protein?

d. At the end of the fractionation, where would the ribosomes be found?

3.

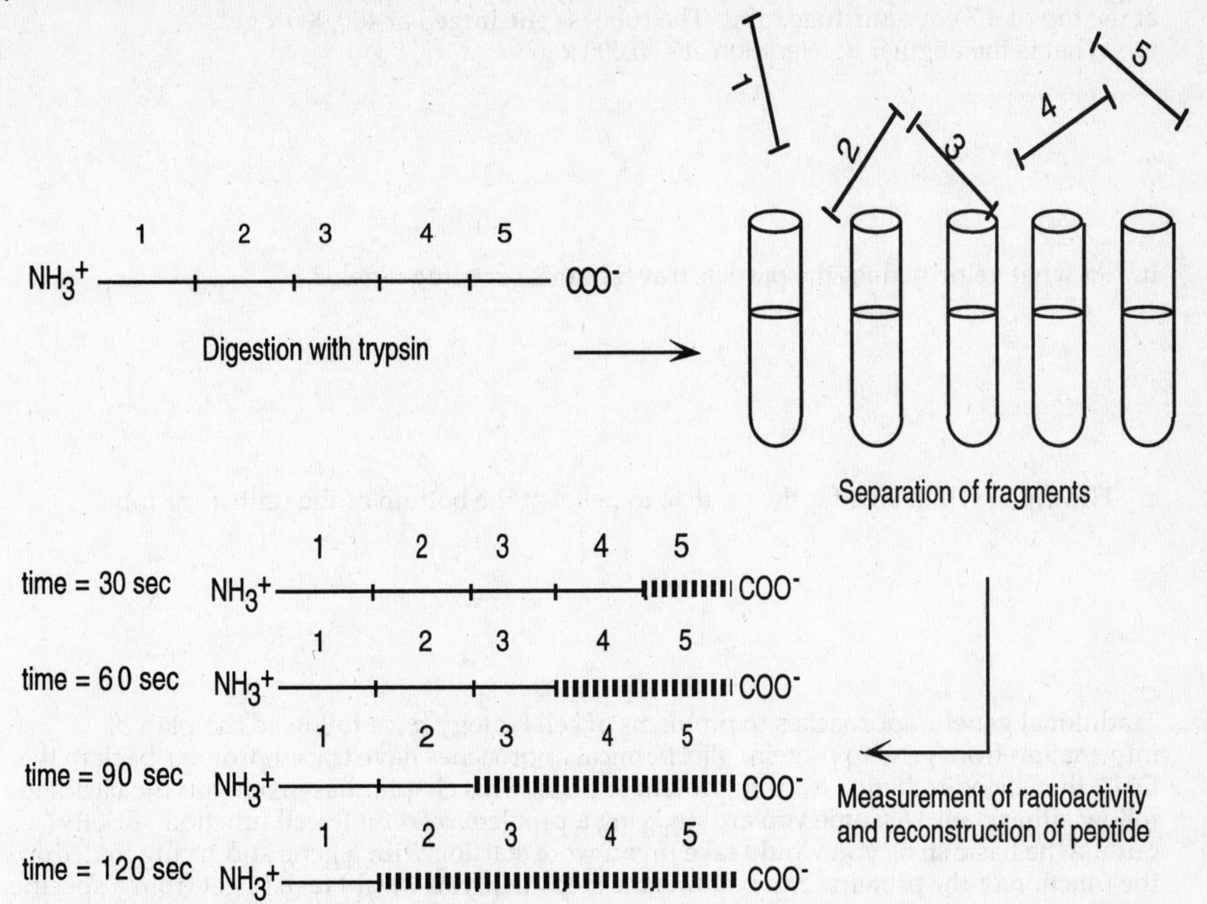

One of the most elegant uses of radiolabeled tracers in cell biology occurred in the experiment done by Howard Dintzis in the early 1960s, where he determined the direction of proteins synthesis. Dintzis cultured reticulocytes actively engaged in synthesizing hemoglobin in the presence of radioactive amino acids. At short intervals, he collected samples of the cells and isolated the hemoglobin. The protein chains were digested with trypsin, a protease enzyme that cleaves polypeptides at the carboxyl sides of both arginine and lysine residues. For hemoglobin, this treatment yields a characteristic set of five fragments. labeled 1 through 5 in the figure. The amino acid sequence of hemoglobin was known, and hence the sequence of each trypsin fragment could be derived. After the fragments were separated, the radioactivity in each was measured. The diagram above summarizes Dintzis' experimental technique and his results. The hatched lines represent parts of the peptide that were radioactive at each of the time intervals. Based on these data, what did Dintzis conclude about the direction in which proteins are synthesized? Why?

4. Suppose you wish to pellet an 18S particle using ultracentrifugation. Assume the particle is at the top of a 7 cm centrifuge tube. The tube is centrifuged at 400,000 x g.
   a. What is the angular acceleration at 400,000 x g?

   b. At what velocity does the particle travel in this centrifugal field?

   c. How long will it take for the particle to pellet at the bottom of the centrifuge tube?

5. Traditional genetic approaches to problems of cell biology have followed the path of information from gene to protein. Biochemical approaches have traced proteins back to the DNA that encodes them, i.e., from protein to gene. This chapter has given you the basics to follow either path. Assume you are studying a problem relating to cell function. Briefly outline the basic steps you would take if you were starting with a gene and trying to deduce the function of the protein. Now outline the steps that you would take to get from a specific protein to a sequenced, isolated gene.

6. You wish to sequence a short nucleic acid (eight nucleotides) using the Sanger dideoxyribo-nucleoside method. After the fragments in each of your four test tubes have been separated and visualized with autoradiography, the patterns on the right are seen. What is the sequence of nucleotides in your fragment? Which end is the 5′ end and which is the 3′ end?

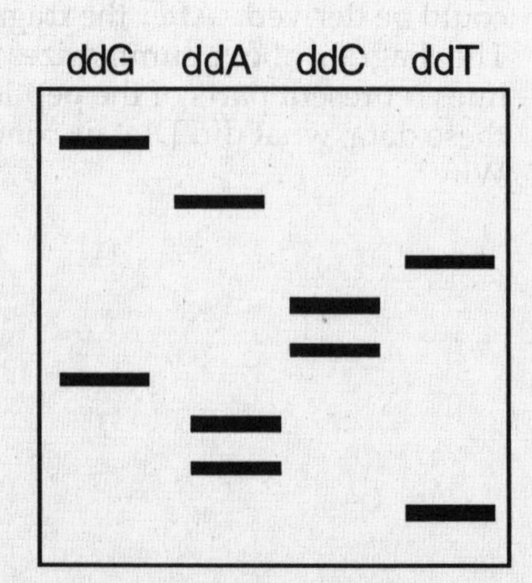

7. Name techniques you might use to answer each of the following questions:
   a. Does the DNA on the hair found at the scene of a crime match that of the defendant in a criminal trial?

   b. Where in a cell are the carbohydrates added to the glycoprotein hormones that are exported?

   c. Is it possible to find a cheap and plentiful source of human growth hormone for treating people with growth hormone deficiencies?

   d. Which amino acids in an enzyme play the most important roles in binding to the substrate?

Notes:

APPENDIX I

Answers to Review Problems

Chapter One

<u>Short Answer</u>

1. Significance
   a. cells: *the smallest units to exhibit the fundamental property of life.*
   b. ATP: *stores chemical energy in cells in a readily available form.*
   c. cell surface receptors: *interact with substances in the environment in highly specific ways, permitting external agents to evoke specific responses in given cells.*
   d. prokaryotes and eukaryotes share the same genetic "language": *they share a common ancestry. In fact, eukaryotes evolved from prokaryotic ancestors.*
   e. nitrogen fixation: *makes otherwise useless nitrogen available for use by all cells by reducing* N_2 *gas to other forms, such as* NH_3.
   f. surface proteins on viruses: *bind to surface components of host cells. The viral surface proteins determine which cells will be infected by the virus.*
   g. genetic similarity between genes of viruses and their hosts: *provides evidence that viruses arose as bits of genetic material derived from the host genome that acquired some degree of autonomy from the host.*
   h. life cycle of *Dictyostelium*: *provides an excellent model system for studying the process of cellular differentiation in multicellular organisms.*

2. Compare and Contrast:
   a. prokaryote vs. eukaryote
   Together these two groups comprise the living world. They share a genetic language, some common metabolic pathways, and many structural features. Prokaryotes are the more primitive cells, and include the Archaebacteria and the Eubacteria. Eukaryotes comprise everything else, including the protists, fungi, plants, and animals. Prokaryotes are smaller. Prokaryotes have a single chromosome made entirely of DNA. They lack a true nucleus and membrane-delineated organelles, and they divide by simple fission. Eukaryotes have a true, membrane-bound nucleus containing chromosomes that are composed of both protein and DNA. They have membrane-bound organelles, endoplasmic reticulum, and cytoplasmic structural and contractile proteins. They divide by mitosis or meiosis, and undergo sexual reproduction. While both groups may have flagella, the flagella function quite differently in each.
   b. endoplasmic reticulum vs. mesosomes
   Both endoplasmic reticulum and mesosomes are intracellular membrane systems that increase the surface area for membrane-affiliated processes. Endoplasmic reticulum is a complex of channels occurring in eukaryotic cells, and is a site of protein and lipid synthesis. Mesosomes occur in prokaryotic cells, and arise as simple infoldings of the plasma membrane.
   c. virus vs. virion
   "Virus" is a general term that refers to a group of non-cellular obligatory intracellular pathogens composed of either DNA or RNA, and a capsid, with or without a lipid membrane. "Virion" is a term describing a virus when it is outside the host.
   d. lytic infection vs. integrative infection
   These are two different mechanisms of viral infection. When a virus is lytic, the metabolism of the host cell is redirected to manufacture new viruses, and assemble new virions. An integrative infection occurs when a virus integrates its DNA into the chromosome of the host, and either

remains dormant or directs the cell to produce progeny virions without causing lysis of the host cell.

Multiple Choice

1. a	2. b	3. c	4. b
5. a	6. d	7. d	8. c
9. d	10. a	11. d	

Problems and Essays

1. *These organelles cannot be maintained indefinitely when isolated from their cells of origin. They are incapable of reproduction outside of the cell, and rely upon the nucleus of the cell for many of the proteins that enable them to undergo respiration and photosynthesis. Therefore, they are not considered alive.*

2. *The process of differentiation, in which genetically identical cells develop into highly specialized cells, is poorly understood. The developmental pathway a cell takes depends upon signals it receives from the environment, and that, in turn, depends upon the position of that cell during the process of differentiation. External chemical signals can selectively activate some genes, while others remain dormant, thereby committing a cell to a particular developmental pathway.*

3 a. *Structures that increase surface area, such as infoldings or microvilli, and many mitochondria to fuel the active transport of substances across cell membranes.*

b. *Structures that provide the equipment for protein synthesis, such as ribosomes and endoplasmic reticulum, and the vesicles that carry proteins from the site of synthesis to the membrane for release, such as Golgi and secretory vesicles.*

c. *Many plastids, both for photosynthesis (chloroplasts) and for storing the products of photosynthesis.*

d. *Prominant intracellular contractile proteins, and intracellular structure organized to direct the force of contraction in a specific direction.*

4. *Viruses are obligate intracellular parasites, and, as such, could not possibly have evolved before their hosts. Bacteria arrived on the scene about two billion years before the first eukaryotic cells, and over three billion years before humans. It is safe to say that the viruses that infect bacteria must have arisen long before those that infect humans. Further evidence that bacterial viruses are older than human viruses is found by comparing their structures. Bacterial viruses are typically more complex than human viruses, indicating that they have had more time to evolve.*

5 a. *The surface area of the cell is the area of each side (length x width) times the number of sides (= 6). SA = 100 μm x 100 μm x 6 = 60,000 μm^2. The volume of the cell is the length x width x height, or 100^3 = 1,000,000 μm^3. Thus, the SA/volume ratio =*

$$SA/vol = 60,000\ \mu m^2 / 1,000,000\ \mu m^3 = 0.06\ \mu m^{-1}$$

b. *No. One hundred twnty-five cubic cells would have a total surface area of 300,000 μm^2. The SA/volume would be 0.3. It would require dividing this cell into 125,000 individual cells, each 2 μm on a side to get a total surface area of 3,000,000 μm^2.*

6. *The common cold is caused by a virus. Unlike bacterial pathogens, viruses do not have their own machinery for replicating their genetic material. They rely entirely upon the host for replication. Thus*

drugs aimed at the bacterial replication process have no effect on viral replication. Drugs aimed at viral replication would likely have deleterious effects on the host organism.

7. *The lipid envelope of the HIV virus is derived from the cell membrane of the host cell. The host cell produces new virions, which are released by budding and take some of the host membrane with them.*

8. *Photosynthesis in both cyanobacteria and green plants occurs by splitting water molecules, and releasing free oxygen. Photosynthesis in other prokaryotes occurs by an entirely different mechanism.*

9. a. *(1) The size of mitochondria and chloroplasts is on the order of 1 to 5 μm, similar to the size of prokaryotic cells. (2) The photosynthetic mechanism of cyanobacteria, involving splitting water molecules, is the same mechanism used by chloroplasts in green plants. (3) The membranes of some prokaryotes are convoluted into mesosomes, much like the membranes of mitochondria, which have similar infoldings.*

   b. *Look for other features of mitochondria and chloroplasts that resemble those of prokaryotic cells. For example: (1) Is there DNA in the organelles? If so, does it exist as a naked coil as it does in prokaryotes? (2) Does the composition of the membrane surrounding the organelles resemble that of the plasma membrane of the cell from which it came, or does it resemble the composition of prokaryotic cell membranes? (3) Are there other metabolic similarities between the organelles and prokaryotes?*

Chapter Two

<u>Short Answer</u>

1. Significance
   a. electronegativity: *describes how strongly a nucleus will attract electrons. In covalent bonds, the nucleus with the highest electronegativity attracts electrons more strongly, resulting in a polar bond.*
   b. free radicals: *implicated in both the cellular deterioration that accompanies aging, and the development of some diseases, including amyotrophic lateral sclerosis, cancer, and atherosclerosis.*
   c. carbon: *the atomic basis of life. Carbon's ability to form four covalent bonds with other carbons and/or atoms has resulted in the evolution of hundreds of thousands of organic molecules.*
   d. functional groups: *responsible for the physical properties and chemical reactivity of organic molecules.*
   e. ß(1 $\longrightarrow$ 4) glycosidic linkages in cellulose: *a covalent bond that cannot be broken by the digestive enzymes of most animals. Cellulose is an abundant, energy rich molecule, but cannot be used as a source of energy by most animals.*
   f. amphipathic nature of phospholipids: *responsible for spontaneous bilayer formation making phospholipids ideal for their structural role as cell membranes.*
   g. protein primary structure: *ultimately determines all the properties of the protein, including its higher orders of structure, solubility characteristics, and function.*
   h. site-directed mutagenesis: *a technique that enables researchers to learn the roles of single amino acids in protein structure and function. This technique could ultimately be used to engineer or repair defective genes.*
   i. Anfinsen's experiments with ribonuclease: *the first series of experiments to show that the secondary and tertiary structure of the protein, ribonuclease, and its enzymatic activity, are based solely upon the primary structure.*
   j. molecular chaperones: *prevent hydrophobic interactions between both newly formed and heat-denatured proteins. If allowed to aggregate, proteins in these vulnerable states precipitate out of solution and become nonfunctional.*

2. Compare and Contrast:

a. K_{eq} of water vs. K_W

Both of these are constants that describe the extent of dissociation of the water molecule under well-defined conditions. K_{eq}, as is true of any equilibrium constant, is the ratio of the concentrations of dissociation products to the concentration of undissociated water, namely $[H^+][OH^-]/[H_2O]$. K_w is a simplification of K_{eq}. Because the concentration of water in water is always a constant in itself, 55.51 M, both sides of the K_{eq} equation are multiplied by 55.1 M and the result is still a constant, albeit a different one (K_w). K_w is equal to 10^{-14}.

b. glycogen vs. starch

These are two polymers of glucose, both of which function as nutritional polysaccharides. Glycogen is stored in certain animal cells, e.g., the liver and muscle, and occurs in only one form: a highly branched, linear polymer. There are two forms of starch, the plant storage polysaccharide: amylose is an unbranched, helical molecule and amylopectin is a moderately branched linear molecule.

c. saturated vs. unsaturated hydrocarbons

In saturated hydrocarbons, every carbon is covalently bound to the maximum number of hydrogen atoms; there are no double or triple carbon-to-carbon bonds. Unsaturated hydrocarbons are characterized by the presence of multiple bonds and thus are not fully reduced. Multiple bonds, unlike single bonds, restrict the free rotation of atoms. They also have higher bond energies.

d. fat vs. oil

At the chemical level, these two lipids are essentially the same molecule; both are forms of triacylglycerols. In fact, an oil is actually a type of fat that is liquid at room temperature. The fatty acids that make up the triacylglycerols of oils have many cis *double bonds. Double bonds introduce kinks into the linear structure of fatty acids and prevent them from packing together, thus lowering their melting point.*

e. purines vs. pyrimidines

These are the two classes of nitrogenous bases that occur in nucleotides. Purines are large, two-ringed structures that include the bases adenine and guanine. Pyrimidines are smaller, single-ringed structures that include the bases cytosine, uracil and thymine.

Multiple Choice

1. a	2. c	3. a	4. b
5. d	6. d	7. a	8. b
9. d	10. d	11. a	12. c
13. b	14. a	15. d	16. b
17. d	18. a	19. c	20. a

Problems and Essays

1. *Solar radiation includes a broad range of wavelengths, including those at the low end of the electromagnetic spectrum. Most UV wavelengths have energy high enough to exceed the approximately 100 kcal required to break covalent bonds. Prolonged exposure to the full range of sunlight, including the UV component, causes tissue damage resulting from the breakage of covalent bonds. If the damage involves non-genetic biochemicals, the cells may die and be sloughed as in a bad sunburn. If cellular DNA is damaged but the cell s do not die, errors in replication may occur when these cells divide, possibly leading to cancer. It is best and healthiest to avoid prolonged exposure to these high energy photons.*

2. *It takes one calorie of energy to raise one gram of water one degree Celsius, which, relative to other liquids, is actually quite high. Much of the energy that is absorbed is used to disrupt the intermolecular hydrogen bonds that form as a result of the polar, asymmetric nature of the molecule. Because energy is absorbed breaking weak bonds, the temperature of the water does not rise as much as that of another liquid might. In this sense, the temperature of a cell is buffered from changes in the temperature of the environment.*

3 a. *The description of three secondary structures and the fact that these proteins all bind calcium are characteristic of a motif.*

b. *The long stretch of amino acids in these proteins, and their functional similarity describe a domain. Domains are often attached to the rest of the protein by means of a "hinge" region.*

c. *Again, the description of similar function with no mention of secondary structure would characterize the membrane-anchor region of these proteins as a domain.*

d. *The zinc finger is a clearly defined set of secondary structures. This, and the common ability to bind DNA, would identify the zinc finger as a motif.*

4. (a) *E;* (b) *A and C;* (c) *A;* (d) *B;* (e) *B and F;* (f) *C, D, and E;* (g) *A is oleic acid, a fatty acid. B is α–D-glucose, a monosaccharide. C is a triacylglycerol, a fat. D is acetic acid, an acid. E is serine, an amino acid. F is adenosine monophosphate, a nucleotide.*

5.

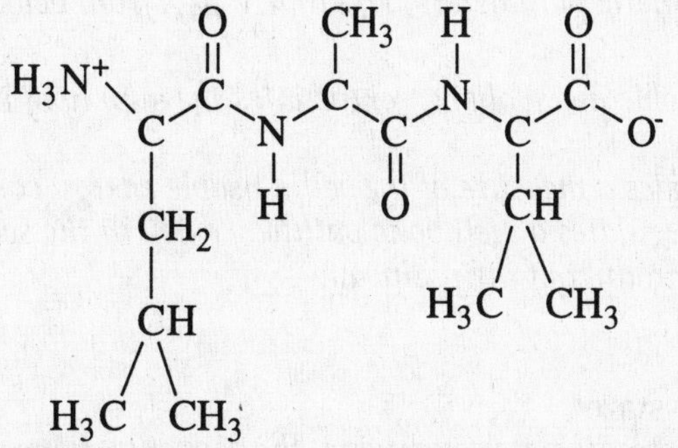

```
             O        CH3   H        O
             ||        |    |        ||
  H3N+      C        C     N        C
       \   / \      / \   / \      / \
        C     N    C     C     C      O-
        |     |    ||          |
       CH2    H    O          CH
        |                    /  \
        CH                 H3C   CH3
       /  \
     H3C   CH3
```

This is just one example of a tripeptide composed of three hydrophobic amino acids. This one, H3N–Leu–Ala–Val–COOH, would partition itself into the membrane, assuming the polar parts of the peptide were shielded from the lipid core. Any tripeptide with alanine, valine, leucine, isoleucine, tryptophan, phenylalanine, and methionine might partition itself into the membrane.

6 a. *Hydrogen bonds have bond energies in the range of 2 to 5 kcal/mole. Let's choose an average value of 3.5 kcal/mole. A mole of G–C pairs would have 3.5 x 3 = 10.5 kcal, and a mole of A–T pairs would have 3.5 x 2 = 7 kcal.*

b. *The G–C rich strand would require more energy to denature, and therefore would be more stable.*

7. *Half of the amino acids comprising one complete turn of the helix must be hydrophobic, and half hydrophilic. Thus 1.8 amino acids of each 3.6 must be hydrophobic. But amino acids don't occur in fractions, so the actual protein would probably have about two hydrophobic acids, followed by about two hydrophilic acids, etc., with a few deviations from that pattern. An example of a decapeptide might look like this:*

Ser–Asp–**Val–Leu**–Gln–Lys–**Tyr–Phe**–Asn–**Phe**

with the hydrophobic acids in bold. Many other sequences are possible.

Chapter Three

Short Answer

1. Significance
   a. standard free energy difference, $\Delta G°'$: *provides a way of comparing the free energy changes of different reactions that do not naturally occur under the same conditions.*
   b. activation energy: *acts as a barrier preventing thermodynamically unstable reactants from reaching equilibrium. Makes possible kinetic stability, even in the face of thermodynamic instability.*
   c. kinetic stability: *enables substances to endure, although they may not be at their lowest energy level. Kinetic stability is made possible by the activation energy barrier.*
   d. structural homology between a substrate and an enzyme inhibitor: *characteristic of competitive inhibitors that bind to the active sites of enzymes.*
   e. transfer potential: *predicts which species will be the donor of an electron, atom or functional group (e.g.. phosphate) and which will be the acceptor.*
   f. enzyme specificity: *ensures that only the appropriate substrate will bind to an enzyme, and only an appropriate metabolic product will be made.*
   g. reciprocal plots of enzyme velocity vs. substrate concentration: *allows for more accurate estimates of important kinetic parameters,* K_M *and* V_{max}, *from velocity vs. substrate concentration data.*
   h. fermentation: *in anaerobic metabolism, rejuvenates the supply of* NAD^+ *for the continuation of glycolysis.*
   i. reducing power: *provides a measure of the cell's usable energy content.*
   j. feedback inhibition: *regulates a metabolic pathway early in the sequence of steps and according to levels of the ultimate product of the pathway.*

2. Compare and Contrast:
   a. equilibrium vs. steady state
   In both equilibrium and steady state conditions, the concentrations of reactants and products remain stable over time. The difference is that at equilibrium, stable concentrations represent identical levels of free energy ($\Delta G = 0$) and equal reaction rates in both the forward and reverse directions. In a steady state, there are differences in free energy levels ($\Delta G \neq 0$), forward and reverse reaction rates are not equal, but concentrations remain the same because reactants are constantly being added and products are constantly being removed. No energy is required to maintain equilibrium, whereas aconstant input of energy is required to maintain a steady state.
   b. kinetic stability vs. thermodynamic stability
   Substances that are kinetically and thermodynamically stable appear to remain unchanged over time. Thermodynamically stable substances areat equilibrium and, indeed, not changing. Kinetically stable substances are not at equilibrium and hence are changing, but the rate of change is so slow that it is imperceptible. The activation energy barrier enables thermodynamically unstable substances to remain kinetically stable.
   c. theoretically reversible reactions vs. essentially irreversible reactions
   All reactions are theoretically reversible; that is, they can proceed in either a forward or reverse direction according to the prevailing conditions under which they occur. Some reactions, however, are essentially irreversible; they only proceed in one direction because the conditions required to reverse them cannot or do not occur under natural conditions.
   d. NAD^+/NADH vs. $NADP^+$/NADPH
   These nicotinamide adenine dinucleotide coenzymes both function as electron "shuttles" in cells, alternately accepting electrons from or donating electrons to metabolic intermediates. The NAD^+*/NADH redox pair functions primarily in catabolic pathways that yield ATP. The* $NADP^+$*/NADPH redox pair functions mainly to donate electrons to anabolic pathways.*

e. competitive vs. noncompetitive inhibition

Both types of enzyme inhibition result in lowered rates of enzyme activity at subsaturating levels of substrate. Competitive inhibitors are substrate analogs that bind to the active site of the enzyme, thus competing with the real substrate and interfering with substrate binding (K_M). The effects of a competitive inhibitor can be overcome by adding high levels of substrate and raising the substrate/inhibitor ratio. Thus the V_{max} of an enzyme in the presence of a competitive inhibitor is unchanged. A noncompetitive inhibitor binds to a site other than the active site. It probably looks nothing like the substrate, and although it may have little effect on substrate binding (and hence K_M), the effect of the inhibitor cannot be overcome by adding saturating levels of substrate or raising the substrate/inhibitor ratio. Thus the V_{max} of the enzyme is lowered.

Multiple Choice

1. b	2. a. **I**	2. b. **I**	2. c. **D**
2. d. **D**	2. e. **I**	3. c	4. a
5. d	6. b	7. a	8. a
9. d	10. a	11. d	12. d
13. b	14. a	15. b	16. a
17. c	18. b	19. d	20. c

Problems and Essays

1. *For the reaction to proceed as written, ΔG′ must be a negative number. Solve for the product/reactant ratio by setting ΔG′ < 0 as follows:*

$$\Delta G' < 0$$

$$\Delta G^{o\prime} + 2.303\, RT \log \frac{[glyceraldehyde\ 3-phosphate]}{[dihydroxyacetone\ phosphate]} < 0$$

$$1800\ cal/mole + (2.303)(1.987\ cal/mol \bullet {}^{o}K)(298^{o}K) \log \frac{[glyceraldehyde\ 3-phosphate]}{[dihydroxyacetone\ phosphate]} < 0$$

$$1364 \log \frac{[glyceraldehyde\ 3-phosphate]}{[dihydroxyacetone\ phosphate]} < -1800\ cal/mol$$

$$\log \frac{[glyceraldehyde\ 3-phosphate]}{[dihydroxyacetone\ phosphate]} < -\frac{1800}{1364}\ cal/mol$$

$$\log \frac{[glyceraldehyde\ 3-phosphate]}{[dihydroxyacetone\ phosphate]} < -1.32\ cal/mol$$

$$\frac{[glyceraldehyde\ 3-phosphate]}{[dihydroxyacetone\ phosphate]} < 0.048\ cal/mol$$

2. *Solve for ΔG°′ as follows:*

$$\begin{aligned} \Delta G^{\circ\prime} &= -(2.303)\, RT \log K'_{eq} \qquad where\ \ K'_{eq} = 0.2818 \\ &= -(2.303)(1.987\ cal/mol \cdot {}^{\circ}K)(298^{\circ}K) \log 0.2818 \\ &= -(1364\ cal/mol) \log 0.2818 \\ &= -(1364\ cal/mol)(-0.5501) \\ &= 750\ cal/mol \qquad or \qquad 0.750\ kcal/mol \end{aligned}$$

3. *Solving this problem requires the use of simultaneous equations. First, use the relationship between equilibrium and standard free energy to find the ratio of products to reactants at equilibrium. Next, you know that for every molecule of glucose 6-phosphate that is produce, one molecule of fructose 6-phosphate must be consumed. Therefore, the total concentration of both, at any stage of the reaction (equilibrium included), must be [glucose 6-phosphate] + [fructose 6-phosphate] = 0.50 M. Solve the problem using this information:*

$\Delta G°' = -400\ cal/mol$

$-400\ cal/mol = -2.303\ RT \log K'_{eq}$

$-400 = -(2.303)(1.987\ cal/mol{\cdot}°K)(298°K)\log K'_{eq}$

$\frac{-400}{-1364} = \log K'_{eq} \qquad 0.2933 = \log K'_{eq} \qquad K'_{eq} = 1.96$

also $K'_{eq} = \frac{[glucose\ 6-phosphate]}{[fructose\ 6-phosphate]}$ *at equilibrium*

and $[glucose\ 6-phosphate]+[fructose\ 6-phosphate] = 0.50M$

so $[glucose\ 6-phosphate] = 0.50M - [fructose\ 6-phosphate]$

thus $1.964 = \frac{0.50-[fructose\ 6-phosphate]}{[fructose\ 6-phosphate]}$

$1.96\,[fructose\ 6-phosphate] = 0.50M - [fructose\ 6-phosphate]$

$2.96\,[fructose\ 6-phosphate] = 0.50M$

$[fructose\ 6-phosphate] = \frac{0.50M}{2.96}$

$[fructose\ 6-phosphate] = 0.17\,M$

$[glucose\ 6-phosphate] = 0.50 - 0.17$

$[glucose\ 6-phosphate] = 0.33\,M$

4. a. *Using the same reasoning as in number (3) above, solve for [fumarate] as follows:*

$\Delta G'° = 0 = -2.303\ RT\log K'_{eq}$

$0 = \log K'_{eq}$

$K'_{eq} = 1.0$

thus $\frac{[fumarate][FADH_2]}{[succinate][FAD]} = 1.0$ *at equilibrium*

therefore $[fumarate] = 0.005M$

b. *It will not change the answer at all. Enzymes have no effect on the equilibrium concentrations of products and reactants.*

5.

$$\frac{-686.0\ kcal/mole\ glucose}{7.3\ kcal/mole\ ATP} \times 0.4 = 37.6\ mole\ ATP/mole\ glucose$$

6.

$$\frac{-2385\ kcal\ /\ mole\ palmitic\ acid}{7.3\ kcal\ /\ mole\ ATP} \times 0.4 = 131\ mole\ ATP\ /\ mole\ palmitic\ acid$$

7. *Fat has over three times more calories per mole than sugar. To gain weight, it is better to eat fats. To lose weight, it is best to stay away from fats.*

8. a. ***B*** b. ***C*** c. ***N*** d. ***C*** e. ***N*** f. ***C***

9. a. *Lineweaver-Burk Plot of Kinetic Data:*

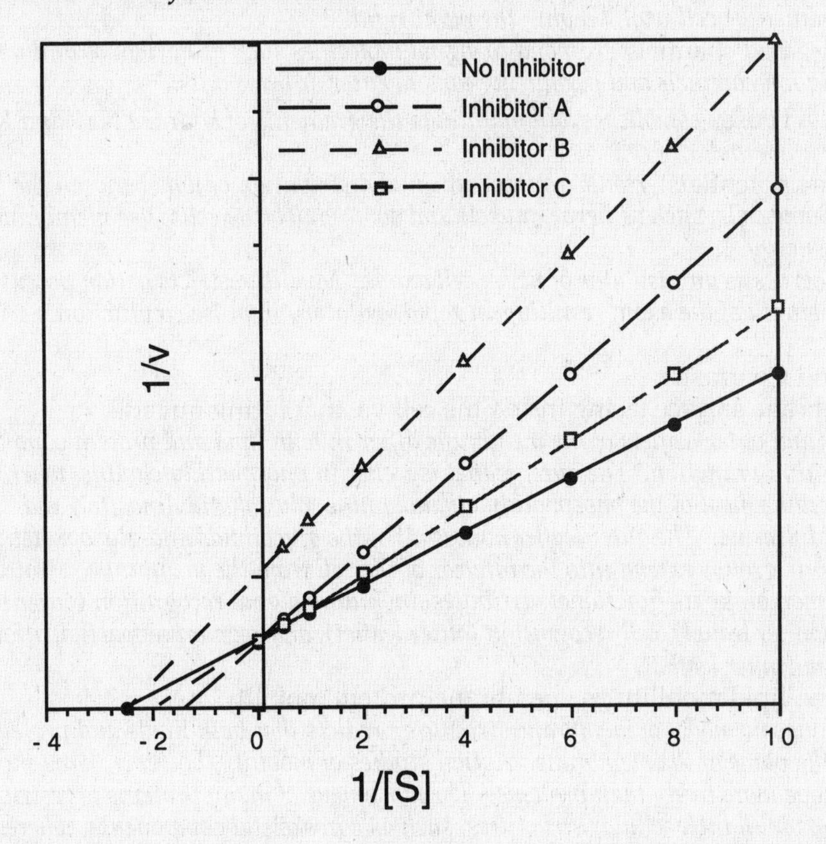

b. and c.

Kinetic Parameters	No Inhibitor	Inhibitor A	Inhibitor B	Inhibitor C
K_M	*0.40*	*0.68*	*0.40*	*0.50*
V_{max}	*2.00*	*2.00*	*1.00*	*2.00*
type of inhibitor	*none*	*competitive*	*noncompetitive*	*competitive*
why?		*V_{max} unchanged*	*K_M unchanged*	*V_{max} unchanged*

Chapter Four

Short Answer

1. Significance

   a. membrane leaflet asymmetry: *results in different functional properties of the inside and outside surfaces of the membrane. For example, the outside surface can recognize signal molecules via specific receptors. The inside surface may have enzymes, or proteins involved in anchoring cytoskeletal elements.*

b. partition coefficient: *a measure of the solubility of a substance in both polar and nonpolar solvents. Substances with high partition coefficients readily enter and leave cells by simple diffusion.*
c. flippase enzymes: *provide a mechanism for the exchange of phospholipids between the inner and outer leaflet of the cell membrane.*
d. freeze-fracture and freeze-etching: *techniques that show the arrangement of membrane lipids and proteins, including microdomains.*
e. detergents: *amphipathic substances that permit otherwise insoluble proteins, such as integral membrane proteins, to be solubilized in aqueous solutions, without denaturing.*
f. turgor pressure in plant cells: *provides support for nonwoody plants. Osmotic water pressure pushes against the cell wall, keeping the plant rigid.*
g. chemical-gated channels: *respond to signal molecules such as hormones and neurotransmitters by opening ion channels and permitting ions to enter or leave cells.*
h. Na^+-K^+, ATPase: *establishes and maintains the gradients of both the* Na^+ *and* K^+ *ions between the inside and outside of cells.*
i. membrane potentials: *result from the unequal distribution of ions between the inside and outside of cells. Some cells, such as nerve, muscle and some endocrine cells, use membrane potentials as a source of energy.*
j. myelin: *serves as an insulator of nerve cell axons. Myelin restricts action potentials to nodes along the length of the axon, resulting in rapid, saltatory impulse conduction.*

2. Compare and Contrast:
a. the membrane surface facing inside the cell vs. that facing outside
The inner and outer leaflets of the membrane differ in both lipid and protein composition, and consequently in function. The outer leaflet is richer in phosphatidylcholine, whereas the inner leaflet contains most of the phosphatidylethanolamine, phosphatidylinositol, and phosphatidylserine. The outer leaflet houses all of the glycolipids and glycoproteins, whereas no carbohydrate groups extend into the interior of the cell from the membrane. Protein asymmetry gives the membrane its functional attributes, including signal recognition (outer leaflet) and response (inner leaflet), cell recognition (outer leaflet), and membrane participation in cellular metabolism (inner leaflet).
b. membrane-lipid mobility vs. membrane-protein mobility
The fluid-mosaic model of membrane structure predicts that both lipids and proteins will move freely in the plane of the membrane. Actual studies of mobility, however, have shown that lipids tend to move more freely than proteins. The movement of many proteins is restricted by connections with intracellular structures, such as cytoskeletal components, whereas the movement of most lipids has no such restrictions.
c. diffusion vs. facilitated diffusion
Both these terms are used to describe the movement of substances across cell membranes down a concentration gradient– a process that does not require an input of energy. Diffusion implies movement through the bilayer without the help of a carrier, whereas facilitated diffusion requires a carrier molecule (usually a protein) that aids the substance in crossing the membrane barrier.
d. membrane potential vs. resting potential
These terms describe the voltage that can be measured between the inside and outside of a cell. Membrane potential is the voltage that occurs across the membrane of a nonexcitable cell. The term resting potential applies to excitable cells, and implies that the cell can undergo changes in that potential related to the work of the cell; e.g., the action potential of nerve and muscle cells.
e. myelinated nerve cells vs. large-diameter nerve cells
Myelination and large diameters in nerve cells are different evolutionary strategies that increase the rate at which nerve cells carry impulses. Invertebrates have adopted the large-diameter strategy, but there are limits to the degree to which increasing diameter can speed up impulses. Vertebrates have evolved a myelin sheath that surrounds nerve cells– an adaptation that increases impulse conduction velocity by as much as twenty-fold.

Multiple Choice

1. c	2. b	3. d	4. b
5. a	6. b	7. a	8. a
9. c	10. b	11. a	12. b
13. a	14. d	15. a	16. a
17. d	18. a	19. b	20. a

Problems and Essays

1. *(a) The regulation of metabolic pathways described in Chapter 2 relies partly on the maintenance of intracellular areas of high substrate and cofactor concentrations. Compartmentalization makes that possible by preventing the substrates and cofactors of specific metabolic pathways from being diluted by mixing with the entire cell's cytoplasm. (b) Resting potentials that are used by excitable cells to effect action potentials depend upon regions of highly regulated ion concentrations. Cellular compartmentalization makes that possible. (c) Throughout this book you will learn about cellular processes made possible by intracellular compartmentalization. Aerobic metabolism, described in Chapter 7, and photosynthesis characteristic of green plants, described in Chapter 8, are two such processes.*

2. *Identify a tissue that would serve as a good source of the enzyme of interest, e.g., the plasma membranes of nerve tissue. Solubilize the Na^+-K^+ ATPase from its native membrane using detergents. Reconstitute the enzyme into liposomes of simple and well-defined composition, then assay the enzyme for activity. By synthesizing a variety of liposomes with different amounts of saturated and unsaturated fatty acids, the fluidity of the liposome can be altered, and the enzyme can be tested under conditions of different fluidity. (To measure the fluidity of the liposomes directly, without having to infer it from the fatty acid composition, either FRAP or SPT techniques can be applied.)*

3. a. ***C*** b. ***C*** c. ***N*** d. ***C*** e. ***N***

4. a. ***P*** b. ***I*** c. ***LA*** d. ***I*** e. ***I*** f. ***I***

5. *The equation that describes the energy required to move a substance into a cell is:*

$$\Delta G = RT \ln \frac{[C_i]}{[C_o]}$$

but in this case we are interested in moving a substance out of a cell, against a concentration gradient. Therefore we must invert the concentration ratio:

$$\Delta G = RT \ln \frac{[C_o]}{[C_i]}$$

Now substitute for the concentration terms:

$$\Delta G = 2.303 RT \log \frac{3.00}{0.04}$$

$\Delta G = 2.62$ *kcal/mole*

6. *The equilibrium distribution of molecules across a membrane results from the forces of both concentration gradients and electrical gradients (measured as a membrane potential). For uncharged molecules, electrical gradients have no effect. For ions, however, membrane potentials may act in either the same direction as concentration gradients, or the opposite direction. If an ion is distributed across a cell membrane such that the concentration gradient and the electrical gradient are equal in magnitude but opposite in direction, then the equilibrium distribution of the ion may be represented by different concentrations across the membrane.*

7. *First calculate the resting potential of the cell under three different conditions, where each of the three ions is assumed to be at equilibrium:*

ion:	Ψ^+	Ξ^+	Θ^-
inside cells	*180*	*6*	*34*
outside cells	*21*	*244*	*51*

$$E_\Psi = 2.303\frac{RT}{zF}\log_{10}\frac{[\Psi_o]}{[\Psi_i]} \qquad E_\Xi = 2.303\frac{RT}{zF}\log_{10}\frac{[\Xi_o]}{[\Xi_i]} \qquad E_\Theta = 2.303\frac{RT}{zF}\log_{10}\frac{[\Theta_i]}{[\Theta_o]}$$

$$E_\Psi = 2.303\frac{RT}{zF}\log_{10}\frac{[21]}{[180]} \qquad E_\Xi = 2.303\frac{RT}{zF}\log_{10}\frac{[244]}{[6]} \qquad E_\Theta = 2.303\frac{RT}{zF}\log_{10}\frac{[34]}{[51]}$$

$$E_\Psi = -55mV \qquad E_\Xi = 95mV \qquad E_\Theta = -10mV$$

Now compare these calculated values with the actual resting potential of +100 mV. The ion Ξ^+ is the one that is closest to its equilibrium potential, and hence probably the one to which the membrane is most permeable. Thus the answers to both (a) and (b) are Ξ^+. (c) The ion Ψ^+ is furthest from equilibrium and, if it were allowed to cross the membrane, would result in the greatest change in membrane potential. Thus the Loligotian action potential probably results from opening voltage-gated Ψ^+ channels.

Chapter Five

Short Answer

1. Significance
   a. evolutionary success of the cyanobacteria: *responsible for the accumulation of oxygen in the atmosphere, and the eventual success of organisms that could utilize oxygen as a final electron acceptor in metabolism.*
   b. cristae: *increase the inner mitochondrial membrane surface area, and thus the number of oxidative phosphorylation sites.*
   c. porins: *create pathways for all but the largest cytoplasmic components to enter the outer mitochondrial membrane.*
   d. spatial arrangement of carriers in the electron-transport chain: *expedites the transfer of electrons between components of the chain by minimizing the distance they must travel within the inner mitochondrial membrane.*
   e. only 20% of Δp is ΔpH: *permits the accumulation of a large electrochemical gradient without lowering the cytoplasmic pH to a level of acidity that would be detrimental to cytoplasmic proteins.*
   f. inner mitochondrial membrane is impermeable to H^+: *prevents the proton gradient established by electron transport from dissipating by diffusion back into the mitochondrial matrix.*
   g. levels of ADP in the mitochondria: *regulate the rate of oxidative metabolism.*
   h. fermentation: *regenerates NAD^+ in the absence of oxygen so that glycolysis can continue.*
   i. catalase in the peroxisomes: *rapidly breaks down potentially toxic H_2O_2 that results from many of the oxidase enzymes residing in these organelles.*
   j. Na^+,K^+ ATPase from red blood cells can be made to synthesize ATP: *This observation, made under highly contrived conditions, provided evidence that an enzyme that catalyzes a reaction with a highly negative standard free-energy change can be made to catalyze the reverse reaction under the right cellular conditions. More specifically, an ATPase can be made to synthesize ATP from ADP and unbound P_i.*

2. Compare and Contrast:
   a. inner vs. outer mitochondrial membranes
   Both of these membranes surround the mitochondrion, but they differ in nearly every aspect of their structure, function, and origin. The outer membrane is highly permeable to even substances of large molecular weight (due to the presence of porins); the inner membrane is highly impermeable to most cellular molecules. The lipid composition of the outer membrane resembles that of the plasma membrane, whereas the inner membrane lipids are more characteristic of bacterial plasma membranes from which they are believed to have evolved. The outer membrane is involved in fatty acid, tryptophan, and epinephrine metabolism (among other things), and in recognizing proteins to be imported into the mitochondria. The inner membrane is involved in electron transport and the synthesis of ATP.
   b. "high energy" vs. "low energy" electrons
   All electrons are affiliated with some atom or molecule. A "high energy" electron is affiliated with a species that is a good reducing agent– one that readily transfers the electron to another compound or atom with the concurrent release of free energy. A low energy electron is affiliated with a compound or atom that is has a low electron-transfer potential, and hence cannot readily be tapped as a source of free energy by a redox reaction with another species.
   c. substrate-level vs. oxidative phosphorylation
   Both of these terms describe the synthesis of ATP from ADP and a phosphate group. In substrate-level phosphorylation, the phosphate group is covalently bound to a donor, and that bond is broken when it is transferred to ADP. In oxidative phosphorylation, there is no phosphate donor. An ATP synthase catalyzes the synthesis of ATP from ADP and free inorganic phosphate using energy derived from a proton gradient . The term "oxidative" refers to the source of energy for making the proton gradient, i.e., the reduction of oxygen by organic reductants in the electron-transport chain.
   d. chemical vs. electrical gradient
   A chemical gradient is based upon a concentration difference between compartments, and the tendency for diffusion to lessen the difference. An electrical gradient occurs where there is a separation of charge between compartments. Like charges repel each other. Protons distributed across the inner mitochondrial membrane are subject to both types of gradients
   e. redox loop vs. proton pump
   These are two possible mechanisms that have been proposed for the translocation of protons across the inner mitochondrial membrane during electron transport. A redox loop occurs when an electron carrier picks up both protons and electrons, passes the electron on to a subsequent carrier, and dumps the proton into the intermembrane space. Redox loops require an alternation of proton-transporting and non-proton-transporting electron carriers. A proton pump moves protons across the membrane using the energy of electron transfer, and may undergo conformational changes to accomplish the task.

Multiple Choice

1. d	2. b	3. c	4. a
5. d	6. a	7. b	8. a
9. c	10. a	11. a	12. a
13. c	14. d	15. c	16. b
17. d	18. b	19. a	20. b

Problems and Essays

1. *(a) Size. The size of the typical mitochondrion is comparable to that of a typical bacterium.*
   (b) Membrane-lipid composition. The inner mitochondrial membrane is rich in cardiolipin and devoid of cholesterol– features of many bacterial plasma membranes.
   (c) The genome, and the transcriptional and translational machinery.
   These features suggest that mitochondria were once free-living. In addition, the circular DNA and the ribosomes resemble prokaryotic features more so than eukaryotic ones.

2. a. acetyl CoA + $2H_2O$ + $3NAD^+$ + FAD + ADP + P_i $\rightleftharpoons$

$2CO_2$ + 3NADH + $3H^+$ + $FADH_2$ + CoASH + ATP

b. *Because the TCA cycle is, indeed, a cycle, for each acetyl CoA that consumes an oxaloacetate, there is one oxaloacetate that is produced from malate.*

3. a. ubiquinone$_{red}$ + 2 cytochrome c_{ox} $\rightleftharpoons$ ubiquinone$_{ox}$ + 2 cytochrome c_{red}

b. $\Delta G^{\circ\prime} = -nF\,\Delta E'_0$

$= -(2)\,(23.063\ kcal/V{\bullet}mol^{-1})\,(0.254\ V - 0.100\ V)$

$= -7.103\ kcal/mol$

c. *No. The 7.1 kilocalories released in the redox reaction are just shy of the 7.3 kilocalories required to synthesize one mole of ATP under standard conditions.*

4. a. *Under standard conditions, ubiquinone would act as the oxidant and cytochrome* b *would act as the reductant, donating electrons to ubiquinone. Ubiquinone has a higher redox potential than cytochrome* b.

b. *Yes, the roles of oxidant and reductant could be reversed by changing the relative concentrations of the oxidized and reduced forms of one or both species. Increasing the concentrations of reduced ubiquinone and/or oxidized cytochrome* b *is one strategy. Decreasing the concentrations of oxidized ubiquinone or reduced cytochrome* b *is another.*

5. a. $NADH + H^+ + 1/2\ O_2 \rightleftharpoons NAD^+ + H_2O$

$\Delta G^{\circ\prime} = -nF\,\Delta E_0'$

$= -(2)\,(23.063)\,(0.82 - (-0.32))$

$= 52.6\ kcal/mol$

b. *On the cellular scale, 52.6 kcal for each mole of NADH is a very large package of energy, about equal to the energy in an entire Oreo cookie. In a single mitochondrion, releasing that amount of energy would be akin to exploding a small stick of dynamite. By releasing the energy in smaller packets, using the carriers of the electron-transport chain, the cell can capture the free energy without suffering damage.*

6. $\Delta p = \Psi - 2.3\frac{RT}{F}\Delta pH$

$230\ mV = 0.8(230mV) - 2.3\frac{RT}{F}\Delta pH$

$46 mV = -59\Delta pH$

$-0.78 = \Delta pH$

where $\Psi = 0.8\Delta p$

and $\Delta p = 230\ mV$

and $2.3\frac{RT}{F} = 59\ mV$

7. a. When inside out submitochondrial particles are engaged in electron transport, the interior of the vesicles *becomes acidic (accumulates protons)*.

b. When electron transport is blocked in intact mitochondria, and an artificial pH gradient is imposed by lowering the pH of the medium, the synthesis of ATP *continues*.

c. FCCP is a drug that partitions itself into membranes and opens a channel specific for the passage of protons. When FCCP is added to a suspension of mitochondria, electron transport___*continues*___, and ATP synthes___*stops*___.

d. Under anaerobic conditions, the pH gradient___*decreases*___relative to aerobic conditions.

Chapter Six

Short Answer

1. Significance
   a. the switch from H_2S to H_2O as a source of electrons for photosynthesis: *required that the photosynthetic process generate a very strong oxidizing agent using the energy of light. Photosynthetic organisms with this ability could exploit many more habitats. Their evolutionary success raised the level of O_2 in the atmosphere, paving the way for the evolution of organisms with aerobic capacity.*
   b. antenna: *increase the variety of wavelengths that can be exploited for photosynthesis, and also the efficiency of photosynthesis, by increasing the number of electrons that can be transferred by the reaction center pigment per unit of time.*
   c. fluidity of the thylakoid membranes: *high fluidity increases the rate of diffusion of the mobile electron carriers plastoquinone and plastocyanin between components of the photosynthetic apparatus.*
   d. carotenoids: *most importantly, protect the photosynthetic apparatus from oxidation by reactive species of oxygen; they also absorb the blue and green wavelengths of light for photosynthesis.*
   e. two photosystems, PSI and PSII: *essential for boosting the energy of electrons from water to a level sufficient to reduce $NADP^+$. Neither photosystem can alone provide sufficient energy to accomplish this highly endergonic task.*
   f. alternate routes that electrons can take from ferredoxin (other than to NADPH): *provide a mechanism for photoautotrophs to reduce highly oxidized species, such as nitrate and sulfate, converting inorganic wastes into reduced molecules that can be incorporated into cellular macromolecules.*
   g. thylakoid membranes are impermeable to protons: *prevents the proton gradient established by photosynthesis from dissipating by diffusion back into the lumen of the thylakoid.*
   h. Rubisco binds to RuBP, not CO_2 or O_2: *leaves the substrate RuBP open to attack by either CO_2 or O_2, only one of which, CO_2, results in productive carbon fixation. This lowers the efficiency of photosynthesis. Most enzymes exhibit high specificity for the most appropriate substrates. The absence of a binding site for CO_2 prevents the enzyme from expressing this level of specificity.*
   i. peroxisomes are closely associated with chloroplasts: *reflects the interdependence of these organelles in the metabolism of glycolate.*
   j. C_4 and CAM pathways: *provide certain plants a means to insulate their Rubisco from O_2, minimizing photorespiration and increasing the carbon-fixing efficiency of the plants.*

2. Compare and Contrast:
   a. chemoautotrophs vs. photoautotrophs
   Both types of organisms can reduce CO_2 and incorporate the carbon into organic molecules. Chemoautotrophs use the oxidation potential of simple inorganic molecules such as ammonia, hydrogen sulfide or nitrites as a source of energy, whereas photoautotrophs use the radiant energy of sunlight to accomplish this task.
   b. grana thylakoids vs. stroma thylakoids
   Together these form the internal membrane system of the chloroplast. Grana thylakoids are stacked pouches, whereas stromal thylakoids are flattened pouches that connect the grana. Many of the protein complexes involved in photosynthesis, including the cytochrome b_6f complex and PSII, are

found in both types of membranes. Stromal thylakoids and the nonappressed membranes of the grana thylakoids are rich in PS I, ATP synthase molecules, and the loosely affiliated Rubisco molecules. There are differences in lipid composition as well. The stromal membranes and the nonappressed grana membranes have about equimolar amounts of MGDG and DGDG, whereas the ratio of MGDG to DGDG in appressed membranes is about three.

c. light-dependent vs. light-independent reactions
Both of these pathways are necessary steps in the de novo synthesis of organic compounds from CO_2. Light-dependent reactions include photolysis, reduction of $NADP^+$, and generation of a proton gradient that is used in the production of ATP. Light-independent reactions include the Calvin cycle– the pathway in which the energy of ATP and NADPH is used to fix carbon into organic compounds.

d. cyclic vs. noncyclic photophosphorylation
In noncyclic photophosphorylation electrons move from H_2O to $NADP^+$, generating a proton gradient that is used to synthesize ATP. The process requires two photosystems, PSII and PSI, for a double boost of light energy. In cyclic photophosphorylation, electrons removed from chlorophyll are replaced not by electrons derived from water, but by the same electrons that were removed– electrons move in a cycle from chlorophyll back to chlorophyll. Only one photosystem, PSI, is required. The energy that boosts the electrons from chlorophyll at the start is used to generate a proton gradient (just as in the noncyclic pathway) which is again used to make ATP.

e. mesophyll cells vs. bundle sheath cells
These cell types are characteristic of C_4 plants that spatially separate the processes of ATP synthesis and $NADP^+$ reduction from sugar production. The cell types are arranged in concentric cylinders, with mesophyll cells in the outer cylinder and bundle sheath cells forming the inner cylinder. ATP and NADPH are produced in the mesophyll cells, where CO_2 is fixed into the four-carbon compound, malate. Malate is transported into the bundle sheath cells where the CO_2 is released from malate, and serves as a substrate for Rubisco and eventual fixation into sugar. The inner cell layer is shielded from atmospheric O_2, and photorespiration is minimized.

Multiple Choice

1. c	2. b	3. a	4. d
5. a	6. b	7. c	8. a
9. a	10. d	11. a	12. a
13. c	14. c	15. d	16. b
17. b	18. a	19. c	20. b

Problems and Essays

1. a. *For H_2O:*

$$\Delta G^{o}{}' = -nF\Delta E'_{o}$$
$$= -(2)(23.063\ kcal / V \cdot mol^{-1})(-0.320 - 0.816V)$$
$$= 52.4\ kcal / mole$$

For H_2S:

$$\Delta G^{o}{}' = -nF\Delta E'_{o}$$
$$= -(2)(23.063 kcal / V \cdot mol^{-1})(-0.320 - (-0.25)V)$$
$$= 3.23\ kcal / mole$$

b. *At 680 nm:*

$$E = \frac{(1.58\times10^{-34}\ cal\cdot sec)(3.0\times10^{8}\ m/sec)}{680\ nm}\cdot(10^{9} nm/m)$$

$$= 6.97\times10^{-20}\ cal/photon\cdot(6.02\times10^{23}\ photons/mol)$$

$$= 4.20\times10^{4}\ cal/mole\ or\ 42.0\ kcal/mole$$

At 485 nm:

$$E = \frac{(1.58\times10^{-34}\ cal\cdot sec)(3.0\times10^{8}\ m/sec)}{485\ nm}\cdot(10^{9} nm/m)$$

$$= 9.77\times10^{-20}\ cal/photon\cdot(6.02\times10^{23}\ photons/mol)$$

$$= 5.88\times10^{4}\ cal/mole\ or\ 58.8\ kcal/mole$$

c. *At 680 nm, the 42 kcal is below the 54.2 kcal required for $NADP^+$ oxidize one mole H_2O. However the energy at 485 nm, 58.8 kcal, is sufficient for $NADP^+$ to oxidize H_2O.*

d.

$$\frac{52.4\ kcal/mole\ NADP^{+}\ reduced}{42.0\ kcal/mole\ photons} = 1.25 mole\ photons/mole\ NADP$$

2. *For each six-carbon sugar, 18 ATPs are consumed, and 12 NADPHs are oxidized. Thus the ratio is three ATP to two NADPH. Plants rely on cyclic photophosphorylation for the generation of a proton gradient to synthesize ATP. $NADP^+$ is not reduced to NADPH during cyclic photophosphorylation.*

3. ___*RL*___ photolysis

   ___*PT*___ QH_2

   ___*PT*___ cytochrome b_6f complex

   ___*SC*___ ferredoxin-$NADP^+$ reductase

4. a. *C_4 and CAM plants;* b. *C_3 plants;* c. *CAM plants;* d. *C_4 plants*

5. a. *PSI: carotenes* → *Chl* b → *Chl* a *660* → *Chl* a *670* → *Chl* a *678* → *Chl* a *685* → *Chl* a *690* → *Chl* a *700: Chlorophyll* a *700 is acting as the reaction center in this photosynthetic unit.*
PSII: xanthophylls → *Chl* b → *Chl* a *660* → *Chl* a *670* → *Chl* a *678* → *Chl* a *680* → *Chl* a *685: Chlorophyll* a *685 is acting as the reaction center in this photosynthetic unit.*

   b. *Light of 700 nm wavelength is absorbed by the reaction-center chlorophyll of PSI, but by itself, PSI cannot split water, and 700 nm light does not have enough energy to activate PSII. Without a source of electrons, noncyclic photosynthesis ceases. Adding a beam at 680 activates PSII, which can then split water and send electrons to PSI. Noncyclic photosynthesis occurs.*

6. *(1) The thylakoid membranes serve as a scaffolding for the photosystems I and II, and the electron carriers. The membranes keep the components of the photosystems in the proper orientation such that the energy of light is funneled into the photoreaction center, and the electrons are passed from donor to acceptor in the proper sequence. (2) The thylakoid membranes separate protons such that a gradient can be established and exploited in the synthesis of ATP. The chloroplast ATP synthase is held in place with the proton channel CF_o open to the proton-rich lumen and the catalytic CF_1 oriented in the stroma for ATP production. (3) The inner and to outer chloroplast membranes separate the reactions*

of carbon fixation from the catalytic reactions of the cytoplasm. These two pathways share many intermediates, and intermixing of the two would result in breakdown of the intermediates before they could be incorporated into sugars.

Chapter Seven

Short Answer

1. Significance
   a. glycocalyx: *mediates cell-to-cell interactions, provides mechanical protection to the cell, and serves as a barrier to particles reaching the plasma membrane.*
   b. basement membrane: *may maintain the polarity of the cell, determine the path of cellular migration, separate adjacent tissues, and act as a barrier to the passage of macromolecules.*
   c. integrins: *attach cells to elements of their microenvironment, including basement membranes and artificial substrates (in cultured cells), and, in some cases to other cells (via binding to IgSF proteins on adjacent cells).*
   d. RGD sequence: *is recognized as a ligand by many integrins. This sequence occurs on several important ECM proteins, including fibronectin, laminin and collagen, as well as on other extracellular proteins.*
   e. hemidesmosomes: *anchor cells to basement membranes. These cell-to-cell connections are especially important in epithelial tissues, such as the epidermis.*
   f. cadherins: *serve as the major protein connecting like cells of a tissue to one another.*
   g. tight junctions: *prevent leakage between cells of a tissue. This is especially important in cells that line body cavities and must maintain different environments on either side of the lining.*
   h. gap junctions: *allow communication (GJIC) between adjacent cells. This is especially important in tissue cells that have to act as a coordinated unit, such as those of the heart.*
   i. plasmodesmata: *permit intercellular communication and the exhange of small-molecular-weight solutes between adjacent cells of plant tissues.*
   j. pectins: *form extensive hydrated gels that act as molecular sieves determining the size of solutes that can penetrate plant cell walls and reach the plasma membrane. Pectins can also fragment when the plant is attacked by pathogens, stimulating a defensive response. In addition, pectins cement the cell walls of adjacent plants together.*

2. Compare and Contrast:
   a. glycocalyx vs. extracellular matrix
   These are both extracellular structures of cellular origin. The glycocalyx is closely affiliated with the cell membrane; it comprises carbohydrate chains of both glycoproteins and glycolipids of the cell membrane. Some cell types have secreted extracellular materials as part of the glycocalyx. The extracellular matrix is a network of extracellular materials that lies beyond the immediate vicinity of the cell membrane. In some tissues, such as bone and cartilage, the ECM is the dominant feature.
   b. fibronectin vs. integrin
   These proteins both participate in cell adhesion, but one is an extracellular protein of the ECM (fibronectin), and the other is an integral membrane protein (integrin). Fibronectin has binding sites for other components of the ECM and for proteins of the cell surface, including integrin. Integrins are membrane-spanning heterodimers with binding sites for fibronectin, other RGD-containing proteins, and other proteins of the ECM. In some cells, fibronectin can act as an integrin-specific ligand, and integrin, as a fibronectin receptor.
   c. focal contacts vs. hemidesmosomes
   These two types of cellular adhesive structures occur in different settings. Focal contacts bind cells to artificial substrates in vitro, whereas hemidesmosomes bind cells to basement membranes in vivo. There are structural differences, notably that focal contacts are characterized by intracellular actin filaments and hemidesmosomes are characterized by intracellular keratin filaments.
   d. tight junctions vs. gap junctions
   These two types of cell-to-cell connections differ in both structure and function. Tight junctions form "belts" that surround the periphery of the cell and prevent leakage of solutes between cells of a

tissue. They are most prevalent in epithelial tissues defining a body compartment and maintaining compositional differences between compartments. Gap junctions connect adjacent cells and permit the diffusion of small-molecular-weight substance between cells. They are most prevalent in tissues that must act in synchrony, such as cardiac muscle and smooth muscle.

e. primary cell walls vs. secondary cell walls
These two types of plant cell walls characterize plant cells at different stages of maturity. Primary cell walls surround growing plant cells and allow some extensibility accommodating further growth. Secondary cell walls occur in more mature plant cells, and are more rigid. They contain more cellulose and more than primary walls. Secondary walls are superior support structures.

Multiple Choice

1. c	2. b	3. d	4. a
5. a	6. d	7. b	8. c
9. d	10. b	11. d	12. a
13. b	14. c	15 c	16. a
17. a	18. b	19. c	20. b

Problems and Essays

1. *One way to determine which ECM protein is acting as an attractant for growing axons is to coat the bottom of a petri dish with alternating strips of collagen and laminin. Introduce cultured embryonic nerve cells and watch for outgrowth from the cell bodies. This experiment was done by R. W. Gunderson (1987.* Devel. Biol. *121:423) and the axon outgrowth was restricted to the areas of the dish coated with laminin. Other experiments might be equally convincing.*

2.

__*EX*__ laminin

__*EX*__ collagen

__*MS*__ heparan sulfate proteoglycans (HSPG)

__*MS*__ connexin

__*EX*__ fibronectin

__*IN*__ keratins affiliated with hemidesmosomes

__*MS*__ integrin

__*MS*__ selectin

__*MS*__ NCAM

__*IN*__ catenins

__*IN*__ actin

3. *(1) The lower layer of the epidermis attaches to its basement membrane via hemidesmosomes that involve integrin-RGD binding. Artificial RGD peptides would compete for these binding sites on integrin. (2) Some cell-to-cell adhesions, particulary between integrins and NCAMs or VCAMs, involve integrin-RGD binding. Artificial peptides would compete for these binding sites. (3) Integrin binding onto epithelial cells of the venules plays a role in leukocyte extravasation during inflammation. This is another potential site of side effects from artificial RGD peptides.*

4.

Cells in suspension (columns) / Cells in monolayer (rows)

	1	2	3	4	5
1	+	-	+	+	-
2		+	+	+	-
3			+	+	-
4				+	-
5					+

5.

cell surface protein	expression increased or decreased in cancer cell	contribution to cell's malignant character
laminin receptors	*increased*	*Increases binding of metastatic cells to laminin in basement membranes of blood vessel walls; increases ability of the cell to invade tissues.*
fibronectin receptors	*decreased*	*Cells can escape from their tissue of origin by sliding over ECM, without sticking.*
integrins	*decreased*	*Because integrins act as fibronectin receptors, the contributions listed above apply here, also; i.e., cells escape from tissues by sliding over the ECM.*
cadherins	*decreased*	*Cells do not recognize or bind like types and can easily break away from their tissue of origin.*
gap junctions	*decreased*	*GJIC is reduced; cell growth is not tightly regulated.*

6. a. *Gap junctions facilitate intercellular communication enabling the tissue can act as a unit.*
   b. *Tight junctions prevent secreted enzymes from diffusing back into tissues of the pancreas.*
   c. *Desmosomes give this tissue its strength and ability to be stretched. Some gap junctions may coordinate contractions during labor.*
   d. *Tight junctions prevent reverse diffusion of salts that are pumped against a gradient.*
   e. *Gap junctions keep rings of contractile tissue coordinated so that contractions are coordinated and productive.*

Chapter Eight

Short Answer

1. Significance
   a. regulated secretion: *one way in which a cell responds to specific stimuli, either from the environment or from within the cell.*
   b. cell fractionation: *a technique used for in vitro study whereby subcellular fractions or structures are isolaated based on differences in physical properties.*
   c. signal sequence: *targets the nascent protein with this stretch of amino acids (the signal sequence) for a specific intracellular or secretory destination.*
   d. *trans* Golgi network: *serves as the site for sorting different vesicles (and their protein cargos) that arise from the Golgi.*
   e. Rothman and Orci experiments on VSV-infected CHO cells: *provided compelling evidence that newly synthesized proteins move from cis to* trans *Golgi cisternae via vesicles that bud from one compartment and fuse with the next.*

f. phosphorylated mannose residues on TGN glycoproteins: *target trans Golgi network (TNG) proteins for the lysosomes.*
g. lysosomes: act as cellular digestive organelles. *Lysosomes digest both exogenous particles that are taken into the cell by phagocytosis and cellular organelles that are damaged or aged.*
h. acid hydrolases: *the digestive machinery of lysosomes, these enzymes have low pH optima and catalyze the hydrolytic breakdown of all types of biologic macromolecules.*
i. receptor-mediated endocytosis (RME): *bulk-uptake mechanism in response to specific extracellular solutes or particles.*
j. coated pits: *result from receptor-mediated endocytosis and form the cell surface precursor to endocytic vesicles, with specific cargo being brought into the cell.*

2. Compare and Contrast:
   a. pulse vs. chase
   These two techniques are used together as a method of tracing the incorporation of precursors into the macromolecules of various cellular organelles. During the pulse phase, cells are given radiolabeled building-block molecules, such as amino acids. After a brief time period, the excess radiolabeled precursors are washed away, and those that have been incorporated into macromolecules are "chased" by measuring which cellular fractions or organelles have become radiolabeled.
   b. protein synthesis on membrane-bound vs. free ribosomes
   Proteins that are synthesized on ribosomes affiliated with the endoplasmic reticular membrane (rough ER) contain specific signal sequences that bind to signal recognition particles (SRPs) recognized by SRP receptors on the ER. These proteins are destined for secretion, specific membrane-bond organelles, or service as integral membrane proteins. Proteins lacking the signal sequence are synthesized on free ribosomes, and comprise a group destined for the cytosol, the nucleus, the mitochonria or chloroplasts.
   c. synthesis of N-linked vs. O-linked oligosaccharides
   The saccharide groups that are added to proteins synthesized on the rough ER can be connected in two ways: either via the nitrogen atom of an asparagine residue (N-linked), or via the oxygen atom of a serine or threonine residue (O-linked). The initial steps in the synthesis of N-linked glycoproteins occurs in the rough ER, where a core segment of the carbohydrate chain is transferred from a lipid carrier, dolichol phosphate, to the nascent peptide. The core sequence of sugars is modified and elaborated in the Golgi. O-linked carbohydrates are added to proteins entirely in the Golgi.
   d. autophagy vs. phagocytosis
   Both of these processes involve intracellular digestion. In phagocytosis, exogenous particles are taken into the cell in endocytic vesicles which fuse with endosomes and lysosomes where digestion occurs. In autophagy, old or damaged organelles are surrounded by membrane derived from the endoplasmic reticulum, and the resulting vesicle fuses with a lysosome initiating digestion.
   e. LDL vs. HDL
   These are particulate protein-lipid complexes that travel via the bloodstream, transporting cholesterol between organs. Low-density lipoproteins (LDLs) carry cholesterol from the liver (which gets it from the digestive tract, where it is absorbed from food) to the cells of the tissues. High-density lipoproteins carry excess cholesterol from the tissues back to the liver where it is secreted as part of the bile. Both are taken into cells via receptor-mediated endocytosis. High blood levels of LDL are correlated with atherosclerosis and heart disease. High blood levels of HDL are associated with a decreased risk of heart disease.

Multiple Choice

1. b	2. a	3. d	4. c
5. b	6. d	7. b	8. a
9. c	10. a	11. d	12. b
13. a	14. b	15. d	16. c
17. a	18. a	19. a	20. b

Problems and Essays

1. ___*RER*___ fibronectin

___*FR*___ cytoskeletal proteins such as actin

___*RER*___ acid hydrolases

___*RER*___ collagen

___*FR*___ triskelion

___*FR*___ the enzymes of glycolysis

___*RER*___ adaptin

___*RER*___ LDL receptor

___*RER*___ signal peptidase

___*RER*___ protein disulfide isomerase

___*FR*___ mitochondrial proteins coded for by nuclear DNA

2. *The bacterial cell is engulfed by the macrophage and radiolabel first appears in the cell in a large endocytic vesicle. The endocytic vesicle then fuses with an early endosome, near the cell periphery. Gradually, as the endosome moves toward the cell interior, the radiolabel has more internal presence. Lysosomes fuse with the late endosome, and the proteins of the bacterium are digested. Radiolabeled amino acids are exported from the late endosome and incorporated into cellular proteins of all types.*

3. a. *The proteolytic cleavage of proinsulin to insulin occurs during the maturation of secretory vesicles, after they leave the* trans *Golgi network.*

b.

	anti P	anti I	anti clathrin
mature secretory vesicles	-	+	-
trans Golgi network	+	-	+
cis and *medial* Golgi cisternae	+	-	-

4. *The C- and N-termini must be directed toward the lumen of the ER, the Golgi, and the secretory vesicles. The side of the membrane facing the cytoplasm will always face the cytoplasm.*

5.

Experimental Conditions	completely synthesized?	signal sequence?	translocated across a membrane?
a. mRNA codes for a cytosolic protein; no SRP present; no SRP receptor present:; no microsomes present	*yes*	*no*	*no*
b. mRNA codes for a secretory protein; no SRP present; no SRP receptor present; no microsomes present	*yes*	*yes*	*no*
c. mRNA codes for a secretory protein; exogenous SRP added; no SRP receptor present; no microsomes present	*no*	*no*	*no*
d. mRNA codes for a secretory protein; exogenous SRP added; free (not membrane-bound) SPR receptor added; no microsomes	*yes*	*yes*	*no*
e. mRNA codes for a secretory protein; exogenous SRP added; exogenous SRP receptor added; microsomes added	*yes*	*no*	*yes*

6. *Ferrotransferrin binds to the transferrin receptor on cell membranes, and a coated pit is formed. A clathrin-coated endocytic vesicle enters the cell, containing a membrane-bound transferrin receptor-ferrotransferrin complex. Clathrin dissociates from the vesicle, which then fuses with an early endosome. As lysosomes bind to the endosome, the pH of the vesicle decreases, and iron diffuses away from the ferrotransferrin. The ferrotransferrin, however, stays bound to the membrane-bound receptor. Iron is transported out of the late endosome and into the cytosol, where it is used in the synthesis of iron-affiliated proteins. The late endosome migrates to the cell surface and fuses with the plasma membrane, where the transferrin-transferrin receptor complex becomes part of the plasma membrane. Transferrin dissociates from the transferrin receptor at the neutral pH of the cell exterior and enters circulation, where it can participate in another round of iron uptake.*

Chapter Nine

Short Answer

1. Significance
   a. overlapping functions of cytoskeletal proteins: *provide redundancy that protects the cell from the consequences of a defective protein.*
   b. microtubule-organizing centers MTOC: *act as sites of origin for microtubule assembly.*
   c. the high metabolic cost of microtubule polymerization: *allows a cell to independently control the rates of assembly and disassembly. Tubulin dimers bound to GTP at the end of a microtubule favor polymerization, whereas those bound to GDP have less affinity for each other and are more likely to disassemble.*
   d. γ–tubulin: *may serve as the nucleating tubulin upon which each new microtubule is assembled.*
   e. intermediate filaments occur only in animal cells: *may reflect the absence of cell walls in animal cells. Cells from all other kingdoms are protected and supported by cell walls, a job that is done by intermediate filaments in animal cells.*
   f. high degree of conservation in the structure of actin proteins: *evidence that virtually every amino acid participates in some essential activity of the protein. This is a reflection actin's many roles in cells.*

g. experiments of A. Huxley, Niedergerke, H. Huxley and Hanson: *provided the first evidence that the shortening of a skeletal muscle cell during contraction is due not to the shortening of individual proteins, but rather to the sliding of interdigitated filaments past one another in a regulated manner.*
h. actin-binding proteins: *determine the organization, assembly, physical properties, and interactions of actin molecules within cells.*
i. contractile stress fibers: *may maintain tension between a cell and its substrate by isometric contraction, the generation of tension without a change in length.*
j. contact inhibition of movement: *is partially responsible for the aggregation of cells, particularly epthelial cells, into tissues.*

2. Compare and Contrast:
a. microtubules vs. microfilaments
These are two types of cytoskeletal elements that occur in virtually all types of eukaryotic cells. Both play important roles in cell motility, cell division, and the maintenance of cell shape. Both, too, can be in a state of dynamic instability in cells, with relative rates of assembly and disassembly reflecting the needs of the cell at any given time. They differ in size, occurrence, the motor proteins with which they affiliate, the subunits of which they are composed, the manner in which they are regulated, and the specific roles they play within cells.
b. dynein vs. kinesin
Both dynein and kinesin are large motor proteins that convert the chemical energy of ATP into movement. Both are found affiliated with microtubules, although only dynein occurs on the microtubules of cilia and flagella. Kinesin is a plus-end directed microtubular motor, and dynein, among its other roles, is a minus-end directed microtubular motor. In spite of their similarities in function, they are not homologous proteins, and they assume quite different three-dimensional shapes. They are not members of a protein family.
c. basal bodies vs. centrioles
Both of these intracellular structures serve as microtubule-organizing centers in cells, and, along with their affiliated proteins, are the sites of microtubular nucleation. They share identical structures of nine evenly spaced fibrils each comprised of three microtubules. Centrioles occur near the nuclei of some but not all eukaryotic cells. During interphase, they are the sites of microtubular convergence, probably a reflection of their roles in initiation. During cell division, the mitotic spindle originates from the centrioles. Basal bodies occur near the base of cilia and flagella, where they give rise to the microtubular axoneme.
d. muscle fibers vs. myofibrils
The term muscle fiber is used to refer to a single skeletal muscle cell, but each cell is actually the multinucleated product of fusion of many embryonic myoblasts. Myofibrils are thinner cylindrical strands, hundreds of which may occur in a single muscle fiber. Myofibrils are composed of a linear array of sarcomeres.
e. pseudopodia vs. lamellipodia
Single cells have the ability to "crawl" over their respective substrates by extending cytoplasmic projections from the leading edge. Cells that move by amoeboid crawling extend broad, rounded protrusions called pseudopodia. Cytoplasm streams into these protrusions from the interior of the cell affecting forward movement. Cells that are normally part of a tissue, such as cultures fibroblasts, send out broad, flattened, sheetlike projections called lamellipodia. Lamellipodia form temporary adhesions with the substrate over which the cell crawls. Both types of cytoplasmic projections involve actin.

Multiple Choice

1. c	2. c	3. a	4. d
5. d	6. d	7. b	8. a
9. c	10. a	11. b	12. a
13. a	14. b	15. d	16. a
17. a	18. b	19. a	20. d

Problems and Essays

1. a. *The negative slopes of these lines are greater than the positve slopes, indicating that disassembly is more rapid than assembly.*
   b. *Line A probably represents activity at the plus end. In general, the plus end is more active than the minus end, and both assembly and disassembly at the plus end are more rapid. The slopes of the A line in both positive and negative directions are greater than those of the B line.*
   c. *dynamic instability*
   d. *Changes in either Ca^{2+} or cAMP levels have been shown to influence the dynamic instability of microtubules.*
   e. *Assuming that the microtubule you have been watching is not involved in a spindle, it would probably disassemble entirely, and the tubulin subunits would be used to build a spindle.*

2. *AMP-PNP, colchicine* ______ axonal transport

   colchicine ______ mitotic spindle formation

   taxol ______ mitotic spindle disassembly

   AMP-PNP, cytochalasin ______ acrosomal reaction

   AMP-PNP, cytochalasin ______ phagocytosis

   AMP-PNP, cytochalasin ______ cytokinesis

3. a. *The presence of both types of filaments indicates that this is a cross section at the A band, but not in the region of the H zone.*
   b.

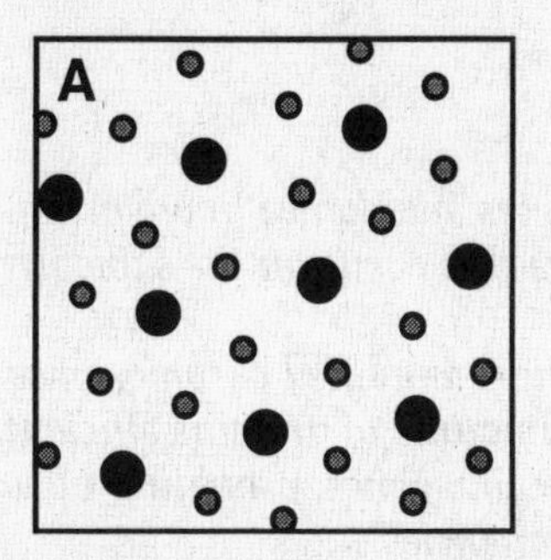

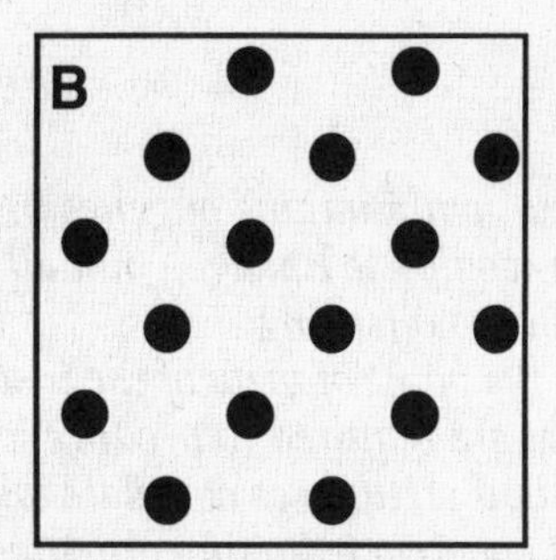

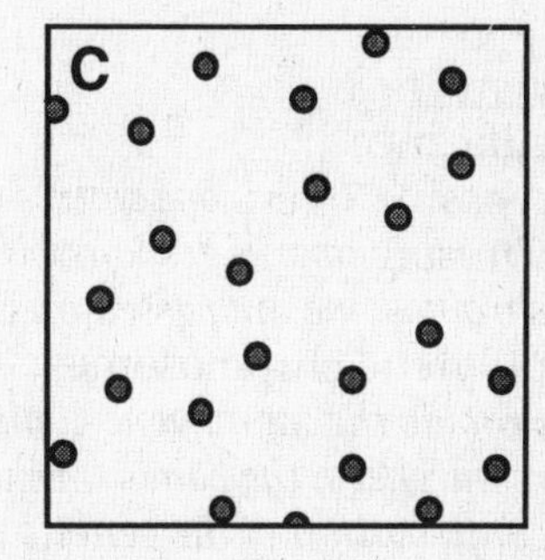

   c. *It might be possible to see some engaged cross-bridges in contracted muscle; these are not visible in relaxed muscle. They would appear as spokes radiating from the thick filaments and contacting the thin filaments.*

4. *The microtubules that form the mitotic spindle are in a state of dynamic assembly and disassembly, both of which are necessary for proper function. The alignment of chromosomes at the equator during metaphase, for example, occurs by the lengthening and shortening of the spindle microtubules. Most importantly, the separation of chromosomes during anaphase results from the shortening of the spindle fibers, a process that is inhibited by taxol.*

5. *Without either the nexin links between adjacent microtubules or the radial spokes that constrain their movements, the ATP-driven action of dynein slides the fibrils relative to one another until they telescope outward from their original positions. It is possible for each fibril to slide completely off both of its adjacent fibrils until they are lined up end to end. If this happens, the length of the axoneme will be equal to the length of all the fibrils in the ring, or nine times its original length.*

6.

NONE	tropomyosin
PM	acetylcholine receptor
SR	Ca^{2+}-ATPase
NONE	titin
SR	Ca^{2+}-release channels

7.

	Ciliary Movement	Skeletal Muscle Movement
Filament Type	*microtubule*	*microfilaments and myosin*
Motor Protein	*dynein and kinesin*	*myosin*
Fuel Molecule	*ATP*	*ATP*
Name of the Functional Unit of Movement	*axoneme*	*sarcomere*

Chapter Ten

Short Answer

1. Significance
   a. crossing over: *one source of new combinations of genes in gametes, and hence in offspring. Crossing over is responsible for incomplete linkage– not all traits that occur on the same parental chromosome are passed on together to offspring.*
   b. polytene chromosomes: *from the point of view of the insect, these amplified chromosomes provide the many gene copies that are required to produce large amounts of proteins efficiently. For researchers, the bands on these giant structures correlate with specific genes, providing visual confirmation of the genetic maps constructed using crossover frequencies.*
   c. topoisomerases: *catalyze the essential first step in DNA replication or transcription– unwinding supercoiled DNA.*
   d. T. H. Morgan's experiments with *Drosophila*: *established that linkage is incomplete, and that the reason for this is that pieces of homologous pairs of chromosome cross over. Morgan used crossover frequencies to construct the first genetic map.*
   e. DNA supercoiling: *packages long stretches of DNA into small, compact units.*
   f. Barbara McClintock's experiments with maize: *the first to show that the genome of a eukaryote is not immutable. McClintock established that some genetic units, which she named transposable elements, can move within the genome.*
   g. restriction fragment length polymorphisms: *used as genetic markers, identifying differences in DNA sequences by means of differences in the fragments resulting from digestion with specific restriction endonucleases. For certain regions of the genome, the RFLPs are so highly individualized that they can serve as positive identifying features. RFLPs have been used to construct physical maps of genomes, and also to find the genes for some diseases including cystic fibrosis.*

h. Avery, Macleod and McCarty's experiments with pneumococcus: *the first to establish that the "transforming principle," the substance that could make a nonvirulent pneumococcus become virulent, is nucleic acid. By inference, it was determined that the genetic material is DNA.*

2. Compare and Contrast:
a. gene vs. allele
A gene is that segment of DNA that codes for a protein. Genes occur at particular positions on chromosomes, called loci. An allele is the exact form of a gene at a given locus. One gene may have several different alleles, all residing at the same locus on either different members of homologous pairs or chromosomes from different individuals. Thus, while any one individual cannot have more than two alleles for a gene, a group of individuals may harbor many different alleles for that gene.
b. Mendel's law of segregation vs. Mendel's law of independent assortment
The law of segregation refers to the manner in which members of homologous pairs of chromosomes, and the genes that reside on them, segregate. One member of each pair, or, as Mendel saw it, one form of each trait, ends up in each gamete. Independent assortment says that the manner in which members of any one pair of chromosomes segregate is independent of how the members of any other pair segregate, or, as Mendel saw it, different traits move into the next generation independently of each other.
c. genomic complexity of prokaryotes vs. genomic complexity of eukaryotes
Prokaryotes have one kind or class of DNA: nonrepeating, coding DNA. Eukaryotes have three general classes of DNA, distinguished by the degree of repetition: nonrepeating DNA, moderately repetitive DNA, and highly repetitive DNA. The differences are best illustrated using reannealing curves. Prokaryotes have simple sigmoidal curves, whereas eukaryotic curves have three plateaus corresponding to the three classes of DNA.
d. genetic map vs. physical map
Both types of chromosomal maps identify the positions of genes on the chromosomes. Genetic maps rely on the frequency of crossover events to derive the relative positions of genes on chromosomes. Physical maps are constructed by first using the nucleotide sequences of overlapping restriction fragments, and then properly ordering the fragments to get the sequence of nucleotides in a chromosome or an entire genome (as in the Human Genome Project).
e. gene therapy using somatic cells vs. gene therapy using germ cells
Both types of manipulation involve introducing new DNA, or changing the DNA of the target cell or tissue. When somatic tissue cells are the target, the changes can only affect that individual. When germ cells are the target, the changes influence the offspring of the germ cell donor, and hence the next generation. Most researchers and practitioners agree that due to the moral implications of altering the next generation, germ cell therapy is unacceptable.

Multiple Choice

1. b	2. a	3. c	4. d
5. c	6. d	7. c	8. d
9. c	10. a	11. c	12. a
13. d	14. b	15. d	16. c
17. d	18. a	19. c	20. b

Problems and Essays

1.

$$1.16\ ribosomes / \sec \cdot 60 \sec / min \cdot 60 min / hr \cdot 24\ \ hr = 100{,}224\ \ ribosomes / 24 hr$$

$$100{,}224\ ribosomes/\ 24\ hr\ /\ gene \times x\, genes = 10 \times 10^{6}\, ribosomes$$

$$x = 99.78\ \ genes$$

In fact, there are between 200 and 2000 copies of the genes for the various ribosomal RNAs in the human genome.

2. *The first* Alu *sequence probably arose by a duplication of the gene for the 7SL RNA, an event that most likely occurred sometime early in the evolution of mammals. Since* Alu *has the features of a transposable element (i.e., terminal inverted repeats), the duplication mechanism may have involved reverse transcription of the 7SL RNA, followed by insertion of the new DNA at a different place in the genome. Over time, the* Alu *sequence underwent many such duplications, and the nucleotide sequences of the duplicated DNA changed by mutation eventually becoming a very large "family" of SINEs.*

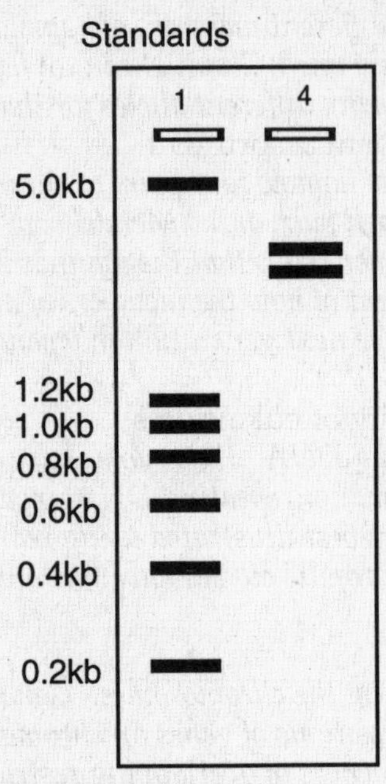

3. a. *First, note from the standards that the relationship between how far a fragment will move on a gel and its size is a logrithmic one, where the smallest fragments of DNA move the farthest toward the cathode. The fragments that resulted from digesting the plasmid in tube 1 are about 1.1 kilobase pairs and about 3.4 kilobase pairs. This adds up to 4.5 kilobase pairs, or 4500 base pairs, corresponding to pFIRST. Give your professor tube 1.*

3. b. *EcoR1 and* HindIII *would cut pSECOND into two pieces of 2900 base pairs and 2700 base pairs. These fragments would migrate nearly together on an agarose gel.*

4. a. $\frac{100{,}000\ genes}{23\ chromosomes} = 4348\ genes\ /\ chromosome$

b. $\frac{3\times10^9\ base\ \ pairs}{1\times10^5\ genes} = 30{,}000\ base\ pairs\ /\ gene$

c. $\frac{3\times10^9\ base\ \ pairs\cdot 0.05}{1\times10^5\ genes} = 1500\ base\ pairs\ /\ gene$

5.

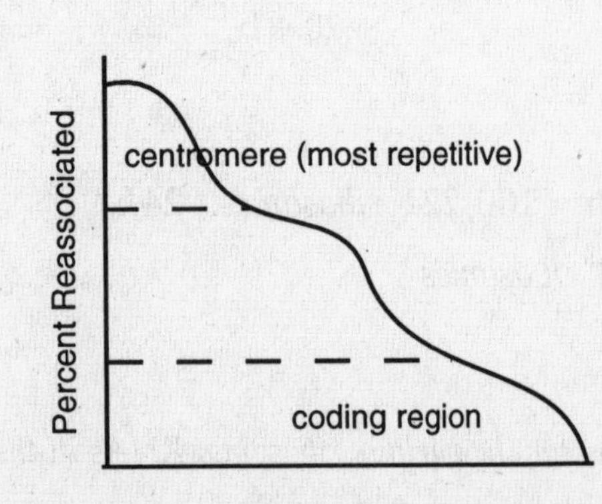

The most repetitive DNA, much of which resides near the centromeres, reanneals first because there are so many copies of it that each strand rapidly finds a complementary strand. The coding regions are usually the least repetitive and reanneal last. These regions are slow to reanneal because each single strand must find its one and only complementary strand.

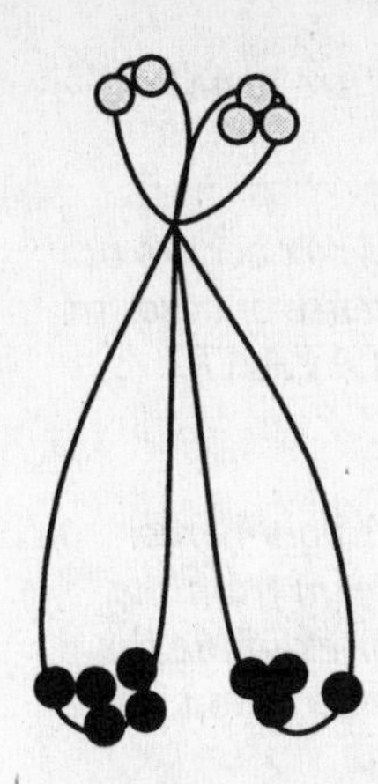

6. *While on the same chromosome, these two genes are far enough apart that they are separated by crossover events at a very high frequency; they appear to act independently. This diagram shows one possible fluorescence pattern that might result from the FISH technique where the distance between the two genes on the chromosome is maximized. Other patterns are possible.*

7. *Band 3 is the only band that differs in a consistent way between those with and those without the disease. This consistency means that the DNA fragment is probably close to, if not part of, the gene for the disease. Band 1 is highly conserved, and band 2 is highly polymorphic between individuals.*

Chapter Eleven

Short Answer

1. Significance
   a. mRNA as an intermediary between DNA and protein: *This indirect system of protein synthesis enables amplification of the amount of protein that can be produced from any one gene. One gene can be transcribed many times, and each RNA can be used many times as a template for protein synthesis. This may evidence to suggest that the first macromolecules in cellular evolution were RNA.*
   b. the highly exergonic nature of RNA polymerization: *decreases the likelihood that RNA synthesis can proceed in the reverse direction by the same mechanism as polymerization.*
   c. slight differences in the sequences of nucleotides in promoter regions: *deviations from the "consensus sequence" change the binding affinity for the RNA polymerase, and hence the likelihood that a gene will be transcribed. This may be a mechanism for regulating which genes are transcribed, and how often.*
   d. methylguanosine cap: *prevents digestion of the 5′ end by nucleases, aids in the transport of mRNA out of the nucleus, and plays a role in initiating translation.*
   e. alternate splicing: *the same gene can code for several different polypeptides.*
   f. Ames test: *rapidly and inexpensively identifies potential mutagens and carcinogens.*
   g. Shine-Dalgarno sequence: *In prokaryotic cells, this sequence orients the mRNA on the small ribosomal subunit and identifies the initial AUG codon.*
   h. our changing concept of the gene: *reflects our ever changing understanding of the complexity of information transfer in cells.*

2. Compare and Contrast:
   a. "one gene – one enzyme" vs. "one gene – one transcription unit"
   These phrases reflect our understanding of the relationship between genes and traits, at two different times in the history of cell biology. The first phrase, derived from the work of Beadle and Tatum, described the finding that one gene carried the information to construct one enzyme. The second phrase expanded Beadle and Tatum's understanding, describing the observation that genes carry elements other than just the information for constructing an enzyme These elemnts include promoter sequences, introns, etc.
   b. higher-order structure of proteins vs. higher-order structure of RNA
   Secondary and tertiary structure are essential to the roles of both proteins and RNA. In proteins, higher-order structure is mainly a result of hydrophobic amino acids that interact better with each other than with the aqueous medium, hydrogen bonding that stabilizes helices and ß sheets, and the

formation of covalent bonds between cysteine residues. In RNA, higher-order structure results primarily from complementary base pairing between nucleotides.

c. Pribnow box vs. TATA box
These are both recognition sites for RNA polymerases found on DNA. The Pribnow box occurs in prokaryotic genes, whereas the TATA box is a feature of eukaryotic genes. The sequence of bases in these two promoters is very similar. The Pribnow box most often has the sequence TATAAT, whereas the consensus sequence of the TATA box is TATAA.

d. upstream promoters vs. internal promoters
The promoter region on DNA is the sequence of bases that is recognized by the RNA polymerase prior to initiation of transcription. In most genes, the promoter region occurs upstream from the template region. However, for genes that are transcribed by RNA polymerase III, for example the genes for the 5S ribosomal RNA, the RNA polymerase binds downstream of the transcription initiation site, to a sequence within the coding DNA.

e. frameshift mutations vs. substitution mutations
A frameshift mutation results when one or two nucleotides are erroneously deleted from or added to a gene. When the gene is transcribed and translated, the reading frame of the message is shifted by one or two nucleotides, and all the subsequent triplet codons are misread. A substitution mutation occurs when one nucleotide base is substituted for another. The result can be harmful, but often it is inconsequential, especially if the substitution occurs in the third base of a triplet coding for an amino acid. Even if the substitution occurs in one of the first two bases, only one amino acid on the protein is altered; the ability of the protein to function may be unaffected.

Multiple Choice

1. b	2. c	3. b	4. a
5. d	6. c	7. b	8. c
9. d	10. b	11. a	12. b
13. d	14. a	15. a	16. c
17. d	18. a	19. d	20. d

Problems and Essays

1. *The sequence TATAA occurs on the complementary strand rather than the actual template strand of the DNA. Recall that regions of DNA are named according to the RNA that is transcribed rather than the template. Recall also that the textbook specifically indicates that the TATA box has the sequence 5′-TATAAA-3′. The 5′ and 3′ are written at the beginning and the end of the sequence respectively– the way that RNA is synthesized. Prokaryotic chromosomes are circular and lack free ends, and eukaryotic genes do not occur at the ends of DNA strands, so the designations 3′ and 5′, which refer only to free ends, are not relevant to the DNA.*

2. *nucleus* hnRNA — *cytoplasm* tRNA

cytoplasm mRNA — *nucleus* pre-RNA

nucleus snRNA — *nucleus* nascent rRNA

cytoplasm rRNA — *nucleus* 45S pre-rRNA

3. *The cDNA is complementary to the mRNA, which is comprised only of the exons from the original gene. It is 5.5 kilobases, and retains only one of the EcoR1 restriction sites. Two fragments result from digestion of the cDNA with EcoR1, and they are 3.0 kb and 2.5 kb long.*

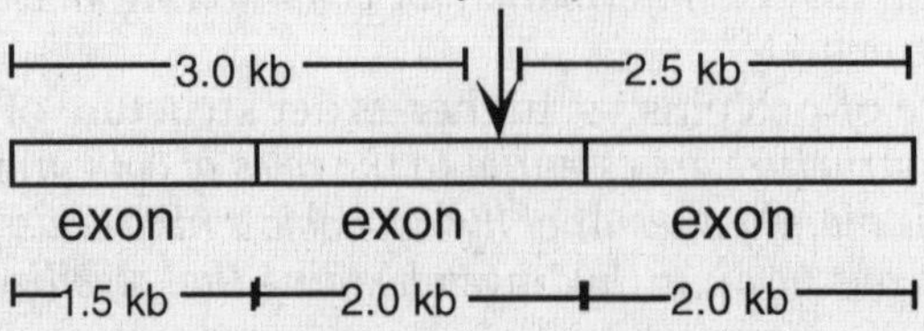

3.0 kb | 2.5 kb

exon exon exon

1.5 kb 2.0 kb 2.0 kb

4. a. *The most abundant codons that result are:* CCC CCU UCC CUC
They code for these amino acids: PRO PRO SER LEU

Other possible(but less likely) codons are: UUC CUU UCU UUU
They code for these amino acids: PHE LEU SER PHE

b. *Proline would be most abundant.*

c. *Phenylalanine would be least abundant.*

5. *The RNA polymerase needs 1 NTP per base x 3 bases per amino acid x 8 amino acids = 24 NTP.*
The aminoacyl tRNA synthetases use 1 ATP per amino acid x 8 amino acids = 8 ATP.
Translocation requires 2 (+) GTP per amino acid x 8 amino acids = 16 GTP(+).
Finally, release factor requires 1 GTP per peptide 1 GTP.
The total number of NTPs required is at least 49 NTP.

Aside from the higher level of activity in children relative to adults, the process of growth involves extensive protein synthesis as a part of both cell division and cell growth. The energy demands of a growing child are higher per pound that that of an individual who has ceased growing.

6. *The amino acid sequence in the protein:*
MET-VAL-HIS-LYS-ARG-THR-LEU-VAL stop
was translated from the mRNA:
*AUG GUU CAU AAA **A**GA ACC UUA GUU UAA*
or some similar sequence.
The mutant protein:
MET-VAL-HIS-LYS-GLU-PRO
is truncated and must have resulted from a frameshift mutation that inserted a premature stop codon. One possible mutation could result in the following RNA:
*AUG GUU CAU AAA *GAA CCU **UAG** UUU AA*
stop
where the asterisk represents the loss of an adenine. Other deletions might give the same mutant protein.

7. *Using the enzyme polynucleotide phosphorylase and a mixture of all the nucleotide triphosphates, make a mixture of synthetic RNAs that differ in random ways in both their base compositions and lengths. Pour this mixture over a column of inert particles to which ATP has been covalently bound. Discard the RNAs that do not adhere to the column particles. Dislodge the RNAs that adhere by using a highly-concentrated salt solution. Using a combination of reverse transcriptase, followed by RNA polymerase and the appropriate precursors and cofactors, replicate the RNAs that adhere, and repeat the column selection step under more stringent conditions, e.g., in the presence of a mild salt solution that requires the RNAs to have a higher affinity for the ATP in order to stick to the column. Repeat this process until you have a population of RNAs with a very high affinity for ATP.*

Chapter Twelve

Short Answer

1. Significance
   a. nuclear pore complex: *regulates the movement of proteins and RNAs between the nucleus and the cytoplasm by recognizing and transporting these molecules in the appropriate direction.*
   b. nuclear localization signal: *enables proteins synthesized in the cytoplasm to be recognized and transported through the nuclear pores into the nucleus.*

c. evolutionary conservation of histone proteins: *evidence that every amino acid in these proteins is functionally important.*
d. H1 histone: *maintains the 30 nm filament structure of chromatin. Without the H1 histone, the 30 nm filament uncoils to form "beads on a string" nucleosomes.*
e. nuclear matrix: *maintains the shape of the nucleus; organizes the loops of chromatin; guides RNAs to nuclear pores.*
f. selective gene amplification: *increases the number of copies of certain genes, particularly some ribosomal RNA genes, so that the expression of those genes is increased.*
g. DNA rearrangement of antibody genes: *provides one mechanism for the synthesis of a diverse variety of antibodies from a limited number of antibody genes.*
h. heterodimerization of transcription factors: *expands the diversity of transcription factors that can be generated from a limited number of peptides, and allows one protein to alter the DNA binding properties of another.*
i. DNase I hypersensitive sites: *identify genes in the process of active transcription.*
j. acetylated histones: *disrupt the structure of the nucleosome and partially dislodge the DNA from the histone proteins, making the DNA more likely to be transcribed.*

2. Compare and Contrast:
a. heterochromatin vs. euchromatin
These are forms of interphase chromatin and represent two different levels of condensation. Heterochromatin remains highly condensed and is not actively transcribed, whereas euchromatin is dispersed and can serve as a template for transcription.
b. constitutive vs. facultative heterochromatin
These are both forms of highly condensed interphase chromatin. Constitutive heterochromatin remains condensed throughout the cell cycle and is never transcribed. It mostly consists of highly repetitive, noncoding DNA and is located around the centromere region of the chromosome. Facultative heterochromatin is condensed and transcriptionally silent during certain phases of the life cycle, but may be dispersed and transcribed during other phases. The inactive X chromosome in females, the Barr body, is an example of facultative heterochromatin.
c. repressible vs. inducible operons
Operons are prokaryotic transcriptional units, including their regulatory sequences, that code for one or several proteins with related functions. An inducible operon is one whose structural genes will be transcribed in response to some key metabolic substance that acts as an inducer. A repressible operon is one whose structural genes will shut off in response to a key metabolic substance, in this case called a co-repressor. Both function by the reversible binding of a repressor to a DNA sequence that lies between the promoter region and the transcriptional start site. In an inducible operon, the inducer binds a repressor, rendering it incapable of binding to the operator; in a repressible operon, the corepressor binds the repressor, enhancing its ability to bind to the operon.
d. DNA methylation in prokaryotes vs. DNA methylation in eukaryotes
The addition of a methyl group to carbon number five of cytosine serves to protect the DNA of prokaryotes from attack by the cell's own restriction enzymes, present to degrade foreign DNA. In eukaryotes, DNA methylation serves to silence transcription of the methylated regions. Methylation of eukaryotic DNA occurs primarily in gene regulatory regions.
e. liver fibronectin vs. fibroblast fibronectin
These two proteins are actually products of the same gene. They differ in the way that the transcript is processed in the nucleus. Fibronectin from fibroblasts has two peptides missing from the liver protein. These regions of the pre-mRNA are retained in fibroblasts but excised from the transcript in liver cells.

Multiple Choice

1. a	2. c	3. b	4. c
5. a	6. d	7. d	8. b
9. c	10. d	11. b	12. a
13. b	14. a	15. d	16. c
17. b	18. a	19. c	20. d

Problems and Essays

1. a.

Genotype	With Lactose ß-gal	With Lactose gal permease	With Lactose transacet	Without Lactose ß-gal	Without Lactose gal permease	Without Lactose transacet
$i^+ P^+ 0^+ z^+ y^+ a^-$	yes	yes	no	no	no	no
$i^+ P^+ 0^+ z^- y^+ a^+$	*no*	*yes*	*yes*	*no*	*no*	*no*
$i^- P^+ 0^+ z^+ y^+ a^+$	*yes*	*yes*	*yes*	*yes*	*yes*	*yes*
$i^+ P^- 0^+ z^+ y^+ a^+$	*no*	*no*	*no*	*no*	*no*	*no*
$i^+ P^+ 0^- z^+ y^+ a^+$	*yes*	*yes*	*yes*	*yes*	*yes*	*yes*

b. *The repressor would not be transcribed due to a mutation in the binding site of RNA polymerase. Thus the structural genes would be expressed whether or not lactose was present.*

2. a. *Band 1. The DNA of the fragment in Band 1 is not sensitive to DNase I a sign that it is not being actively transcribed in either tissue.*

   b. *Gel B. The globin gene (and its regulatory regions) are present in Band 2, and sensitive to increasing concentrations of DNase I only in Gel B. This gene is actively transcribed in red blood cells of birds but not in brain tissue.*

3. *Translation provides a mechanism for amplification of a gene product, because a single transcript can be translated many times. RNA that is not translated, such as rRNA, misses out on that amplification step. In cells with a high demand for ribosomes, such as amphibian oocytes, the DNA that codes for all but one of the rRNA subunits is amplified by selective replication. Thousands of transcriptionally active copies of the genes are produced. The single rRNA subunit that is not selectively amplified, that of the 5S rRNA, is present in the genome in 20,000 copies.*

4.

__*HLH motif*__ The transcription factor occurs as a heterodimer with two alpha-helical segments separated by a loop. The alpha helices are perpendicular to one another.

__*HLH motif*__ MyoD

__*zinc finger motif*__ This transcription factor has several zinc ions affiliated with it.

__*HMG-box motif*__ The transcription factor bends DNA resulting in a spatial arrangement that enhances its access to other regulatory factors.

__*zinc finger motif*__ TFIIIA

__*HMG-box motif*__ UBF

5.

__*active*__ Core histones are heavily acetylated.

__*inactive*__ Chromatin is tightly affiliated with histones.

__*active*__ Chromatin is hypersensitive to DNase I.

_________*inactive*_________ Chromatin is methylated.

_________*active*_________ Chromatin is affiliated with histones, but the histone octamers are disrupted.

_________*inactive*_________ Chromatin is in the form of constitutive heterochromatin.

Chapter Thirteen

Short Answer

1. Significance
   a. Meselson and Stahl experiment: *established that DNA replication in prokaryotes is semiconservative .*
   b. temperature-sensitive mutants: *have enabled researchers to study nearly every type of physiological process, including those* whose absence would be lethal (such as DNA synthesis).
   c. theta structure of bacterial chromosomes: *enables visualization of the replicating bacterial chromosomes. These structures also confirmed the circular nature of bacterial chromosomes.*
   d. RNA primers: *provide a starting point for DNA polymerases that cannot initiate replication, but elongate only existing nucleotide strands. The initiation of replication occurs when mismatched bases are most likely to be incorporated. Thus RNA primers may also minimize errors, by initiating the process with a temporary strand that is later removed.*
   e. single-stranded binding proteins: *stabilize the single-stranded DNA molecule so that it can serve as a template for replication.*
   f. oriC site: *the one and only starting point for replication in* E. coli.
   g. autonomous replicating sequences (ARSs): *promote replication of the DNA in which they are contained– either the yeast chromosome where they are occur naturally, or any DNA in which they are experimentally inserted.*
   h. DNA repair enzymes: *prevent DNA damage, occuring as a result of many natural phenomena, from accumulating and causing mutation and cell death.*
   i. intergenic complementation technique: *is used to study mutations that occur in complex, multigenic pathways. By fusing two cells with different mutations in the same pathway, a functional pathway can be reconstructed. Analysis using this technique can establish the number of genes involved in a pathway.*

2. Compare and Contrast:
   a. RNA primers of leading strands vs. RNA primers of lagging strands
   On both the leading and lagging strands, replication depends upon elongation of RNA primers. Due to semidiscontinuous replication, there are many more primers synthesized on the lagging strand than on the leading strand. Also, on the leading strand, RNA synthesis is initiated at the origin of replication by RNA polymerase. On the lagging strand, RNA synthesis begins at the replication fork and is carried out by a different enzyme, called primase.
   b. *ori*C vs. ter
   Both of these are regions on the bacterial chromosome. OriC *is the site of initiation of replication; ter is the site where the two replication forks meet and replication ceases.*
   c. base substitution errors in transcription vs. base substitutions in replication
   If a base substitution occurs during transcription, the RNA that results contains an error, and that error may affect all of the polypeptides that are translated from that transcript. But because transcripts are generally short-lived, the error may have little overall, long-term effect on the cell. Base substitutions that occur during replication, on the other hand, become part of the cell's genome. The error is be passed down to all cells that arise from that cell and is apparent in all the RNA that is transcribed from that DNA. If the error occurs in a germ cell of a eukaryote, it may even be passed on to the next generation.

d. replication origins in prokaryotes vs. replication origins in eukaryotes
Prokaryotes generally have one specific site where replication begins, oriC. *Most eukaryotes have many sites, called autonomous replication sequences. In mammalian cells, the origin sites for replication may depend more on features of chromatin organization than on specific DNA sequences.*

e. autonomous replicating sequence vs. origin recognition complex
These are the two essential components of replication in yeast. The autonomous replicating sequence is the region of DNA that is recognized by a complex of initiation proteins. That protein complex is called the origin recognition complex. Thus it is the binding of the latter to the former that initiates replication.

Multiple Choice

1. a	2. c	3. d	4. c
5. a	6. c	7. d	8. b
9. b	10. a	11. d	12. c
13. b	14. a	15. d	16. a
17. c	18. a	19. b	20. a

Problems and Essays

1.

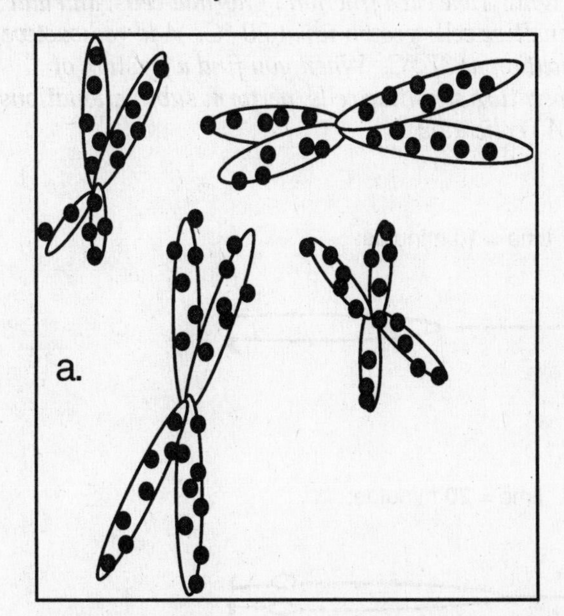

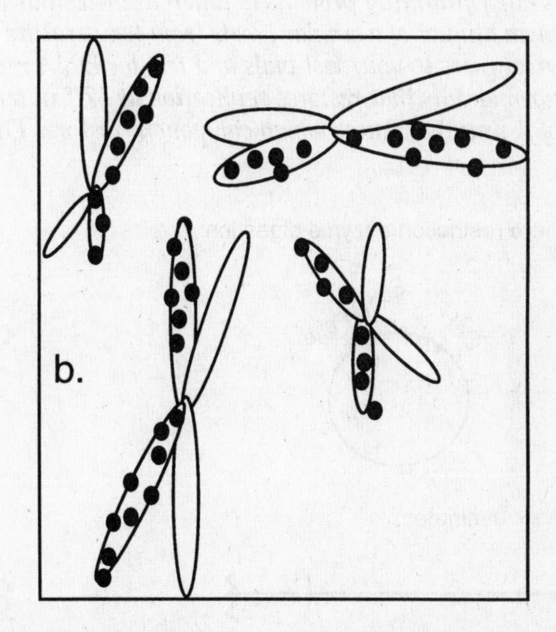

c. *After the first round of DNA replication, each sister chromatid is a combination of one strand from the unlabeled parental DNA duplex, and one newly synthesized strand made with radioactive thymidine. Thus both chromatids will have one labeled strand of DNA. After the second round of replication, only one of the sister chromatids will have a labeled strand of DNA. The second sister chromatid, comprising the unlabeled parental strand from diagram (a) and one newly synthesized strand (also unlabeled), will have no labeled thymidine. Although it is not illustrated here, when experiments like these are actually done, some label appears in the "unlabeled" chromatid during the second cell division. This is because some exchange of genetic material between chromatids occurs during metaphase of mitosis. The corresponding region of the "labeled" chromatid in these cases is free of label.*

2.

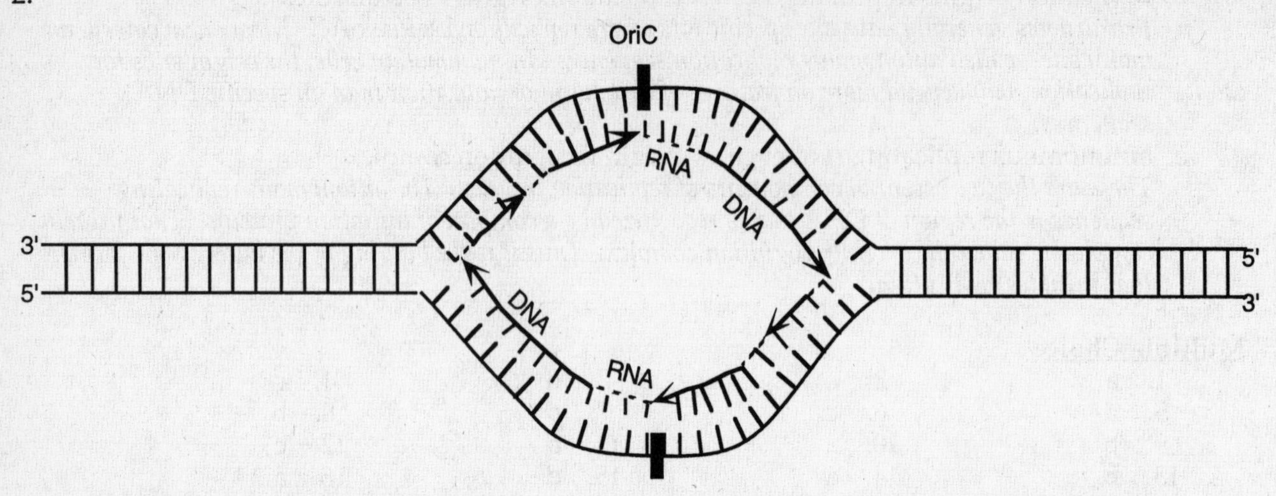

3. *Grow your normal E. coli at 37 °, then lyse the cells. Using biochemical techniques, isolate lysate fractions that represent different cellular proteins. You could, for example, isolate fractions of the lysate containing proteins of different molecular weight. Take each fraction of normal cells, and add it to an aliquot of a similar lysate from temperature sensitive cells grown up at 20 °C. Add radioactive precursors to your test vials and test for DNA replication at 37 °C. When you find a fraction of normal cells that restores replication at 37 ° in temperature sensitive cells, perform subfractionations of it until you know which component restores DNA replication.*

4.

a. Before restriction enzyme digestion:

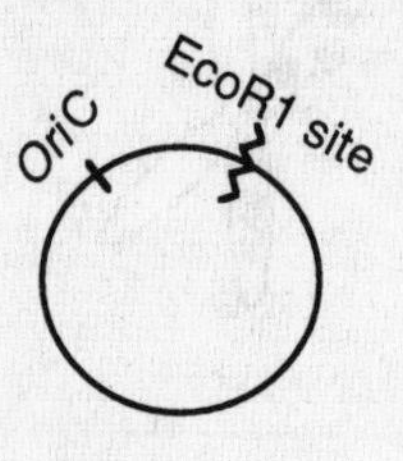

b. time = 0 minutes:

c. time = 5 minutes:

d. time = 10 minutes:

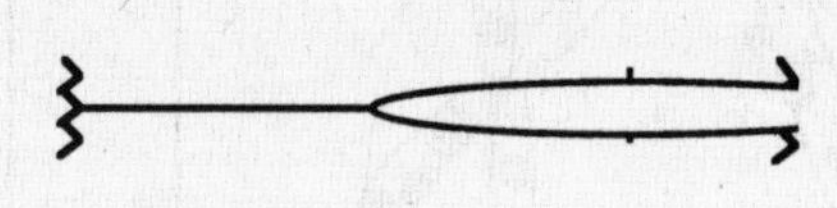

e. time = 20 minutes:

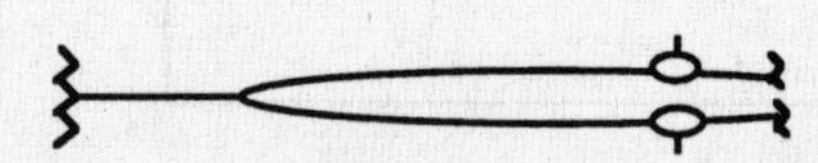

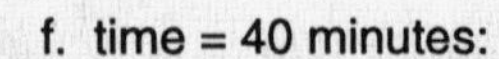

f. time = 40 minutes:

5.

Natural Bases	Deaminated Bases	Natural Bases	Deaminated Bases
cytosine	uracil	guanine	xanthine
thymine	cannot be deaminated	adenine	hypoxanthine

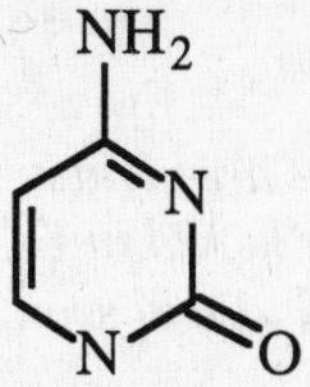

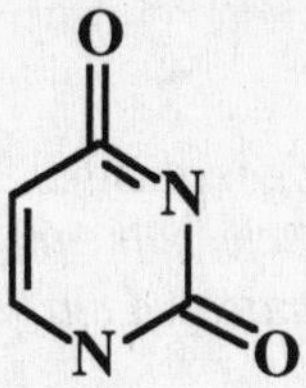

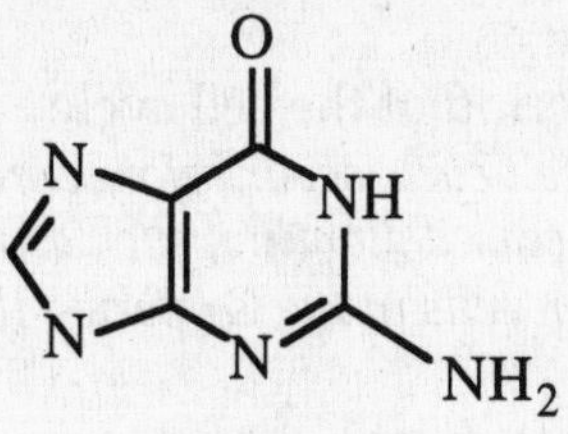

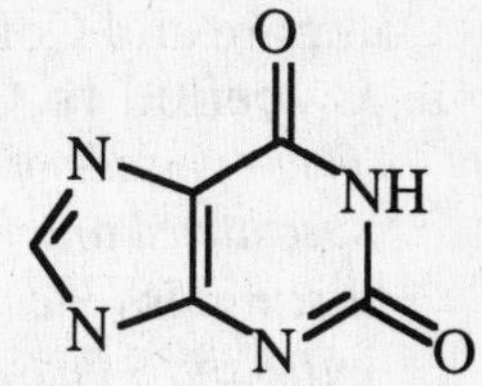

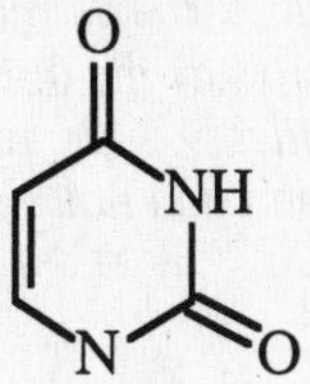

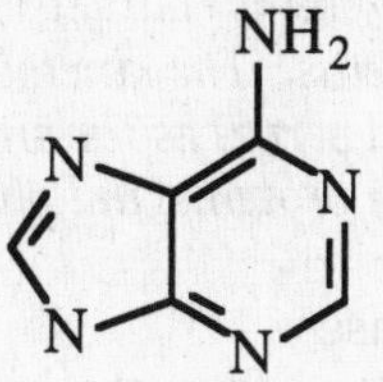

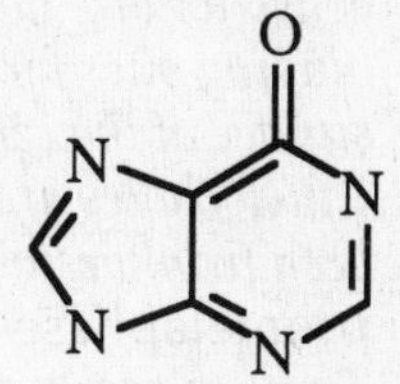

Uracil is the base that occurs in RNA in place of thymine.

6. *If cytosine becomes deaminated as a result of environmental damage, the base that forms is uracil. Because uracil is not a normal component of DNA, it is recognized and removed by a DNA glycosylase. If, in fact, uracil were the normal base that paired with adenine, as is the case in RNA, then DNA glycosylases would not be able to distinguish between uracil that was part of the DNA sequence, and uracil that resulted from deamination of cytosine. Uracil would base pair with adenine, but if that uracil was originally a cytosine, the proper partner would be guanine.*

Chapter Fourteen

Short Answer

1. Significance
   a. MPF (maturation promoting factor): *triggers the transition of a cell from the G_2 phase to the M phase, but only when it is activated by cyclin. MPF acts as a cyclin-dependent protein kinase.*
   b. cell cycle checkpoints: *provide points at which the progress through the cell cycle can be halted if conditions are not appropriate for cell division or DNA replication.*
   c. p53 protein: *this transcription factor helps to prevent a cell from becoming malignant by controlling the synthesis of a cyclin-Cdk inhibitor that is required for a cell to enter the S phase of the cycle.*
   d. formation of the mitotic chromosome: *packages the genetic material in a highly condensed form such that its distribution to daughter cells during mitosis is efficient.*
   e. phosphorylation of histones: *plays a role in chromosome packaging prior to cell division by decreasing the affinity of histones for DNA.*
   f. mitotic spindle: *organizes the dividing cell by establishing the poles (at the positions of the centrosomes) and establishing the positions of the chromosomes both before and after separation.*
   g. fragmentation of the endoplasmic reticulum: *facilitates the partitioning of the endoplasmic reticulum into daughter cells during cell division.*
   h. polar microtubules: *form a structural basket that maintains the integrity of the mitotic spindle during cell division.*

i. slow rate of chromosome movement during anaphase: *ensures that chromosomes segregate accurately, and prevents them from becoming tangled.*
j. chiasmata: *represent positions along the chromosomes where recombination has likely occurred.*

2. Compare and Contrast:
a. G_0 period vs. G_1 period of the cell cycle
These periods of the cell cycle occur before chromosomes have replicated. Cells that are highly specialized and have ceased dividing, such as nerve and muscle cells, are said to be locked in G_0, *to distinguish them from cells in the* G_1 *period that have unreplicated chromosomes but will soon enter the S phase.*
b. centromere vs. kinetochore
Both of these structures occur at the point of attachment of sister chromatids in a replicated, condensed chromosome. The centromere is part of the chromosome, and forms the site of primary constriction. DNA in this region is characterized by highly repeated sequences that serve as the binding sites for specific proteins. The kinetochore is a platelike structure that occurs at the outer surface of the centromere, and serves as the site of attachment of spindle microtubules. Kinetochores are comprised of proteins and possibly DNA, but the exact composition has not yet been firmly established.
c. prometaphase vs. metaphase
These are early stages of mitosis. Prometaphase precedes metaphase, and is characterized by both the attachment of the chromosomes to the spindle fiber microtubules, and the movement of chromosomes toward the equator of the spindle. During metaphase, the chromosomes become fully aligned at the metaphase plate. Metaphase serves as a "checkpoint" for mitosis. The cell will remain at this stage until the chromosomes are properly aligned.
d. anaphase A vs. anaphase B
Anaphase is characterized by two separate but simultaneous events: anaphase A and anaphase B. Anaphase A refers to the movement of chromosomes toward the poles, and anaphase B refers to the movement of the spindle poles away from each other.
e. phragmoplast vs. contractile ring
These are both structures that separate the cytoplasm of dividing cells. The phragmoplast in plant cells is a region where microtubules accumulate and eventually form the cell plate. Animal cells divide their cytoplasm by cinching in the cell membrane via actin-containing contractile filaments that form the contractile ring.

Multiple Choice

1. c	2. b	3. a	4. b
5. d	6. a	7. c	8. a
9. c	10. d	11. c	12. a
13. b	14. b	15. c	16. a
17. b	18. a	19. c	20. c

Problems and Essays

1.

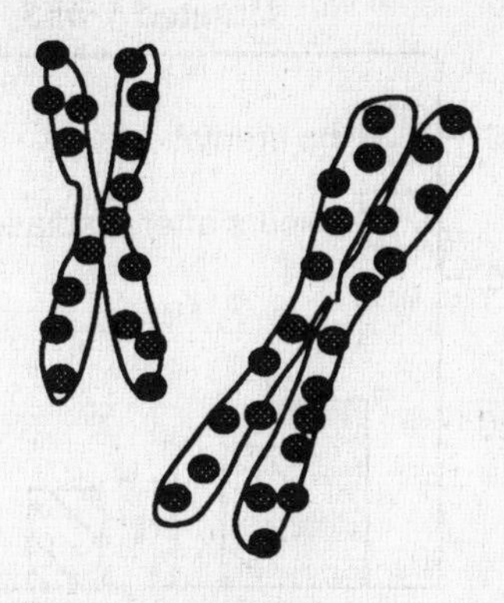

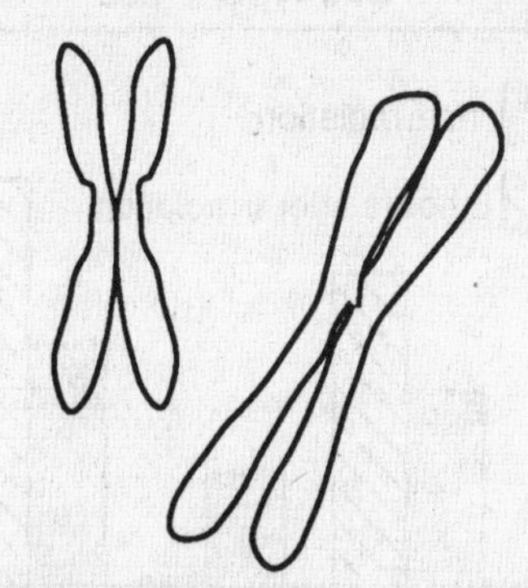

In experiment 1, the active MPF in the cytoplasm of the S cell would induce DNA replication in the G_1 nucleus, and thymidine would be incorporated into the G_1 chromosomes. In experiment 2, the G_2 nuclei could not be induced to undergo a second round of replication, hence no radioactive thymidine would occur in the G_2 chromosomes.

2. *First, isolate the DNA from the human cells, and digest it into fragments, using restriction endonucleases. Separate the fragments, using gel electrophoresis, and then introduce each fragment into a different culture of temperature-sensitive yeast mutants. Try growing your yeast mutant cultures with the human DNA, at both the permissive temperature and the nonpermissive temperature. The yeast cells that are able to grow at the nonpermissive temperature have been "rescued" by the human DNA and must contain a functional gene for the mutated cell cycle protein. Now isolate the DNA from your rescued yeast, and separate the human DNA from those cells. That DNA has an essential cell cycle gene– one that encodes a cell cycle control protein.*

3.

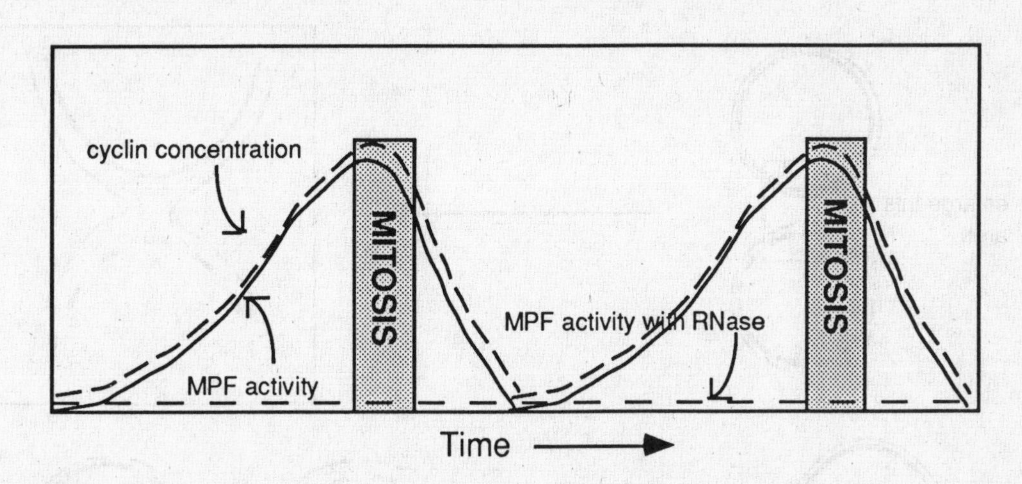

4. a. *Mutant 1 is deficient in the protein coded by the gene* cdc25, *hence it is the* cdc25⁻ *mutant. The cdc25 activates the cdc2 by removing inhibitory phosphates. Cells grow abnormally long if the* cdc25 *gene is not producing a functional phosphatase. Mutant 2, on the other hand, is deficient in the protein coded by the* wee1 *gene.* Wee1 *codes for a kinase that adds the inhibitory phosphates to cdc2 In the absenceof wee1, cells divide prematurely.*

   b. *In both mutants, the duration of the G_2 phase is abnormal. In mutant 1 it is lengthened, and in mutant 2 it is shortened.*

5.

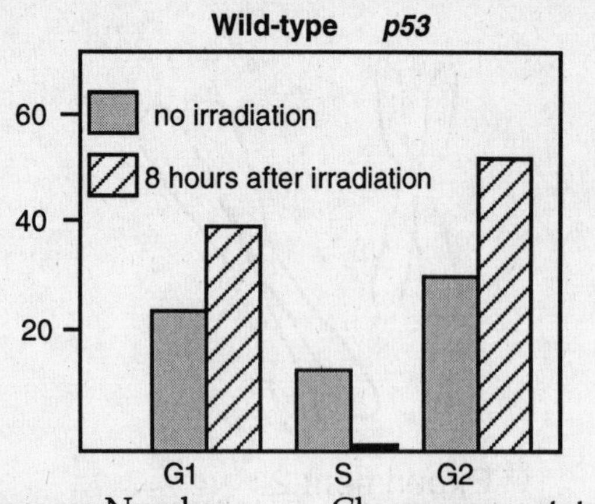

6.

Chromosome Number	Chromosome status	
E	R	a. G_2
D	R	b. meiotic metaphase
E	R	c. mitotic prometaphase
D	R	d. meiotic pachytene
D	R	e. meiotic interkinesis
E	U	f. mitotic cytokinesis
E	U	g. G_o

7. a.

b.

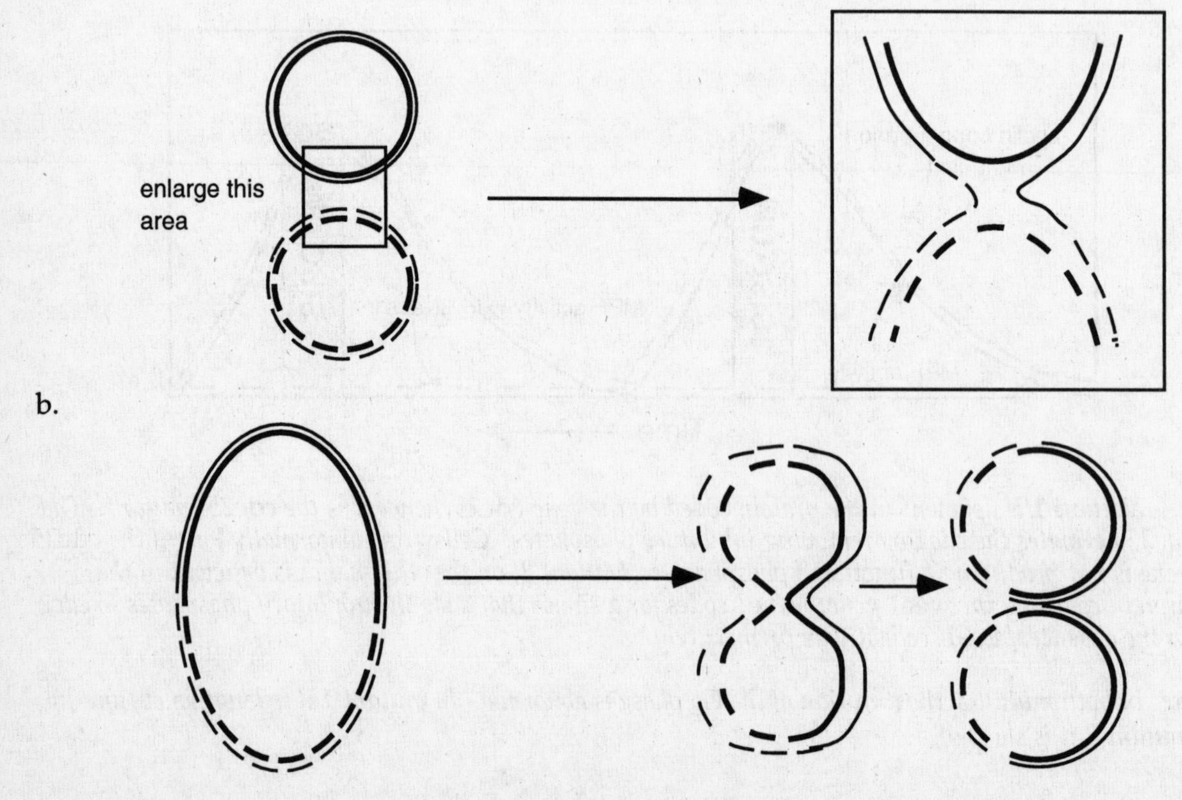

Chapter Fifteen

Short Answer

1. Significance
   a. the many steps of a signal transduction cascade: *provide many potential sites for signal amplification, as well as potential sites for regulation of the signal.*
   b. cAMP: *acts as a second messenger, activating protein kinase A in response to ligand binding to a membrane-bound receptor on the external cell surface.*
   c. phosphatases: *remove phosphates from proteins. Depending on the substrate protein, dephosphorylation may activate or inactivate the substrate, and terminate the intracellular signal.*
   d. calmodulin: *binds up to four calcium ions. In the bound form, calmodulin can activate protein kinase C, phosphodiesterase, or calcium transport.*
   e. SH2 domain: *endows a protein with high affinity for phosphotyrosine motifs on IRSs. Proteins with these domains bind to IRSs, and their activity is modified in a way that alters cell differentiation and growth.*
   f. receptor tyrosine kinases: *phosphorylate tyrosine residues on substrate proteins (IRSs) upon binding their specific ligands at the cell surface. Phosphorylation results in changes in cell growth and differentiation.*
   g. Ras protein: *the protein that acts as the focal point for several important signal cascades involved in cell differentiation and growth. Mutations in the ras gene are implicated in 30 to 40 percent of human cancers.*
   h. MAP kinase cascade: *important (and widespread) signal transduction pathway involved in regulation of transcription, replication, and translation.*
   i. crosstalk: *enables different signal transduction pathways to function in a coordinated and interdependent fashion.*
   j. apoptosis: *rids the organism of cells that are no longer needed or may be potentially harmful to the individual.*

2. Compare and Contrast:
   a. kinase vs. phosphatase
   Both types of enzymes involve phosphate groups, and both play important roles in intracellular signal transduction pathways. Kinases transfer phosphate groups to tyrosine, serine, or threonine residues on target proteins, whereas phosphatases remove phosphate groups from those residues on target proteins.
   b. heterotrimeric G proteins vs. monomeric G proteins
   These proteins both function as signal transduction molecules, relaying information from a ligand-bound receptor at the cell surface to an intracellular effector molecule. Both are bound to GTP in the active form, and both have GTPase activity. Upon hydrolysis, both GDP-bound G proteins are switched off. Heterotrimeric G proteins function by dissociation of the α subunit which interacts with an effector. Monomeric G proteins, such as ras, activate effectors. When a ligand binds to a receptor tyrosine kinase and causes autophosphorylation, SH2 proteins are recruited, and, via an sos intermediate, the G protein releases GDP and binds GTP.
   c. phospholipase C vs. protein kinase C
   Both of these enzymes function in signal transduction pathways, but their substrates and actions are quite different. Phospholipase C, activated by the α subunit of a heterotrimeric G protein following, for example, the binding of acetylcholine to its receptor, removes the inositol phosphate moiety from phosphatidylinositol in the membrane, creating IP_3 and DAG, both of which function as second messengers. Protein kinase C, activated by DAG and sometimes calcium, transfers a phosphate group to serine or threonine residues on target proteins.
   d. tyrosine kinases vs. serine/threonine kinases
   Each of these kinases transfers phosphate groups to target proteins, but to different sites on the target proteins. Most kinases are specific for either tyrosine residues or serine and/or threonine residues.

e. signaling pathways in plants vs. signaling pathways in animals
Basically, members of these two kingdoms use quite similar signal transduction pathways, with a few exceptions. Whereas both groups use changes in intracellular calcium, IP_3, and DAG, only animals use cyclic nucleotide monophosphates as second messengers, and only plants use salicylic acid. Histidine kinases appear to be specific to plants, as well.

Multiple Choice

1. b	2. d	3. b	4. d
5. b	6. b	7. a	8. c
9. d	10. c	11. d	12. b
13. d	14. c	15. b	16. a
17. c	18. b	19. d	20. c

Problems and Essays

1.

Treatment	Additions	DNA Replication
a. none	PDGF or EGF	yes
b. microinjection (intracellular) of constitutively active Ras protein	none	*yes*
c. microinjection of saline ("mock" injection)	PDGF or EGF	*yes*
d. microinjection of saline ("mock" injection)	none	*no*
e. microinjection of anti-Ras antibodies	PDGF or EGF	*no*
f. microinjection of anti-Ras antibodies	none	*no*

2. a. √
b.
c.
d. √
e. √
f.

3. *The Ras protein is a small monomeric G protein that is turned on by binding GTP and turned off by hydrolyzing GTP to GDP. If the* ras *gene is mutated such that the GTPase activity is lost (in this case due to a single amino acid substitution), Ras is never inactivated, and the signal cascade is always on. Transcription, translation, replication, and hence growth and division proceed unregulated. The result is malignancy.*

4. *Because four moles of GppNp bind to 0.8/100 moles of receptors, then 500 GppNp bind per receptor. Thus the extra 500-fold amplification must occur at the level of the receptor-G protein interaction.*

5. *The serotonin receptor is a membrane-bound, heterotrimeric G-protein coupled receptor whose effector is adenylyl cyclase.*

Chapter Sixteen

Short Answer

1. Significance
   a. aneuploidy in transformed cells: *a sign that the genome of cancer cells is not being carefully maintained, and that the growth of cancer cells is not as dependent upon a normal chromosome complement as the growth of normal cells.*
   b. Percival Pott's studies of British chimney sweeps: *the first studies to link cancer to environmental factors– in this case, soot.*
   c. tumor-suppressor genes: *code for proteins that keep the cell cycle in check. Mutations in both copies of these important genes result in uncontrolled cell growth and reproduction.*

d. proto-oncogenes: *code for proteins involved in many aspects of the cell cycle, including the signal transduction pathways that initiate cellular reproduction. Malfunction of either allele for these genes, as a result of mutation, inappropriate transcriptional control, or infection by viruses, transforms them into oncogenes.*

e. chromosomal deletion in children with retinoblastoma: *occurs in the region of the chromosome containing a gene, RB, that plays an important role in down-regulating the cell cycle. This is a clear indication that the condition can be inherited.*

f. the oncogene, sis, of the simian sarcoma virus: *the first oncogene to be identified as coding for a growth factor, namely PDGF.*

2. Compare and Contrast:

a. benign vs metastatic tumor
Both types of tumors result from uncontrolled cell growth, but benign tumors lack the ability to metastasize to distant sites in the body and are therefore contained and treatable by surgery. Metastatic tumors occur when cells break away from the original tumor and invade adjacent or distant sites, growing uncontrollably at these new sites. Metastatic tumors often begin as benign tumors.

b. tumor virus genes vs. oncogenes
Both of these stretches of DNA code for proteins that cause the cell to lose control of growth and reproduction. Tumor virus genes enter the genome of the host cell by viral infection, inserting the oncogenic DNA into the host DNA, transforming the host cell to a cancer cell. The term oncogene is a more general term, referring to any gene (whether it be of viral origin, or a mutated or malfunctioning part of the normal cell genome) that transforms a cell from normal to malignant.

c. initiators vs. promoters
These terms both refer to environmental agents that play a part in transforming cells from normal to malignant. The initiator causes the original mutation that results in the creation of an oncogene. The promoter is the agent that causes the cancer to be expressed, either by creating conditions of cell growth that favor the development of other mutations or via other epigenetic mechanisms.

d. tumors in children with familial retinoblastoma vs. those with sporadic retinoblastoma
Both of these cancers result from the malfunction of the product of the RB gene. If the condition is familial, i.e., inherited as a deletion in one member of chromosome 13, the tumors tend to be multiple and in both eyes. The other member of the missing allele on chromosome is invariably mutated. If the condition arises sporadically, it is usually characterized by one tumor in one eye, and while there is no chromosomal deletion, both alleles of the RB gene are found to be mutated.

e. oncogenes that encode growth factors vs. those that encode growth-factor receptors
Any oncogene will result in a cell that has lost normal controls over its growth and reproduction. When a gene coding for a growth factor is overexpressed, uncontrolled growth results from an overabundance of the growth factor in the cell's environment. If the gene coding for the receptor is mutated, the growth-factor signal may be constitutively "on," regardless of whether the growth factor is there to turn it on.

<u>Multiple Choice</u>

1. d	2. a	3. b	4. d
5. b	6. c	7. b	8. b
9. d	10. b	11. a	12. d
13. c	14. a	15. d	16. d
17. b	18. b	19. c	20. d

Problems and Essays

1. a. chemotherapy: *Treating cancer patients with high doses of toxic chemicals is designed to selectively kill cancer cells by exploiting their higher sensitivity to these toxins. For many cancers, doses that are sufficient to eradicate the cancer are too toxic for the survival of normal cells, and either the cancer is not eliminated, or the patient dies.*
   b. gene therapy: *Introducing functioning genes into cells that have lost control of growth and reproduction is a possible therapy because in cancer patients, proteins that control these functions are coded by mutated genes. While this approach may someday be the treatment of choice, so far it has been impossible for the new and functioning bits of DNA to reach billions of cancer cells buried deep within tissues.*
   c. surgery: *Removing cancer cells from the body surgically would cure the disease if the cancer were caught before it had a chance to metastasize. Unfortunately, that is not always the case, and it is not always possible to determine if a cancer has metastasized.*
   d. stimulating the immune system: *This approach, which recruits the body's own immune surveillance system into detecting and destroying cancer cells, was a source of great optimism in the early 1980s, but has not been as successful as was hoped.*

2. $$\frac{5\times10^{9}\ \textit{people}\ \times\ 100\times10^{12}\ \textit{cells / person}}{50\ \textit{years}}\times\frac{1}{3}=3.33\times10^{21}\ \textit{malignant cells per year}$$

3. a. *Isolate the RNA from your strain of normal, nonmutated tumor virus. Using reverse transcriptase and radiolabeled deoxyribonucleotide triphosphates, synthesize complementary DNA (cDNA) in vitro. Isolate the RNA from the mutated tumor virus, and hybridize it to the cDNA from the normal virus. Pass the hybridized nucleic acid over a hydroxyappatite column to separate the DNA-RNA hybrids from the unhybridized DNA. The unhybridized DNA represents part or all of the gene that is responsible for causing murine leukemia.*

   b. *The first step is to determine the sequence of the cDNA involved in the cancer, and therefore the amino acid sequence of the gene product. One good strategy for determining function involves comparing this sequence to those of genes of known function, and looking for particular amino acid domains or motifs that occur in an oncogene product whose function can be identified. For example, does the protein contain a zinc finger motif? How about a tyrosine kinase domain? Comparisons of this kind give clues about the function of the protein. The sequences of nucleic acids and amino acids for many different proteins are avialable on data bases accessible to students and researchers in this field. Ask if your university can access these gene bank data bases.*

4. a and b:

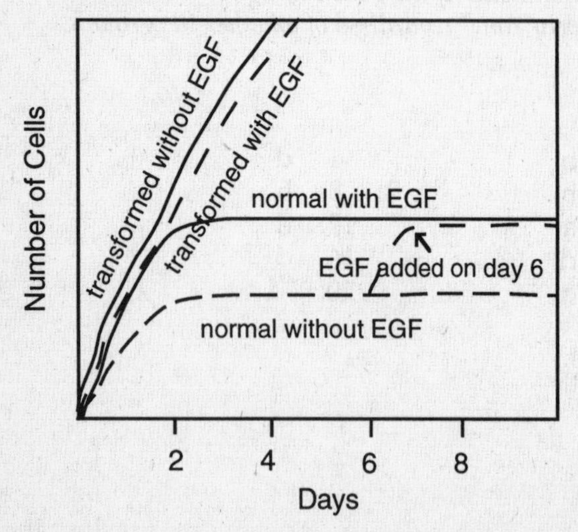

   c. *Cells in the normal +EGF and -EGF cultures are arrested in the Go phase of the cell cycle after day four.*

5. *If a mutation occurs in one allele of a tumor-*

5. *If a mutation occurs in one allele of a tumor-suppressor gene, the product of that gene will not function properly, but the malfunctioning protein, by itself, does not cause malignancy. The cause of cancer is the absence of a properly functioning tumor-suppressor protein. The gene product of the other allele, which is not mutated, can assume the role in cell cycle regulation of that locus. Thus a mutation in a tumor-suppressor gene acts as a recessive. In the case of an oncogene, the presence of the malfunctioning protein is enough to cause cancer. The action of the gene product is to stimulate cell growth and division, not down-regulate it. Thus even in the presence of a properly functioning proto-oncogene product, which would be produced from the allele that did not undergo mutation, the malignant phenotype would be expressed. The gene would act as a dominant.*

6. *A cancer initiator causes the first (and sometimes only) mutation in a sequence of events that leads to cancer. Mutations are permanent features of the genome; they do not fade or heal with time. Tumor promotion usually results from stimulated growth or some other condition whereby the likelihood of a second mutation (and full-blown cancer) is increased. The effects are cumulative over time, and further, the passage of time cannot mitigate the permanent nature of initiation.*

7. a. *Cell line A is transformed. The* p53 *gene is probably involved in the transformation.*
   b. *Cell line B is normal.*
   c. *Cell line C is transformed. Based on the descrption, it is not possible to identify the oncogene, but it is probably not the* p53 *gene because apoptosis relies on a functioning p53 protein.*
   d. *Cell line D is transformed, and* p53 *is undoubtedly the oncogene involved.*

Chapter Seventeen

Short Answer

1. Significance
   a. microscopic resolution: *determines the fine detail that can be seen by a given microscope under given conditions.*
   b. interference microscopy: *makes transparent objects more visible. This technique can be used to examine unstained living cells that would be impossible to see in bright-field microscopes.*
   c. electron microscopy: *provides vastly greater resolution than any form of light microscopy by using a beam of electrons instead of a beam of light. This technique was instrumental in determining the fine structure of cells*
   d. autoradiography: *when it is applied to whole cells, identifies the intracellular location of radiolabeled components. When it is applied to gels or agar plates, this technique identifies the bands or colonies that are radioactive.*
   e. cell culture: *removes cells from the complex factors that influence them as part of multicelluar organisms, and allows them to be studied under simplified conditions. This technique has proven to be one of the most successful and important advances in cell biology.*
   f. two-dimensional gel electrophoresis: *separates complex mixtures of proteins on the basis of both molecular weight and isoelectric point. This high-resolution technique can separate thousands of proteins from the same mixture.*
   g. nucleic acid hybridization: *is used to isolate or identify specific DNA fragments by base-parings between complementary DNA or RNA probes and the DNA of interest. This technique can also be used to measure the similarity between two DNA sequences.*
   h. yeast artificial chromosomes: *a technique for cloning and storing DNA fragments that are larger (100 to 1000 kb) than any that can be cloned using plasmids. This has been instrumental in the success of the Human Genome Project thus far.*
   i. animal models for human disease: *provide a laboratory system for studying aspects of human disease that would be impossible or unethical to study in humans.*
   j. polymerase chain reaction: *rapidly amplifies minute quantities of DNA. This technique is more time- and cost-effective than cloning, and it has virtually revolutionized DNA technology.*

2. Compare and Contrast:

a. magnification vs. resolution
These terms both apply to the ability of the microscope to make small specimens visible. Magnification refers to the ratio of the size of the image formed by the microscope to the actual size of the specimen. Resolution is the distance between two points that can be seen as distinct entities. Magnification aids in resolution, but does not solely determine it. Both are necessary to see subcellular structure.

b. transmission electron microscopy vs. scanning electron microscopy
For magnification and resolution of small structures, hese two techniques depend upon a beam of electrons altered by a specimen. TEMs form images from electrons that are transmitted through a specimen and then focused on a screen or photographic plate. SEMs depend on the image produced when electrons bounceg off the surface of a specimen and then onto a screen or photographic plate. TEMs give high resolution of small intracellular structures seen through thinly sectioned specimens. SEMs reveal surface features of unsectioned specimens.

c. differential centrifugation vs. density-gradient centrifugation
Both techniques separate and resolve particles in suspension by subjecting cellular homogenates to centrifugal force. Differential centrifugation is usually used to separate organelles and large subcellular particles, all of which are more dense than the medium. Density-gradient centrifugation can also be used to separate organelles and large particles, but is often applied to small particles or macromolecules. In density gradient centrifugation, the medium forms a gradient, and the particles have densities that are less than the medium at the bottom of the tube. Particles sediment from the top of the gradient to the region of equal density within the medium.

d. cloning eukaryotic DNA in bacterial plasmids vs. in phage genomes
Both of these are techniques for amplifying and storing DNA fragments from eukaryotic sources. Bacterial plasmids are commonly used vectors appropriate for cloning tens to hundreds of DNA fragments. For larger DNA samples, viral vectors are better. Bacterial "lawns" on petri plates can accommodate hundreds to thousands of viral plaques, and each one can be screened for the gene of interest.

e. cloning genomic DNA vs. cloning cDNA
Both types of clones can be used to amplify and isolate genes of interest. Genomic DNA includes regulatory sequences and intervening sequences whereas clones using cDNA include coding sequences only. Purified cDNA is good for sequencing a gene, and determining the amino acid sequence of a protein. Clones of genomic DNA can provide information about the evolution of DNA, members of gene families, and the ways in which genes are regulated.

Multiple Choice

1. c	2. b	3. a	4. b
5. a	6. d	7. d	8. b
9. a	10. d	11. c	12. c
13. b	14. c	15. b	16. c
17. d	18. c	19. a	20. d

Problems and Essays

1. a.

$$D = \frac{0.61 \times \lambda}{n \sin \theta}$$
$$D = \frac{0.61 \times 527 \text{ nm}}{1.0 \sin 70°}$$
$$D = 342 \text{ nm}$$

b.

$$D = \frac{0.61 \times \lambda}{n \sin \theta}$$
$$D = \frac{0.61 \times 450 \text{ nm}}{1.0 \sin 70°}$$
$$D = 292 \text{ nm}$$

c.

$$D = \frac{0.61 \times \lambda}{n \sin \theta}$$
$$D = \frac{0.61 \times 450 \text{ nm}}{1.5 \sin 70°}$$
$$D = 195 \text{ nm}$$

The resolution in part (c), 195 nm, is about the best you can expect from a light microscope.

2. a. *Whole homogenate, Supernatant 1, and Pellet 2*
   b. *Pellet 2*
   c. *Whole homogenate*
   d. *Supernatant 3*

3. *Amino acids are linked together starting with the amino end and ending with the carboxyl end. In Dintzis' experiment, the first fragments to incorporate radiolabel were those at the carboxyl terminal. Those peptides were nearly finished when the labeled amino acids were added, and therefore only this last segment of the peptide incorporated labeled amino acids. The last fragments to incorporate the label were those at the amino terminal, indicating that their synthesis had begun after the labeled precursors were added.*

4. a.

$$400{,}000 = \frac{(\omega^2 r)\ \text{cm / sec}^2}{980\ \text{cm / sec}^2}$$

$$(\omega^2 r) = 3.92\times 10^8$$

b.

$$v = s(\omega^2 r)\ \textit{where}\ s = 18\times 10^{-13}\ sec$$

$$v = (18\times 10^{-13}\ sec)\times(3.92\times 10^8)$$

$$v = 7.06\times 10^{-4}\ cm / sec$$

c.

$$\frac{7\ cm}{(7.06\times 10^{-4})(3600 sec / hr)}$$

$$= 2.76\ hr$$

5. *Gene to Protein:*
   a. *Isolate the gene of interest from genomic DNA.*
   b. *Use a clone of the genomic DNA to probe a population of mRNAs.*
   c. *Make a cDNA using the mRNA for the gene of interest.*
   d. *Sequence the cDNA, and use its sequence to deduce the amino acid sequence of the protein.*
   e. *Compare the amino acid sequence with that of other proteins whose functions are known.*
   f. *Express the gene using a plasmid vector in E. coli or some other expression vehicle.*

   Protein to Gene:
   a. *Isolate the protein by using either its molecular weight, its isoelectric point, or its function.*
   b. *Sequence the amino acids of at least part of the protein.*
   c. *Deduce the nucleotide sequence of at least part of the gene that codes for the protein.*
   d. *Synthesize a radiolabeled oligonucleotide to use as a probe to search the genomic DNA for the gene.*
   e. *Isolate the entire gene and determine its sequence (and that of regulatory elements).*

6.

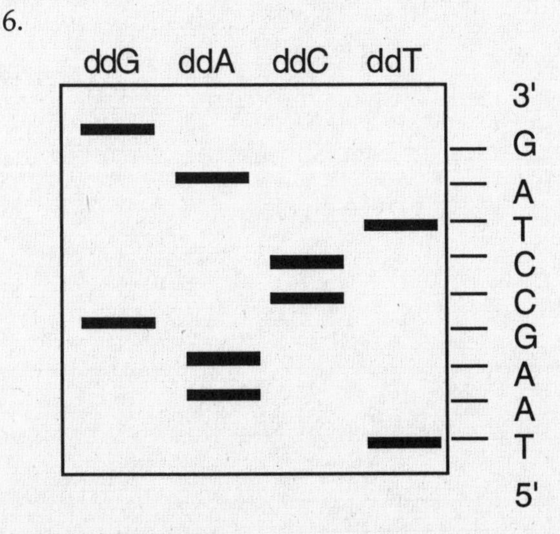

7. a. *polymerase chain reaction followed by restriction enzyme digestion and gel electrophoresis to search for restriction fragment length polymorphisms*
b. *radiolabeled carbohydrate tracers followed by autoradiography*
c. *nucleic acid hybridization between a radiolabeled probe for growth hormone, followed by gene isolation using restriction enzyme digestion, recombinant DNA, and cloning, and then protein purification from cultures of cells genetically engineered to express the hormone*
d. *site-directed mutagenesis followed by enzyme kinetics*

APPENDIX II

Answers to Even-Numbered, End-of-Chapter Questions

Chapter 1

2. *They greatly increase the surface area/volume ratio of the cell, allowing much greater exchange between the cell and the lumen of the intestine. Would be unable to absorb sufficient nutrients from the lumen to survive.*
4. *Chloroplasts provide the machinery for photosynthesis, which allows the plant to produce complex organic materials from inorganic substances. Cell walls provide mechanical support for plants and offer protection to the enclosed cells. The plasmodesma provides a pathway for communication between cells. The large central vacuole has numerous, less-obvious functions described in Chapter 8.*
6. *The Driesch experiment separated the first two cells completely, while the Roux experiment left the dead cell in continued contact with the live one. In the Roux experiment, the living cell presumably recognized the presence of the dead cell, which prevented the living cell from developing into a complete embryo.*
8. *The nuclear envelope that separates the nucleus and cytoplasm in a eukaryotic cell provides the basis for regulating the movement of substances between the two compartments. The DNA of a bacterial cell is presumably much more accessible to cytoplasmic substances than that of a eukaryotic cell.*
10. *A highly flattened cell because it will have a much greater surface area/volume ratio.*
12. *That viruses contain genetic material; that they are capable of producing more of themselves, albeit only inside a host cell; that they contain complex biological macromolecules. These are all important criteria of living organisms.*

Chapter 2

2. *Glucose would be the most soluble and* $CH_3(CH_2)_5CH_3$ *the least.*
4. *The structure of DHAP is on p. 104. It does not have stereoisomers.*
6. *Glycogen.*
8. *2 and 4. A rise in pH, would increase the loss of a proton from a —COOH group and from a* $—NH_3^+$ *group.*
10. *It would be expected to slow the rates of the reaction because the products of the first two reactions would diffuse into the surrounding medium rather than be passed directly into the active sites of the second and third enzymes.*
12. *8,000* (20^3). *4 carboxyl terminals, one per chain.*
14. *Since the pK is very near to physiological pH, the amino acid can shift from the charged to uncharged state with slight changes in conditions that occur during a reaction.*

Chapter 3

2. *Inhibition of the first enzyme shuts down the entire pathway, allowing the substrate to be used for other purposes. If one of the later enzymes were inhibited, energy would be wasted and unusable metabolites might accumulate.*
4. *The substrate concentration is saturating at all three concentrations and thus the enzyme is operating at approximately* V_{max}.

6. *One. Zero.*
8. *Both are derived from ATP.*
10. *greater; greater.*
12. *3) less than one-tenth.*
14. *Probably not, because its formation is too endergonic (+11.8 kcal/mol), which is the reason that it can be used to form ATP by substrate-level phosphorylation (i.e., its phosphate transfer potential is much higher than that of ATP).*

Chapter 4

2. *Since steroids are relatively small hydrophobic compounds, they would be expected to diffuse through the plasma membrane. Accordingly, the receptors for these hormones are found in the cytoplasm. In contrast, receptors for nearly all other types of hormones, including insulin, are found at the surface of the plasma membrane. Being a protein, insulin would not be expected to penetrate the plasma membrane.*
4. *Incorporate a component, such as an antibody, into the lipid bilayer of the liposome that will recognize and bind to a specific component on the outer surface of the target cell membrane.*
6. *See Figures 4.15 and 4.31.*
8. *It would be expected to cause the level of saturation of the fatty acids of the membrane to increase (the number of polyunsaturated fatty acids to decrease), which would raise the transition temperature of the lipid bilayer. The activity of membrane desaturases would be expected to decrease.*
10. *No. It would be more likely that ouabain is inhibiting the Na^+-K^+ ATPase which builds the Na^+ gradient that is required for the secondary transport of glucose.*
12. *It is proposed that the sialic acid at the end of the oligosaccharide chains acts to block recognition sites for the removal of the blood cells. During normal red blood cell aging, these sugar residues may be removed, which exposes the underlying sugars, which targets the cells for removal from the bloodstream by the spleen. Treatment of the cells with neuraminidase would mimic the aging process of the cells, causing their premature recognition and removal.*
14. *Na^+ would be expected to move into the cell more rapidly because its influx is favored by the electrical component of the gradient as well as the difference in concentration. In contrast, K^+ would be moving down its concentration gradient as it exited the cell, but against its electrical gradient (inside negative).*
16. *The intracellular concentrations would be expected to be hypertonic to the extracellular fluids, which would cause the cells to gain water and maintain turgor pressure. The intracellular fluids of animal cells would be expected to have similar solute concentrations to the extracellular fluids so that cells would neither gain nor lose water.*

Chapter 5

2. *1 and 2 are correct. The same ones would be true for submitochondrial particles treated with DNP.*
4. *NADH is the strongest reducing agent; O_2 is the strongest oxidizing agent; and NAD^+ has the greatest affinity for electrons.*
6. *The first major drop in potential would be bypassed and electrons would be fed into coenzyme Q from a donor at roughly the same energy level.*
8. *Malate concentrations must be kept high enough so that the reaction is favorable under conditions that operate in the cell.*
10. *They move through the ATP synthase in the other direction during the phosphorylation of ADP and in the uptake of P_i.*
12. *Isolated mitochondria would acidify the medium, whereas submitochondrial particles (which are inside-out particles) would alkalinize the medium.*

14. *Because the outer membrane is freely permeable to protons, therefore the proton concentration in the intermembrane space would rapidly equilibrate with that of the cytosol, lowering the cytosolic pH.*

Chapter 6

2. *C_4 plants.*
4. *#s 2, 4, and 5.*
6. *When NADPH levels are high.*
8. *In chloroplasts, the internal compartment of the thylakoid can be maintained at a very low pH because there are virtually no enzymes in that compartment. In contrast, the mitochondria eject protons into the medium, which would seriously affect enzymes of the cytosol.*
10. *No. The C_4 plant has to use additional ATP to convert pyruvate to PEP.*
12. *18 ATPs (6 to regenerate RuBP) and 12 NADPHs are presumably required for each hexose produced from CO_2. According to the numbers shown in Figure 6.18, noncyclic electron transport should be able to generate approximately 3 molecules of ATP for pair of NADPHs, which is the ratio required for CO_2 assimilation.*
14. *Manganese is part of the oxygen-evolving complex and acts to store electrons removed from H_2O; Magnesium is part of all the chlorophyll molecules and is involved in light absorption; Iron is part of cytochrome b_6f and the various iron-sulfur proteins and undergoes oxidation and reduction during electron transport.*

Chapter 7

2. *Anti-fibronectin antibodies would disrupt neural crest cell migration; anti-laminin antibodies would disrupt axon outgrowth; and fibroblast adhesion might be disrupted by RGD peptides, which would block formation of focal contacts, or by tenascin, or by various types of antibodies.*
4. *Without an extracellular matrix, cartilage would lack the ability to provide mechanical support, the corneal stroma would lack its durability and transparency, and tendons would lack their tensile strength.*
6. *Both types of adhesive junctions are important in maintaining the integrity of an epithelium. Hemidesmosomes are important in holding the layer to the underlying substrate and desmosomes in holding the cells to one another.*
8. *The follicle cells and cardiac muscle cells form gap junctions with one another. As a result, the FSH would bind to receptors on the follicle cells, stimulating the production of cyclic AMP molecules that could diffuse through the gap junctions into the cardiac muscle cells, causing their contraction.*
10. *Because gap junctions assemble from connexons that diffuse within the plane of the plasma membrane into the region where the gap junction is forming. Once there, connexons from apposing membranes become aligned to form a continuous passageway. A decrease in temperature would be expected to reduce the ability of integral proteins to diffuse through the membrane.*
12. *The cell walls of plants protect the enclosed cell, prevent it from rupturing due to the influx of water, and provide mechanical support for the plant. The cells of an animal are bathed by extracellular fluids whose composition is strictly regulated by various organs, especially the kidneys. Since the osmotic concentration of the extracellular fluids is maintained at the same value as that of the cells, the cells are not faced with the gain or loss of water. In addition, mechanical support is provided by various types of extracellular skeletal materials, which preclude the need for supportive cell walls.*

Chapter 8

2. *The cells would be expected to secrete lysosomal enzymes into the medium. If the cells were incubated with ^{32}P-phosphate, the label would not be expected to appear in lysosomal enzymes as it would in control cells. Golgi fractions prepared from the cells should lack the enzyme N-acetylglucosamine phosphotransferase, which transfers the phosphorylated sugar from UDP to a mannose receptor.*
4. *No. It is synthesized in the Golgi complex.*
6. *The process can be studied in embryos where it occurs for the first time. Alternatively, cells could be stimulated hormonally or with agents that increase intracellular calcium to provoke granule discharge.*
8. *The thicker the section, the poorer the resolution because particles emitted from the source can travel a greater distance through the section before they strike the emulsion. This will allow the silver grains to form farther from the site in the cell where the radiolabel is present.*
10. *Macrophage because a cell engaged heavily in phagocytosis is continually removing membrane from the cell surface that must be replaced.*
12. *Plasma membrane; SER; RER; lysosome; plant vacuole; plasma membrane; TGN (or other membranes of the biosynthetic pathway); cytosol.*
14. *The surface of the RER would be smooth not unlike the SER.*
16. *It would change the conformation of the cytoplasmic domain of the receptor such that the receptor's affinity for adaptins and clathrin would increase, causing the receptors to become trapped in coated pits.*

Chapter 9

2. *The dynein cycle shown in Figure 9.39 is similar to the actomyosin cycle shown in Figure 9.62. ATP is required for dissociation o of the dynein cross-bridge. In the absence of ATP, the dynein arm remains attached to the neighboring B tubule and the axoneme is said to be in rigor.*
4. *9; 11.*
6. *Warm the preparation, add GTP, add EGTA, add more tubulin, add taxol. Chill the preparation, add colchicine or another depolymerizing agent, add Ca^{2+}, place the mixture under hydrostatic pressure.*
8. *No. In fact, all the microtubules of the axon are oriented in the same direction; different motor proteins move vesicles in anterograde and retrograde directions.*
10. *The heads, because they must all have a structure that enables them to move along the same type of microtubular track and all of them require a catalytic site that can hydrolyze ATP and use the energy to drive the conformational changes of the power stroke. In contrast, the structure of the tails would be expected to vary according to the type of cargo being transported.*
12. *Those antibodies that bind to sites on the motor protein that are essential to its motor activity, such as its ATPase active site, would be expected to be much more inhibitory than antibodies that bound to less essential sites.*
14. *Very little, since other kinesin-like proteins may be able to compensate for the loss of kinesin. This does not mean that kinesin normally does not play a major role in cellular activities.*

Chapter 10

2. *Because he would not have been able to distinguish maternal and paternal members of a pair of homologous chromosomes to be able to follow how they assorted.*
4. *Both of them would be linked to one or more other genes that resided somewhere in the middle of the chromosome. If cDNAs could be made for both genes and mixed together, sites at both ends of one chromosome should be labeled.*
6. *No. Most restriction enzymes require sequences between 4-6 nucleotides in length (depending on the enzyme); such sequences would be expected to occur every 4^4-4^6 (256-4056) nucleotides.*

8. *The solution of double-stranded DNA would be considerably more concentrated (e.g., gram/ml) than that of single-stranded DNA because single-stranded DNA has a higher absorbance per unit weight than does double-stranded DNA as evidenced by the hyperchromic shift in Figure 10.16.*
10. *2, 3, 4 are correct. 35 percent.*
12. *No. The T_m is the point at which 50 percent of the base pairs have separated. This could occur in a population of DNA molecules where all of the DNAs remained in a duplex form being held together by the 50 percent of the bases that were still bonded to one another.*
14. *It is random because in any given cell, the rate of amplification of the DHFR gene is no higher than that of other genes. The difference is that amplification of other genes provides no selective advantage for continued survival of the cell, whereas amplification of the DHFR gene does. Consequently, cells continue to accumulate more and more copies of the DHFR gene as other cells die off.*

Chapter 11

2. *In a nonoverlapping code, the entire polypeptide would be nonfunctional past the site of mutation, while in an overlapping code, only three amino acids would be affected.*
4. *The radioactivity profile would follow the absorbance profile very closely, showing peaks at 18S and 28S. There might also be very small peaks of radioactivity in the rRNA precursor size ranges, reflecting the fact that rRNA precursors are still being synthesized with ^{3}H-uridine.*
6. *2^3 or 8 different codons.*
8. *Eight (one for initiation and termination and two for each elongation).*
10. *It strongly suggests that, but is not definitive since the code could have still been overlapping and the other two codons affected were degenerate ones that did not change their coding properties.*
12. *It would take 5 nucleotides (2^5). 2^4 would only be able to code for 16 different amino acids, whereas 2^5 could code for 32 amino acids.*
14. *Because it has such a short half-life. Although rRNAs and tRNAs are synthesized at a much slower rate, they have a much longer lifetime, which causes them to accumulate while the mRNA is being degraded.*

Chapter 12

2. *Objects the size of multiprotein complexes and coated metal particles are capable of passing through the nuclear pores. Such large-sized substances presumably trigger a specific opening response before they are allowed passage. Albumen is not a protein normally found in the nucleus and lacks the signals required for its penetration.*
4. *It shows that the histones provide precise docking sites for each DNA strand along the entire circumference of the core histone particle. Changes in amino acid sequence that affect the nature or positions of any of these sites might be expected to disrupt nucleosome structure.*
6. *A translocation had occurred that moved DNA sequences normally found on one chromosome to another.*
8. *These are sites where the machinery for replication and transcription can be concentrated, which should greatly increase the efficiency of these synthetic activities by decreasing the time required for interactions to occur.*
10. *The stop codon at the end of the coding region of the z gene had been mutated so that it encoded an amino acid rather than a stop message, which allowed the ribosome to read through to the next coding region, forming a fusion protein.*
12. *No. There could be many reasons why such nuclei failed to support development, including the likely fact that nuclei are easily damaged in these experiments, for example, by the cytoplasm in which they are placed. In most cases where many variables occur, negative results leave open numerous possible interpretations.*
14. *DNA rearrangement during antibody synthesis and alternate slicing of primary transcripts.*

Chapter 13

2. *If the replication machinery is tightly associated with the nuclear matrix, then the DNA polymerase is presumably immobilized while the DNA template actually moves relative to the enzyme.*
6. *Without* oriC, *the two strands will not become separated to form a replication fork. The origin of replication contains an AT-rich section that is subject to strand separation.*
8. *The mutant genes of the first group could encode a subunit of the core holoenzyme for DNA polymerase III, which is required for continued incorporation; those of the second group would not be required immediately, such as a helicase or DNA polymerase I, allowing replication to continue for a brief period; those of the third group would likely be required for initiation of replication at the origin.*
10. *It would appear as a solid black line, i.e., a continuous string of silver grains. No. You are only looking at one small part of one chromosome. Other segments of the DNA might not be replicating at all at this time (which is the case).*
12. *The matrix provides a mechanism to spatially organize the various factors involved in replication. The foci allow the proteins required for replication to be concentrated, which would be expected to decrease the time required for specific interactions and thus increase the efficiency of replication.*
14. *The cell in S phase would have a much higher silver grain density over its nucleus, but the nonreplicating cell would still show incorporated label due to DNA synthesis that accompanies DNA repair.*

Chapter 14

2. *Examples would include the synthesis of enzymes involved in production of deoxyribonucleoside triphosphates and the synthesis of cyclins that activate Cdks required for the* G_2*-S transition.*
4. *The minimum amount of time of labeling would be 12 hours. This time would be needed for the cells that had just entered* G_2 *at the start of the labeling period to reenter S phase. You could determine the time it took for the number of cells in the population to double.*
6. *A mutation in* CAK *or* cdc25 *would be expected to stop cell division; a mutation in* wee1 *would be expected to promote cell division.*
8. *By fusion and by mitosis without cytokinesis. That mitosis and cytokinesis are independent processes that do not have to occur together.*
10. *100 percent; 25 percent.*
14. *The fetuses should show evidence of secondary nondisjunction rather than primary nondisjunction, i.e., evidence that the cells contain sister chromatids rather than both homologues from the mother. Evidence contradicts this conclusion since most mistakes occur during meiosis I.*

Chapter 15

2. *Inactivation of phosphatase 2 would lead to the permanent activation of kinase 3, which would lead to the continued unregulated activation of the transcription factor and likely the unregulated growth of the cell. Phosphatase 3 would still, however, be able to inactivate the transcription factor. Inactivation of protein kinase 3 might be expected to stop the cell from progression through the cell cycle because the transcription factor could not be activated.*
4. *All three act by binding to a protein, but* Ca^{2+} *binds to a protein (calmodulin) that activates other proteins,* IP_3 *binds to and directly opens an ion channel, and cAMP binds to and directly activates an enzyme.*
6. *You could see if they competed with one another for the same binding site. This could be determined, for example, by radioactively labeling the two hormones and determining the amount of bound radioactivity when cells were incubated with the two hormones separately or together over a range of hormone concentrations.*
8. *The GTP analog would presumably bind to the G protein and remain unhydrolyzed, causing the stimulation to be prolonged. The liver cell would continue to break down glycogen, while the epithelial*

cell treated with EGF would continue to proliferate. The cholera toxin would be expected to have a similar effect as the GTP analog on the glucagon treated liver cell, but would not have an effect on proliferation of an epithelial cell since this response is not mediated by a heterotrimeric G protein.

10. *Cells possess high concentrations of calcium-binding proteins that are able to "soak up" excess calcium ions, and they also possess large numbers of calcium pumps in their plasma and cytoplasmic membranes that removes the ions from the cytosol. If you injected the calcium solution, you would expect to see a very localized and transient rise in* Ca^{2+} *levels, with a quick return to normal levels.*
12. *No. Calmodulin is a small protein whose structure is highly conserved throughout eukaryotes. It is much more likely that such a molecule would have only one or two recognizable binding sites and that the various effectors have similar sites that are capable of binding to calmodulin.*
14. *Dominantly because mutations in* Ras *that lead to a protein that cannot be inactivated, such as one whose GTPase active site is noncatalytic, will feed continuous signals along the MAP kinase cascade provoking unregulated growth. The presence of normal* Ras *molecules will not be able to prevent the effects of the abnormal ones.*